Characterization of Metallic Materials: Microstructure, Forming and Heat Treatment (Second Edition)

Characterization of Metallic Materials: Microstructure, Forming and Heat Treatment (Second Edition)

Guest Editors

Seong-Ho Ha
Young-Ok Yoon
Dong-Earn Kim
Young-Chul Shin

Basel • Beijing • Wuhan • Barcelona • Belgrade • Novi Sad • Cluj • Manchester

Guest Editors

Seong-Ho Ha
Materials–Supply Chain
R&D Department
Korea Institute of Industrial
Technology
Incheon
Korea, South

Young-Ok Yoon
Materials–Supply Chain
R&D Department
Korea Institute of Industrial
Technology
Incheon
Korea, South

Dong-Earn Kim
Regional Industry Innovation
Department
Korea Institute of Industrial
Technology
Incheon
Korea, South

Young-Chul Shin
Flexible Manufacturing
R&D Department
Korea Institute of Industrial
Technology
Incheon
Korea, South

Editorial Office
MDPI AG
Grosspeteranlage 5
4052 Basel, Switzerland

This is a reprint of the Special Issue, published open access by the journal *Materials* (ISSN 1996-1944), freely accessible at: www.mdpi.com/journal/materials/special_issues/8W6B01TG35.

For citation purposes, cite each article independently as indicated on the article page online and using the guide below:

Lastname, A.A.; Lastname, B.B. Article Title. *Journal Name* **Year**, *Volume Number*, Page Range.

ISBN 978-3-7258-2882-1 (Hbk)
ISBN 978-3-7258-2881-4 (PDF)
https://doi.org/10.3390/books978-3-7258-2881-4

© 2025 by the authors. Articles in this book are Open Access and distributed under the Creative Commons Attribution (CC BY) license. The book as a whole is distributed by MDPI under the terms and conditions of the Creative Commons Attribution-NonCommercial-NoDerivs (CC BY-NC-ND) license (https://creativecommons.org/licenses/by-nc-nd/4.0/).

Contents

Yanchao Wang, Huimin Duan, Zhongna Zhang, Lan Chen and Jingan Li
Research Progress on the Application of Natural Medicines in Biomaterial Coatings
Reprinted from: *Materials* 2024, 17, 5607, https://doi.org/10.3390/ma17225607 1

Ji-Young Kim, Seung-Chae Yoon, Byeong-Keuk Jin, Jin-Hwa Jeon, Joo-Sik Hyun and Myoung-Gyu Lee
Friction Characteristics of Low and High Strength Steels with Galvanized and Galvannealed Zinc Coatings
Reprinted from: *Materials* 2024, 17, 5031, https://doi.org/10.3390/ma17205031 25

Jie Zhang, Haibin Zheng, Chengwei Zeng and Changlong Gu
High-Precision Instance Segmentation Detection of Micrometer-Scale Primary Carbonitrides in Nickel-Based Superalloys for Industrial Applications
Reprinted from: *Materials* 2024, 17, 4679, https://doi.org/10.3390/ma17194679 46

Ziheng Ding, Chaogang Ding, Zhiqin Yang, Hao Zhang, Fanghui Wang and Hushan Li et al.
Ultra-High Strength in FCC+BCC High-Entropy Alloy via Different Gradual Morphology
Reprinted from: *Materials* 2024, 17, 4535, https://doi.org/10.3390/ma17184535 68

Guanwen Luo, Zhiwei Peng, Kangle Gao, Wanlong Fan, Ran Tian and Lingyun Yi et al.
Preparation of Metallized Pellets for Steelmaking by Hydrogen Cooling Reduction with Different Cooling Rates
Reprinted from: *Materials* 2024, 17, 4362, https://doi.org/10.3390/ma17174362 82

Yeong-Maw Hwang, Tsung-Han Ho, Yung-Fa Huang and Ching-Mu Chen
Formability Prediction Using Machine Learning Combined with Process Design for High-Drawing-Ratio Aluminum Alloy Cups
Reprinted from: *Materials* 2024, 17, 3991, https://doi.org/10.3390/ma17163991 101

Wanlong Fan, Zhiwei Peng, Ran Tian, Guanwen Luo, Lingyun Yi and Mingjun Rao
Toward Metallized Pellets for Steelmaking by Hydrogen Cooling Reduction: Effect of Gas Flow Rate
Reprinted from: *Materials* 2024, 17, 3896, https://doi.org/10.3390/ma17163896 123

Xinchen Pang, Guifang Zhang, Peng Yan, Zhixiang Xiao and Xiaoliang Wang
Study on the Migration Patterns of Oxygen Elements during the Refining Process of Ti-48Al Scrap under Electromagnetic Levitation
Reprinted from: *Materials* 2024, 17, 3709, https://doi.org/10.3390/ma17153709 138

Koh-ichi Sugimoto, Shoya Shioiri and Junya Kobayashi
Effects of Mean Normal Stress and Microstructural Properties on Deformation Properties of Ultrahigh-Strength TRIP-Aided Steels with Bainitic Ferrite and/or Martensite Matrix Structure
Reprinted from: *Materials* 2024, 17, 3554, https://doi.org/10.3390/ma17143554 149

Emanuela Cerri and Emanuele Ghio
On Strain-Hardening Behavior and Ductility of Laser Powder Bed-Fused Ti6Al4V Alloy Heat-Treated above and below the β-Transus
Reprinted from: *Materials* 2024, 17, 3401, https://doi.org/10.3390/ma17143401 162

Xiaoliang Wang, Guifang Zhang, Peng Yan, Xinchen Pang and Zhixiang Xiao
Numerical Simulation of Electromagnetic–Thermal–Fluid Coupling for the Deformation Behavior of Titanium–Aluminum Alloy under Electromagnetic Levitation
Reprinted from: *Materials* 2024, 17, 3338, https://doi.org/10.3390/ma17133338 180

Sun-Ki Kim, Seung-Hyun Koo, Hoon Cho and Seong-Ho Ha
Hot Deformation Constitutive Analysis and Processing Maps of Ultrasonic Melt Treated A5052 Alloy
Reprinted from: *Materials* **2024**, *17*, 3182, https://doi.org/10.3390/ma17133182 194

Zichen Qi, Zhengchi Jia, Xiaoqing Wen, Hong Xiao, Xiao Liu and Dawei Gu et al.
Enhanced Mechanical Properties of Ti/Mg Laminated Composites Using a Differential Temperature Rolling Process under a Protective Atmosphere
Reprinted from: *Materials* **2024**, *17*, 2753, https://doi.org/10.3390/ma17112753 211

Ji Li, Yujie Wo, Zhigang Wang, Wenhao Ren, Wei Zhang and Jie Zhang et al.
Research on the Influence of Cold Drawing and Aging Heat Treatment on the Structure and Mechanical Properties of GH3625 Alloy
Reprinted from: *Materials* **2024**, *17*, 2754, https://doi.org/10.3390/ma17112754 226

Daiwei Liu, Guifang Zhang, Jianhua Zeng and Xin Xie
Numerical Simulation of Segregation in Slabs under Different Secondary Cooling Electromagnetic Stirring Modes
Reprinted from: *Materials* **2024**, *17*, 2721, https://doi.org/10.3390/ma17112721 245

Review

Research Progress on the Application of Natural Medicines in Biomaterial Coatings

Yanchao Wang, Huimin Duan, Zhongna Zhang, Lan Chen * and Jingan Li *

School of Materials Science and Engineering, Zhengzhou University, Zhengzhou 450001, China; wangyancaho@gs.zzu.edu.cn (Y.W.); duanhuimin@gs.zzu.edu.cn (H.D.); zhongna_zhang@gs.zzu.edu.cn (Z.Z.)
* Correspondence: chenlan@zzu.edu.cn (L.C.); lijingan@zzu.edu.cn (J.L.); Tel.: +86-185-3995-6211 (J.L.)

Abstract: With the continuous progress of biomedical technology, biomaterial coatings play an important role in improving the performance of medical devices and promoting tissue repair and regeneration. The application of natural medicine to biological materials has become a hot topic due to its diverse biological activity, low toxicity, and wide range of sources. This article introduces the definition and classification of natural medicines, lists some common natural medicines, such as curcumin, allicin, chitosan, tea polyphenols, etc., and lists some biological activities of some common natural medicines, such as antibacterial, antioxidant, antitumor, and other properties. According to the different characteristics of natural medicines, physical adsorption, chemical grafting, layer-by-layer self-assembly, sol–gel and other methods are combined with biomaterials, which can be used for orthopedic implants, cardiovascular and cerebrovascular stents, wound dressings, drug delivery systems, etc., to exert their biological activity. For example, improving antibacterial properties, promoting tissue regeneration, and improving biocompatibility promote the development of medical health. Although the development of biomaterials has been greatly expanded, it still faces some major challenges, such as whether the combination between the coating and the substrate is firm, whether the drug load is released sustainably, whether the dynamic balance will be disrupted, and so on; a series of problems affects the application of natural drugs in biomaterial coatings. In view of these problems, this paper summarizes some suggestions by evaluating the literature, such as optimizing the binding method and release system; carrying out more clinical application research; carrying out multidisciplinary cooperation; broadening the application of natural medicine in biomaterial coatings; and developing safer, more effective and multi-functional natural medicine coatings through continuous research and innovation, so as to contribute to the development of the biomedical field.

Keywords: natural medicines; biomaterials; coatings; applications; orthopedic implants; cardiovascular and cerebrovascular stents; wound dressings

Citation: Wang, Y.; Duan, H.; Zhang, Z.; Chen, L.; Li, J. Research Progress on the Application of Natural Medicines in Biomaterial Coatings. *Materials* **2024**, *17*, 5607. https://doi.org/10.3390/ma17225607

Academic Editors: Csaba Balázsi and Christopher C. Berndt

Received: 1 October 2024
Revised: 29 October 2024
Accepted: 12 November 2024
Published: 16 November 2024

Copyright: © 2024 by the authors. Licensee MDPI, Basel, Switzerland. This article is an open access article distributed under the terms and conditions of the Creative Commons Attribution (CC BY) license (https://creativecommons.org/licenses/by/4.0/).

1. Introduction

Biomaterial coatings are increasingly used in medical and biological tissue engineering, ranging from diagnosis and treatment to tissue repair and organ replacement. In the field of diagnostics, sensors and detection reagents made of biomaterials can quickly and accurately detect biomarkers, providing strong support for the early diagnosis of diseases [1–3]. In terms of treatment, the application of biomaterials such as drug carriers and interventional devices has significantly improved the therapeutic effect and reduced side effects [4,5]. In tissue engineering and regenerative medicine, biomaterials play a key role [6,7]. For example, scaffold materials used for bone repair can guide the growth of bone cells and promote bone tissue regeneration [8–10]. Biomaterials such as artificial blood vessels and heart valves in the cardiovascular field help to restore normal blood circulation [11–14]. In addition, biomaterial coatings have also been used in ophthalmology, stomatology, neurosurgery, and other fields, making great contributions to improving the quality of life of patients [15–19]. Figure 1 illustrates some of the applications. Biomaterial coatings are an effective means to significantly improve the performance and biocompatibility of biomaterials. However, there are still some problems with traditional biomaterial coatings. The risk of implant infection remains, due to limited antimicrobial performance and the difficulty of effectively countering the growing problem of bacterial resistance; the anticoagulant effect is not ideal, and thrombotic complications may still occur during long-term use. In addition, the biocompatibility of some coatings remains to be improved, which may cause a local inflammatory response or immune rejection. In addition, traditional coatings may use toxic chemical agents in the preparation process, posing a potential threat to the environment and human health. In the complex physiological environment of living organisms, the coating may be subject to mechanical stress, chemical corrosion, biodegradation, and other factors, which will lead to the peeling or degradation of the coating, which will affect the performance and service life of the coating [20–22]. For biomaterial coatings with drug release, how to accurately control the release rate and the amount of release is a problem, and a release that is too slow may not achieve therapeutic effect. Releasing it too quickly may lead to increased side effects of the drug [23–25].

In view of the above-mentioned problems in traditional biomaterial coatings, natural medicines are gradually gaining attention as a potential solution. Natural medicine refers to drugs of natural origin, unmodified or synthesized and used under the guidance of modern medical theories, including pharmaceutical ingredients from plants, animals, and microorganisms. Natural medicines are rich in bioactive ingredients and diverse mechanisms of action, such as antibacterial, anti-inflammatory, antioxidant, etc. The application of natural medicines to biomaterial coatings is expected to overcome the limitations of traditional coatings. For example, some natural medicines have potent antimicrobial activity and are able to effectively inhibit the growth of multiple drug-resistant bacteria, providing new strategies for preventing implant infections [26]. At the same time, the anti-inflammatory and antioxidant properties of natural medicines can regulate the body's immune response, reduce inflammatory damage, and improve biocompatibility [27–30]. In addition, natural medicines usually have low toxicity and good biodegradability [31], which is in line with the development concept of green environmental protection. The role of natural medicine is equivalent to building a barrier on the surface of the scaffold to realize the functionalization of the stent. At present, the application of natural medicine in the coating of biological materials has made some progress. Researchers have explored the application of a variety of natural medicines in different types of biomaterial coatings. In terms of antimicrobial coatings, natural medicines such as berberine, curcumin, and allicin have been used to prepare coatings with antimicrobial functions and have shown good results in in vitro experiments and animal models. These coatings can effectively inhibit the attachment and growth of bacteria, reducing the incidence of infection. In terms of anticoagulant coatings, the application research of natural drugs such as salvia extract and hirudin is also constantly advancing, providing a new way to improve the anticoagulant performance of cardiovascular devices. In the coating that promotes tissue regeneration, natural medicines such as collagen and chitosan are used to promote cell adhesion, proliferation and differentiation, and accelerate the process of tissue repair and regeneration. There is also the use of polymer coatings to encapsulate natural medicines to achieve controlled release of drugs under certain conditions.

While the application of natural medicines in the coating of biomaterials is still in the research stage, there are some issues that need to be addressed. For example, the stability and controlled release of natural medicines need to be further optimized, and the preparation process of coatings needs to be perfected to achieve large-scale production and clinical application. However, with the continuous deepening of research, it is believed that natural medicine will have broad application prospects in the field of biomaterial coating, and more new natural medicines will be discovered, and multidisciplinary cross-integration will promote the progress of technology. Also, the specific appropriate drug coating is selected for treatment according to the patient, so as to make a greater contribution to the improvement of medical outcomes and the quality of life.

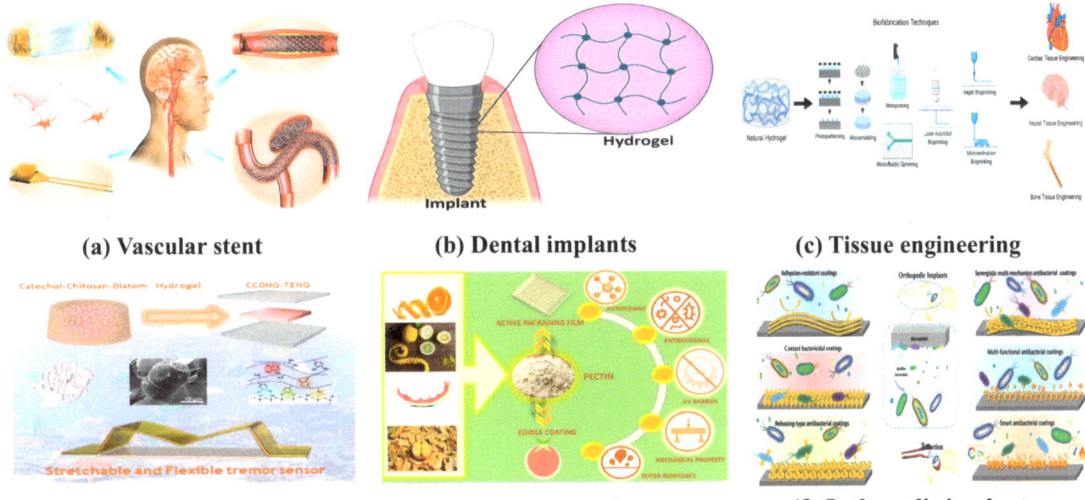

Figure 1. Some applications of biomaterial coatings: (**a**) Mg-based alloys have been used in neuroscience as filaments within nerve conduits to accelerate nerve regeneration, the nerve electrode, devices for neural recording and monitoring, and stents for carotid artery stenosis and aneurysm treatment [32]; (**b**) the incorporation of compounds such as titanium dioxide (TiO$_2$), dopamine, fluorine-substituted hydroxyapatite (FHA), tetraethyl orthosilicate (TEOS), and silica nanoparticles (SNs) into the hydrogel structure can improve the biocompatibility, stability, and peripheral inflammation of implants [33]; (**c**) the prepared hydrogels are used for cardiac, nervous, and bone tissue engineering [34]; (**d**) catechol chitosan diatom hydrogel (CCDHG) was developed for use in TENG electrodes, and m-type defibrillation sensors were developed based on CCDHG-TENG to evaluate low-frequency motion in patients with Parkinson's disease [35]; (**e**) plant-based multi-confectionery gums can be used to produce polymer films for active packaging [36]; (**f**) an antimicrobial coating can be built on the surface of orthopedic implants [37].

2. Preparation Method of Coating

Different natural medicines possess unique properties, functional groups, and mechanisms, while biomaterials exhibit distinct characteristics. How does a natural medicine bind to the surface or within another material? Several common methods are identified through a review of the literature, including physical adsorption, chemical grafting, layer-by-layer self-assembly, the sol–gel method, and electrospinning.

2.1. Physical Adsorption

Principle: The use of intermolecular van der Waals force, electrostatic action, etc., to adsorb natural drugs to the surface of biomaterials. The biomaterial is soaked in a solution containing natural medicines, and after a certain period of incubation, the drugs are adsorbed on the surface of the material. For example, the titanium alloy material is soaked in the solution of tea polyphenols, and the tea polyphenols form an antioxidant coating on the surface of the titanium alloy through physical adsorption [38]. Clindamycin was loaded onto cerium dioxide nanoparticles (CNPs) by the physical adsorption method [39]. Polyacrylamide (PAM) and graphene oxide (GO) were combined by the physical adsorption method, and then loaded with clove essential oil to make composite nanomaterials for antibacterial packaging [40]. Liu Sisi et al. [41] first prepared chitosan/silk fibroin nanofiber multilayer membranes through self-assembly layer by layer, and loaded the antibacterial drug berberine into the multilayer membrane by the physical adsorption method, thereby preparing berberine–chitosan/silk-fibroin nanofiber multilayer membranes. Lubna

Shahzadi et al. [42] prepared chitosan hydrogel by the freezing gel method, and then soaked it in a solution of thyroxine completely adsorbed into the hydrogel, and then dried it to form a hydrogel. Although the physical adsorption method is relatively simple and low-cost, the binding force between the coatings is weak, and the loading of drugs on the coating is limited.

2.2. Chemical Grafting

Natural medicines are covalently bound to active groups on the surfaces of biomaterials through chemical reactions. Covalent bonds are employed to secure the surface of the material, which is first activated to introduce active groups. These groups subsequently react with natural drugs that have undergone specific chemical modifications. However, the reaction conditions can be harsh, and the construction process is often time-consuming and costly, complicating actual production [43]. Additionally, the involvement of certain groups within the drug molecule in the formation of covalent bonds may alter the overall structure of the drug, potentially impacting both the biological activity of the material and the drug itself [44]. Examples of chemical grafting are provided in the literature, as detailed in Table 1.

Table 1. Chemical grafting.

Materials	Reaction Process	Reference
Glycidyl methacrylate (GMA), carbon fiber (CF)	The COOH on the surface of CF would lose electrons and then remove carbon dioxide to generate carbon radicals on the surface of CF. The carbon radical would attack the carbon–carbon double bond in GMA to initiate the radical polymerization of GMA monomers and graft polymers would be formed on the CF surface.	[45]
Waterborne polyurethane coatings (WPU), graphene oxide (GO)	The hydroxyl group on the surface of GO reacts with the amino group of the WPU molecular chain to form an amide bond.	[46]
Graphene oxide (GO), Ethylenediamine (EDA), polyvinylidene fluoride (PVDF)	GO-NH2 obtained by GO and EDA by a coupling reaction was grafted with dehydrofluorinated PVDF (DF-PVDF) in a Michael addition reaction.	[47]
Carboxymethyl cellulose (CMC), epigallocatechin gallate (EGCG-g-CS)	Polyionic complexes were formed by ionic bonding of the amino groups in EGCG-g-CS with the carboxyl groups in CMC.	[48]
Epsilon-poly-l-lysine (EPL), cellulose	Hydroxyl groups on C6 of cellulose were oxidized to carboxyl groups by TEMPO, and a grafting reaction was achieved between newly formed carboxyl groups of cellulose and amino of EPL.	[49]

2.3. Layer-by-Layer Self-Assembly

The layer-by-layer self-assembly technology is based on the interaction force between different molecules, so that the molecules are alternately deposited on the base to form a multi-layer structure. These interaction forces include electrostatic interactions, hydrogen bonds, coordination bonds, van der Waals forces, etc. It is simple, easy to operate, and the thickness can be controlled, with good adaptability. Tan et al. [50] used the electrostatic interaction between sodium alginate (SA) and chitosan (CS) to form a stable modification layer on the surface of liposomes. Zhang et al. [51] used polyurethane as the base material to alternately deposit heparin and chitosan onto a decellularized scaffold (PU/DCS).

2.4. Sol–Gel Method

Principle: The hydrolysis and polycondensation reaction of the precursor in solution are used to form a gel, in which the natural medicine is embedded, and then applied to the surface of the biomaterial. Natural medicines are mixed with sols and subjected to steps such as gelation, drying, and heat treatment to form a coating. Violav et al. [52] used tetraethyl orthosilicate as a precursor of silica by the sol–gel method, in which polyvinyl alcohol was dissolved in ethanol and added to the silicon matrix, and green oxalic acid was slowly added to the mixed solution and, finally, an organic–inorganic material was prepared. The sol–gel method can be used to prepare nanoparticles and encapsulate drugs to achieve targeted drug delivery [53]. The porosity of the coating prepared by this method is controllable, and the coating uniformity is good, but the reaction conditions need to be strictly controlled.

2.5. Electrospinning

In the electrospinning process, a solution or melt containing natural medicines and polymers is placed in a syringe with a metal needle attached to the front end of the syringe to apply a high voltage between the needle and the receiving device. Under the force of an electric field, a polymer solution or melt forms a Taylor cone from the tip of the needle and ejects a charged trickle. As these streams fly towards the receiving device, the solvent volatilizes or the melt cools and solidifies, forming continuous fibers that are deposited on the receiving device, resulting in a fiber coating containing natural medicines [54,55]. For example, curcumin is mixed with a polylactic acid–glycolic acid copolymer (PLGA) to prepare an electrospinning solution [56]. The electrospinning process was used to obtain a coating of PLGA fibers containing curcumin. Curcumin has biological activities, such as anti-inflammatory and antioxidant activities, and this coating can be applied to tissue-engineered scaffolds to promote tissue repair and regeneration.

3. Natural Medicines and Biomaterials Form Coatings and Their Applications

Loading coatings on the surface of biomaterials not only changes the physical and chemical properties of the materials but also affects their biocompatibility and biointegration capabilities. The implantation of biomaterials with surface coatings can improve the compatibility of materials, reduce immune responses, better bind to cells and tissues, and reduce postoperative complications and the need for reoperation. It can also enhance the mechanical properties and corrosion resistance of the material. In addition, surface coatings can endow biomaterials with specific functions, such as using natural medicines with antibacterial, anti-inflammatory, and tissue-regeneration properties to achieve functional therapeutic effects when combined with biomaterials. Here, some common natural medicines that work in combination with biomaterials based on their biological properties are used, such as by forming an antimicrobial coating that inhibits the growth of bacteria and can reduce the risk of infection. Combining these with biomaterials facilitates the reduction of inflammatory responses based on the resistance properties of natural medicines. According to the needs of functionalization, the controlled release of drugs can be realized to achieve effective treatment.

3.1. Curcumin

Curcumin is a natural polyphenolic compound, often found in turmeric rhizomes, the chemical formula of curcumin is $C_{21}H_{20}O_6$. Curcumin can not only be used as a spice in food, but also in medicine and health, because of its multiple biological activities. Curcumin has antioxidant, anti-inflammatory, and antibacterial properties, and has a certain protective effect on Alzheimer's disease and Parkinson's disease [57–61].

Studies have proved that curcumin has broad-spectrum antibacterial activity, and has an antibacterial effect on both Gram-positive and -negative bacteria [62,63]. Curcumin can destroy the structure and permeability of bacterial cell walls and cell membranes, interfere with the synthesis of bacterial nucleic acids, proteins and fatty acids, and inhibit the formation of bacterial biofilms [62–65]. Kandaswamy et al. [66] loaded curcumin and berberine chloride into polymer nanofibers, and curcumin and berberine chloride were able to reduce the production of extracellular polymeric plasma (EPS), which would damage the protective barrier of the biofilm and lead to the instability of the overall structure of the biofilm. Nanofibers have a significant antibacterial effect on methicillin-resistant Staphylococcus aureus (MRSA). W et al. [67] encapsulated montmorillonite (MMT), L-malic acid (LMA) and curcumin (Cur) into a bacterial cellulose (BC) matrix to prepare multifunctional nanofilms; curcumin can interact with bacterial proteins and bind to deoxyribonucleic acid molecules to destroy the cell wall and cell membrane of bacteria [64], thereby destroying the cell membrane and inhibiting the growth of Staphylococcus aureus and *Escherichia coli*. Simin et al. [68] prepared hydroxyapatite–gelatin–curcumin nanofiber composites by electrospinning, which were released from the composites through a two-step mechanism, which would enter the bacteria and cause damage to the bacteria by the interaction between bacterial proteins and DNA. Li [69] uses a single-step precipitation process to load curcumin onto the surface of hydroxyapatite (HA), and curcumin-functionalized HA can inhibit the cell attachment of Staphylococcus aureus and pseudomonas aeruginosa, which can be used as an antimicrobial agent to control the risk of postoperative infection. Curcumin is loaded on a polyvinyl alcohol/cellulose nanocrystal (PVA/CNC) membrane, which is concentration-dependent on curcumin, with decreased activity of breast cancer cells and liver cancer cells at 8 mg/mL. A chitosan/curcumin multilayer coating was loaded on PEEK/BG/h-BN by electrophoretic deposition method, and the antimicrobial activity of the coating was tested according to the optical density, and it was found that the multilayer coating had a good antibacterial effect on Gram-positive and Gram-negative bacteria, because curcumin changed the permeability of the bacterial cell membrane and entered the bacteria to destroy DNA activity, resulting in the death of bacteria [70], so the chitosan/curcumin-PEEK/BG/H-BN coating can be used for antiseptic orthopedic implants [71].

Curcumin's ability to inhibit the production of inflammatory cytokines, such as interleukin-1 (IL-1) and interleukin-1 (IL-6) and TNF-α, as well as the significant inhibition of other inflammatory mediators such as nuclear factor (NF-κB), cyclooxygenase 2 (COX2), and matrix metalloproteinase-9 (MMP 9), can reduce inflammation at the implant site. Li et al. [72] prepared polychin–curcumin-loaded PCL-PEG nanofibers by electrospinning technology, and found that the drug-loaded nanofiber materials affected the expression of IL-6, MMP-2, TIMP-1, TIMP-2 and iNOS genes in multiple stages of wound healing, and could shorten the time of wound healing. Curcumin-loaded lipid–poly(lactic-*co*-glycolic acid) (PLGA; Cur@MPs) hybrid microparticles (MPs) fabricated using an oil-in-water single emulsion method are embedded into gelatin-based scaffolds. It was found that the scaffold can regulate the Nrf2/HO-1 signaling axis, inhibit the secretion of pro-inflammatory factors by macrophages, and inhibit the migration and angiogenesis of vascular endothelial cells, which can be used to prevent complications after corneal transplantation. Li et al. [73], found that the novel material can activate the Nrf2/ARE pathway, inhibit inflammation, reduce reactive oxygen species to reduce cellular oxidative stress, and can be used in wound dressings to promote wound healing. Xian mou Fan et al. [74] prepared a new type of polyvinyl alcohol (PVA)-chitosan (CS)/sodium alginate (SA)-Cur hydrogel (PCSA), which can convert pro-inflammatory M1 macrophages into anti-inflammatory M2 macrophages, and was found to be able to clear ROS, down-regulate IL-1β, and up-regulate CD31 expression in a full-thickness wound model of rat diabetes, which can accelerate wound healing and promote angiogenesis and collagen deposition. The hydrogel can be used for diabetic wound healing. Nanoparticle curcumin was loaded onto the surface of titanium implants for modification, and curcumin was found to enhance the nuclear translocation and DNA binding activities of Runx2 and osterix, which can be used to enhance the expression of osteoblast and bone-matrix protein genes [75]. First, hydroxyapatite was used to modify the orthopedic implants of titanium and alloys to enhance the biological activity, and then curcumin (Cur) and epigallocatechin gallate (EGCG) were added, and it was found that the release of curcumin and EGCG could promote the growth of osteoblasts, and curcumin could enhance the expression of osteocalcin and Runx2 to promote the proliferation and differentiation of osteoblasts, which can be used to treat bone tumors [76]. First, Gao et al. [77] designed a double-layered drug-loaded RSF coating on the PET surface and encapsulated the anti-inflammatory drug (curcumin) and tissue release factor (Zn+) in the REF coating, as shown in Figure 2. The release of curcumin and Zn+ can inhibit the inflammatory response and promote tissue regeneration.

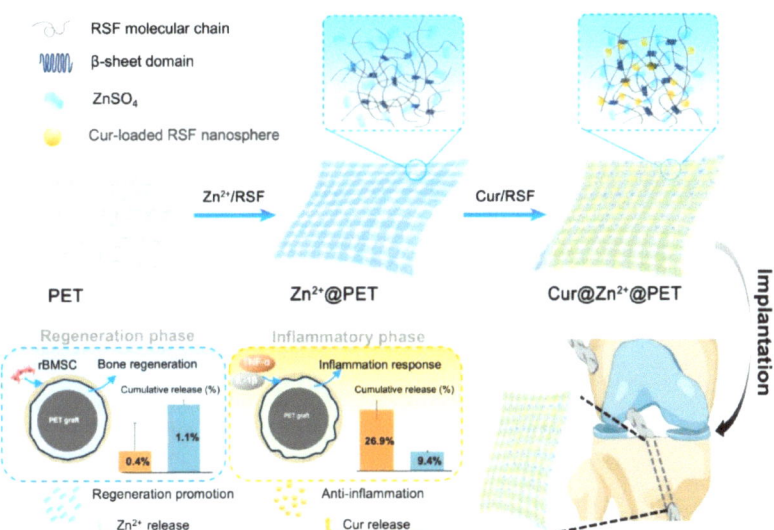

Figure 2. A dual-loaded multi-layered RSF coating with curcumin and Zn^{2+} on PET grafts, which followed a time-programmed pattern of drugs release, could intervene anti-inflammatory and tissue regeneration in a time-matched way, and ultimately improve graft–host integration [77].

Zhang Ling et al. [78] synthesized monocarbonyl curcumin analog A2 and found that curcumin analog A2 can increase the level of ROS in endothelial cells, thereby inducing cell death and exerting antiangiogenic activity. The curcumin analog A13 activates the Nrf2/ARE pathway, thereby reducing oxidative damage to ameliorate myocardial fibrosis in diabetic rats [79]. Chen et al. [80] evaluated the anti-tumor effect of WZ26 in both CCA cells and a CCA xenograft mouse model. WZ26 significantly increased ROS in the CCA cancer cells, thereby inducing mitochondrial apoptosis and inhibiting cancer cell growth both in vitro and in vivo.

3.2. Allicin

Allicin is a sulfur-containing organic matter extracted from garlic, with a strong garlic odor; its chemical name is diallyl thiosulfinate, its chemical formula is $C_6H_{10}S_{20}$, and its functional groups include the sulfide bond (-S-) and sulfonate group (-SO-). Allicin inhibits and kills a variety of bacteria, fungi, and viruses, as well as reducing the risk of atherosclerosis by regulating blood lipids to reduce blood cholesterol and triglyceride levels. It inhibits the aggregation of platelets, reduces blood viscosity, and prevents the formation of thrombosis; it can scavenge free radicals in the body, reduce oxidative stress to cells, and delay aging [81–86].

Allicin has a broad-spectrum antibacterial effect on a variety of bacteria, such as Gram-positive and -negative bacteria, and the application of allicin to wound dressings can stimulate the proliferation of fibroblasts and the synthesis of collagen, and promote the growth of granulation tissue. Pang et al. [87] prepared a multifunctional nanoplatform for Allicin@ZIF-8/AgNPs, and the release of allicin can reduce the level of ROS to inhibit the production of cytokines and promote the production of collagen fibers, stimulate the production of new blood vessels, and improve the speed of wound healing. Wang et al. [88] prepared a micellar composite hydrogel loaded with allicin; the release of allicin can be antibacterial, reduce ROS, and promote cell metabolism. According to the constructed rat full-thickness skin injury model, it was found that the wound healing time was shortened, and the color of granulation tissue was lightened by HE staining. Therefore, allicin-loaded micellar composite hydrogel can be used as a wound dressing to accelerate wound healing. Liu Yongxu et al. [89] added allicin (All) to chitosan (CS)/polyvinyl alcohol

(PVA)/graphene oxide (GO) composites, and used electrospinning technology to prepare nanofiber membranes, which have strong antibacterial effects according to antibacterial experiments and allicin in vitro release surface nanofiber membranes, which can be used for antibacterial wound dressings and tissue engineering.

Allicin can upregulate alkaline phosphatase (ALP) and osteocalcin (BGLAP) to promote the proliferation and differentiation of mesenchymal stem cells (hMSCs), and the release of allicin was found to promote collagen formation in a rat distal femur model [90]. Allicin was applied to the treatment of mouse models of cerebral hemorrhage, and it was found that the injection of allicin had a neuroprotective effect, which could inhibit the accumulation of cells around the hematoma and reduce axonal damage [91]. It also inhibits pro-inflammatory factors in the brain, such as interleukin 6 and C-X-C motif ligand 2 in the brain. Therefore, it can be applied to the treatment of intracerebral hemorrhage disorders. The nanofiber wound dressing was prepared by using polycaprolactone (PLC) and silk fibroin, and the antibacterial effect could be used to treat burns because the water contact angle was reduced to 85.5° and the Young's modulus was increased to 5.12 MPa after the addition of allicin [92]. A hydroxyapatite drug delivery system loaded with allicin was prepared, and polycaprolactone (PCL) increased the kinetics of allicin release, increasing the release of allicin from 35% to 70%, and antibacterial experiments found that allicin had an excellent antibacterial effect on Staphylococcus aureus, and had good compatibility with osteoblasts, and could be used as a bone-tissue engineering scaffold to exert antibacterial, drug release, and compatibility characteristics [92]. In the treatment of a rat diabetic wound model, it was found that the loading of allicin dressing improved the wound closure rate, and the epithelial cell thickness and collagen deposition were higher than those of the non-allicin dressing group, so the clinical application of a allicin/chitosan/polyvinyl alcohol dressing in diabetic wounds has great potential [93]. Using a 3D-printed bone-tissue engineering scaffold to load curcumin and allicin, the drug release can promote the proliferation of osteoblasts, reduce the activity of osteoclasts, and enhance osseointegration, which can be used for low-load bone-tissue engineering and dental applications. Using the porous structure of the white flesh of grapefruit peel as a scaffold, a new type of natural antibacterial patch was prepared by using the antibacterial properties of allicin, and it was found that the natural antibacterial patch could effectively kill Gram-negative and -positive bacteria, and also had anti-skin infection activity [94]. Han et al. [95] prepared C-HA-Cys hydrogel coatings by an amide reaction using catechol hyaluronic acid (C-HA) and cystine (Cys). The H2S-releasing donor allicin is loaded into the hydrogel to form a smart biomimetic coating (this is shown in Figure 3). Loading the surface of the vascular interventional device with a C-HA-Cys hydrogel coating enables drug release and alters the lesion microenvironment, thereby preventing stent restenosis.

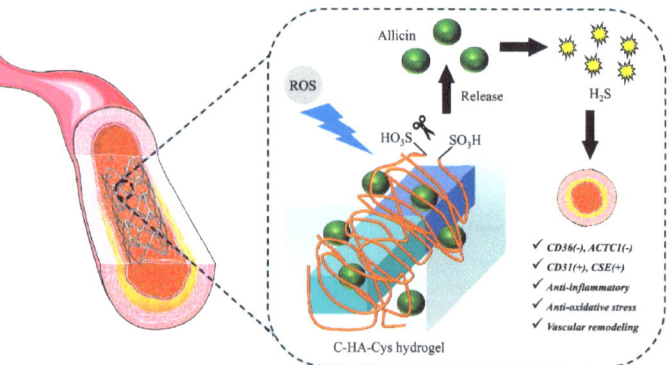

Figure 3. C-HA-Cys hydrogel coatings were prepared by an amide reaction using catechol hyaluronic acid (C-HA) and cystine (Cys). The H$_2$S-releasing donor allicin is loaded into the hydrogel to form a smart biomimetic coating [95].

3.3. Berberine

Berberine is an alkaloid extracted from various plants such as Coptis chinensis and Phellodendron chinensis. The molecular formula is $C_{20}H_{18}NO_4$, containing quaternary ammonium salt functional groups, and the taste is extremely bitter. Berberine can exert neuroprotective effects by upregulating the survival of PI3K/Akt/Bcl-2 cells and the Nrf2/HO-1 antioxidant signaling pathway [96]. Berberine can lower blood glucose, maintain homeostasis, inhibit the release of inflammatory mediators (TNF-α, NF-κB, phosphorylated NF-κB-p65, cox-2 and iNOS) [97], improve atherosclerosis, protect cardiovascular and cerebrovascular vessels [98], and interfere with bacterial metabolic processes, such as destroying the structural integrity of bacteria to inhibit bacterial proliferation.

Berberine has a wide range of pharmacological effects [99–101], such as adding berberine to the surface coating of medical devices (urinary catheters, orthopedic implants), which can effectively inhibit the reproduction of bacteria on the surface of materials. The application of berberine in the coating of tissue-engineered materials can accelerate wound healing and tissue integration; medical devices that come into contact with blood (such as cardiac and cerebrovascular stents) inhibit the aggregation of platelets and the activity of clotting factors, thereby reducing the risk of thrombosis; it can also be used in combination with other bioactive ingredients to prepare multifunctional composite coatings. Zhong Xiaojun et al. [102] isolated Coptisine, an alkaloid from Coptis chinensis, used the fluorescent probe PI to detect it, and found that the PI signal in Cryptococcus cells treated with Coptisine increased, and the surface Coptisine changed the permeability of the bacterial membrane, resulting in the leakage of key substances in the cell and cell death. In addition, Coptisine can also lead to an excessive accumulation of ROS in Cryptococcus cells, and lipid peroxidation targets mitochondria and nuclei, among others, and ultimately leads to cell death. Dong et al. [103] found that berberine could increase the expression of Runx2, activate the typical Wnt/β-catenin signaling pathway, and promote the osteogenic differentiation of MSCs. At the same time, berberine can alleviate hydrogen peroxide (H_2O_2)-induced apoptosis of rat bone marrow mesenchymal stem cells in vitro. Berberine is combined with nano-hydroxyapatite/chitosan bone cement to prepare an antibiotic drug delivery system for bone healing [104]. E et al. [105] loaded berberine into a PCL/Gel scaffold and evaluated the healing process after stent implantation in a rat skull defect model, and found that the hydrophilicity of berberine was improved, which could promote the proliferation of fibroblasts and osteoblasts, and provide the possibility for bone-healing applications. A hydrogel was developed by combining the hypolipidemic properties of silk fibroin and the antioxidant properties of melanin with the therapeutic effect of berberine, which can be used for the wound healing of diabetic wounds [106]. The release of berberine in the hydrogel is able to protect the wound from bacterial infection. The hydrophobic and cationic nature of berberine makes microbial cell membranes permeable and stably intercalated with DNA. It inhibits the synthesis of DNA and the protein of bacteria, and has a good antibacterial effect on bacteria.

Cellulose acetate–hyaluronic acid electrospun fibers loaded with berberine were manufactured by electrospinning [107], which can reduce the production and release of inflammatory factors, block the activation of inflammatory signaling pathways such as NF-kB, and can be used as a wound dressing to improve wound-healing time [108]. By sulfonation on the surface of polyetheretherketone (PEEK) of orthopedic implants to form a spongy three-dimensional structure, osteogenic active Cnidium monnieri nanoparticles were added, and a silk fibroin–berberine coating was applied to the surface to increase antibacterial function. In rat femur implantation experiments, it was found that Ost Ber @SPEEK promoted the activity of osteoblasts and prevented bacterial infection. Functionalized PEEK-based implants were prepared for use in orthopedics to promote osteogenic differentiation (Figure 4) [109]. Titanium and titanium alloys have good corrosion resistance and biocompatibility applied to orthopedic implants, but because titanium and titanium alloys do not have antibacterial and osteogenic properties, the graphene oxide (GO) of berberine (Ber) will be loaded on the titanium surface; that is, Ber-GO@Ti, Ber-GO@Ti has strong antibacterial activity against Staphylococcus aureus [104], which can promote the osteogenic differentiation of MC3T3-E1 cells, and there are no toxic and inflammatory cells in vivo experiments. This versatile coating can be applied to orthopedic implants to function. Two-dimensional Ti_3C_2 and berberine were loaded into the 3D-printed biphasic calcium phosphate (BCP) scaffold [110], both Ti_3C_2 and BBR had good antibacterial properties, Ti_3C_2 was beneficial to the generation of new bone, and the functionalized BCP scaffold was used to treat the problems of bone-defect infection according to the antibacterial and osteogenic properties.

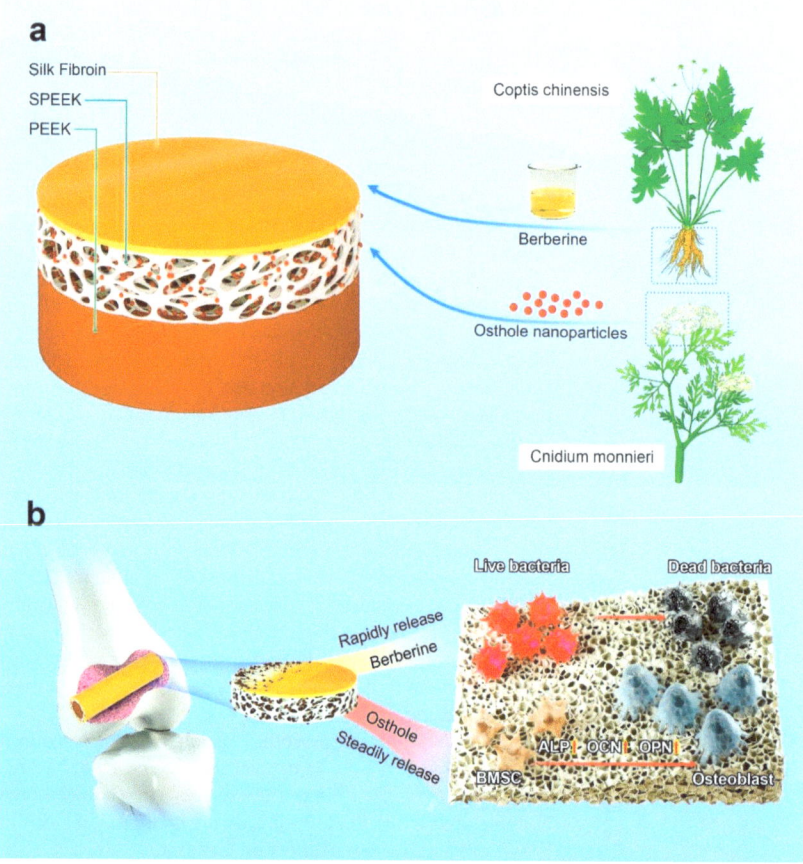

Figure 4. Polyetheretherketone (PEEK), which can be used for orthopedic implants, is selected to form a spongy three-dimensional structure on the surface through a sulfonation reaction and embedded osthole nanoparticles with osteogenic activity. The silk fibroin–berberine coating with antimicrobial function is loaded on the surface of the material [109]. (**a**) Composition of the coating; (**b**) The antibacterial and osteogenic functions of the coating.

3.4. Tea Polyphenols

Tea polyphenols are a general term for polyphenols in tea, the main components of which are catechins, which belong to the flavanols and have a basic structure of 2-phenybenzopyran. Common catechins include epicatechin (EC), epigallocatechin (EGC), epicatechin gallate (ECG), and epigallocatechin gallate (EGCG), among others, as shown in the figure. The hydrogen atom, with multiple phenolic hydroxyl groups, is easy to react with free radicals and terminate the chain reaction of free radicals, as it has strong antioxidant properties. Tea polyphenols can inhibit inflammatory factors (such as tumor necrosis factor-α (TNF-α) and interleukin-6 (IL-6), etc., and reduce the inflammatory response after stent implantation. It can regulate the function of vascular endothelial cells, increase the release of nitric oxide, inhibit the aggregation of platelets and the formation of thrombosis, and reduce the risk of cardiovascular and cerebrovascular diseases [111,112].

Zhang et al. [113] evaluated the leakage of protein, DNA, and K^+, and found that tea polyphenols (TPs) destroyed the cell membrane, resulting in the loss of intracellular components, and finally led to cell death. Using the good antimicrobial activity of tea polyphenols, it is encapsulated into electrospun fibers and can be used for wound healing. As a biodegradable metal material, magnesium alloy is susceptible to corrosion and premature leakage of implants, Zhang Bo et al. [114] loaded the active factor epigallocatechin gallate (EGCG) in green tea onto the magnesium alloy (AZ31) matrix through layer by layer self-assembly method, and the coating provided an anti-corrosion barrier for bare AZ31. After a series of experiments, it was found that the surface of the EGCG/Mg coating has good corrosion resistance and superhydrophilicity, which can inhibit the denaturation of fibrinogen, improve hemocompatibility, prolong coagulation time and reduce thrombosis, and can be applied to cardiovascular implants. Zhang et al. [114] constructed an anodizing/tea polyphenol composite layer on the surface of AZ31 magnesium alloy, and the EGCG selected not only improved the corrosion resistance of the magnesium matrix, but also improved the surface modification of the medical magnesium alloy on the implant according to its antioxidant and anti-inflammatory properties.

Ren et al. [115] co-mixed tea polyphenols with high molecular weight polyethylene for joint replacement prostheses, which use tea polyphenols to produce excess reactive oxygen species (ROS) to destroy the bacterial membrane structure and reduce bacteria-induced inflammation in the body. Li Huimin et al. [116] immersed the porous structure of carboxymethyl chitosan/sodium alginate/tea polyphenol into the impregnation solution $CaCl_2$/glycerol/ethanol to obtain CC/SA/TP compounds with antibacterial, antioxidant, and hemostatic functions, and the porous structure of CC/SA/TP can regulate wound bleeding and may be applied to wound dressings.

Wen et al. [117] used electrospinning technology to prepare a new type of composite fiber using polyurethane (PU)/polyacrylonitrile (PAN)/tea polyphenols (TPs) as raw materials, and found that the 120 μm composite fiber could continuously release TPs, which had good antibacterial properties against Gram-positive and Gram-negative bacteria, and scavenged DPPH free radicals. Based on its excellent performance, it is used in wound dressings and tissue catheters. Wang Chaoyang et al. [118] made a polyacrylic acid/polyacrylamide/tea polyphenol/Yunnan Baiyao gelatinous powder, and the hydrogel had good antibacterial properties against Staphylococcus aureus and *Escherichia coli* in the antibacterial experiment. In in vitro coagulation, a large number of red blood cells accumulate on the surface of the hydrogel to accelerate blood clotting. Based on the anti-inflammatory properties and cell proliferation of tea polyphenols and Yunnan Baiyao, biomaterials can be used to treat acute bleeding and promote wound healing. Jinhua et al. [119] used polylactic acid–glycolic acid (PLGA) nanoparticles to load tea polyphenols, and then coated platelet membranes (PMs) on the surface of nanoparticles, and the nanoparticles were targeted to accumulate in lung and vascular endothelial cells in vivo, which can reduce the secretion of inflammatory cytokines, and this biomimetic nanomaterial is expected to be applied to treat infectious diseases.

3.5. Heparin

Heparin is a mucopolysaccharide sulfate, commonly used as an anticoagulant in clinical practice, which can inhibit the binding of platelet surface receptors to thrombin substances, reduce the aggregation and adhesion of platelets, and reduce the risk of thrombosis. Heparin can bind to antithrombin III, enhance the affinity and inhibition of antithrombin III and coagulation factors, and accelerate the inactivation process of coagulation factors. The grafting of heparin to the surface of hollow fibrous membranes modifies the adsorption of surface proteins, resulting in reduced platelet adsorption [120]. Heparin significantly promotes the inactivation of antithrombin III on thrombin and inactivates FXa, FIIa, and FVIIa to inhibit blood clotting. In addition, it stimulates the release of fibrinolytic and anti-coagulant substances from the vascular endothelium by activating protein C, and regulates tissue plasminogen activators to solubilize the fibrin network. Heparin has been reported to bind to pro-inflammatory molecules on the cell surface, such as P-selectin, intercellular adhesion molecule-1 (ICAM-1), and integrin macrophage-1 antigen (Mac-1) on the cell surface, through electrostatic interactions, thereby inhibiting activated coagulation-assisted inflammation and preventing receptor-mediated pro-inflammation to reduce the risk of thrombosis [121].

Yu Wen et al. [122] deposited chitosan and heparin on the surface of 316L stainless steel through layers of self-assembly technology to form a chitosan/heparin composite coating, which prolonged thrombin time and thromboplastin time to reduce the formation of thrombosis. Liu Xingyu et al. [123] found that a heparin/chitosan coating can reduce the adhesion of Staphylococcus aureus and inhibit the formation of stones in ureteral stents by modifying the ureteral surface and retrograde. Bukola O Awonusi et al. [124] applied heparin nanoparticles to the surface of Zn-Cu alloy for modification, which can be used for urinary system implantation, and reduce bacterial infection caused by implants and reduce the occurrence of inflammation.

Polyetheretherketone (PEEK) can be used in orthopedic and dental alternatives to simulate the mechanical properties of bone, with good biocompatibility and good chemical stability [125]. Zhang Wenning et al. [126] prepared heparinized hydrogels by the free radical cross-linking polymerization (FRCP) of heparin and polydopamine. According to the catechol portion of the PDA, O_2 redox is converted to H_2O_2, and heparin creates acidic conditions that favor the conversion of H_2O_2 to -OH, enhancing the combined sterilization of cations and ROS. Through antimicrobial experiments, it was found that the presence of these two improved the antimicrobial activity of hydrogels, which was applied to implants with antithrombotic and antibacterial properties. Jin Yingying et al. [127] conjugated anticoagulant heparin (Hep) and antibacterial carboxymethyl chitosan (CMCS) to the surface of polydopamine (PDA)-coated polyurethane (PU) membrane, namely PU/PDA-Hep/CMCS, and the amine and carboxyl groups in CMCS can bind to the bacterial surface, destroy the cell membrane, induce the leakage of components in the package, and lead to cell death. HEP can promote the binding of antithrombin and thrombin involved in activating factors to exert antibacterial effects. CMCS implantable materials can significantly reduce the activity of NF-kB and down-regulate inflammatory cytokines in rabbit models. In in vitro antimicrobial experiments, CMCS implants had good antibacterial properties against *Escherichia coli* and Staphylococcus aureus. Wang et al. [128] synthesized a molecular layer composed of sodium carboxymethyl cellulose sulfate (SCMC) and chitosan (CS) on the surface of polylactic acid through layer-by-layer self-assembly method, and the modified polylactic acid membrane has good cytocompatibility, antibacterial and anti-inflammatory abilities, and can be used for the surface modification of cardiovascular implantation. Hua et al. [129] designed a smart biomimetic coating which is modified with heparin and can release NO (Figure 5). Using heparin anticoagulation and NO release to inhibit the proliferation of smooth muscle cells, the biomimetic coating can improve the microenvironment of endothelial cells, and the application to vascular stents can inhibit the formation of thrombosis and resist stenosis.

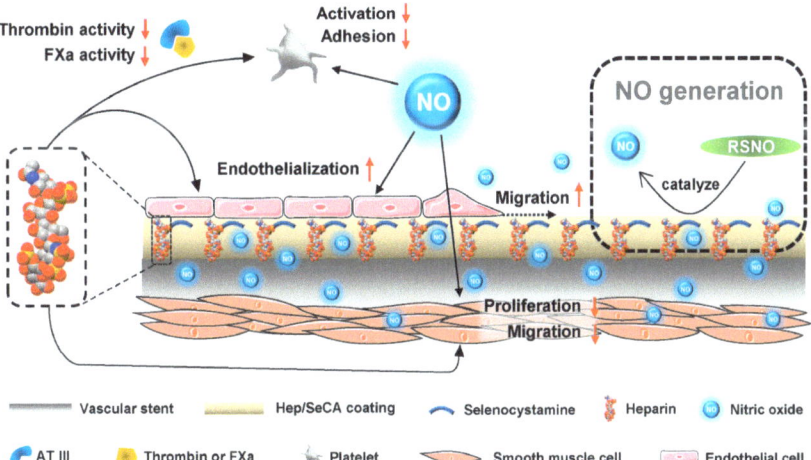

Figure 5. Biomimetic engineering of an endothelium-like coating through the synergic application of bioactive heparin and nitric oxide-generating species. The endothelium–biomimetic coating imparts the modified cardiovascular stent with the ability to combine the physiological capabilities of both heparin and NO, which creates a favorable microenvironment for inhibiting the key components in the coagulation cascade, such as Factor Xa and thrombin (Factor IIa) and platelets, as well as the growth of ECs over SMCs. These features endow the vascular stent with the abilities to impressively improve the antithrombogenicity, induce re-endothelialization, and prevent restenosis in vivo [129].

3.6. Propolis

Propolis is a natural ingredient consisting of a resinous mixture of honey compounds of various plant origins. Propolis contains a large amount of flavonoids, polyphenols, and other organic substances; propolis can change the permeability of cell membranes and lead to the leakage of cell contents, reduce energy metabolism, inhibit RNA polymerase to inhibit protein synthesis [130], and can also inhibit adenosine triphosphate (ATP) to hinder cell activities to exert antibacterial effects [131]. The antibacterial mechanism of propolis is shown in Figure 6.

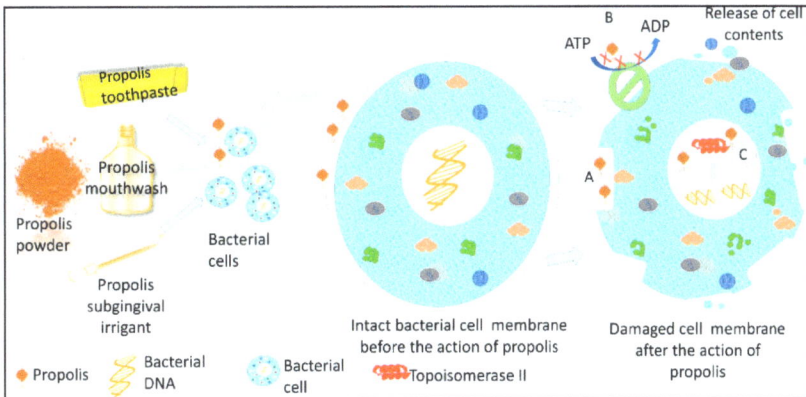

Figure 6. The mechanisms of antibacterial action of propolis—(A) propolis causes damage to the cell membrane, leading the cell contents to leak out, causing cell lysis. (B) Propolis inhibits adenosine triphosphate (ATP) formation, inhibiting mobility and the metabolism of the cell, impeding cell function (C) Propolis inhibits topoisomerase activity, causing DNA damage and mitotic failure [131].

Using the extract of propolis (EEP) as a raw material and mixing with silver nitrate to synthesize silver nanoparticles, sodium alginate was used as a fixative to fix the propolis–silver nanoparticles on the surgical suture line to prevent infection at the surgical site. According to the good antibacterial activity of propolis, propolis can be applied to the surface of porous polyurethane material to form a hydrogel, which can be used as a wound dressing to promote wound healing [132]. Propolis can effectively inhibit the proliferation of bacteria; the surface of wollastonite can produce hydroxyapatite, which can improve the viability of cells, and the use of propolis-modified wollastonite ($CaSiO_3$) 3D-printed scaffolds can be used to repair bone damage [133]. Propolis was loaded onto TiO_2 nanotubes (TNT), and it was found that the TNT loaded with propolis could reduce the expression of inflammatory factors IL-1β and TNF-α, and increase the expression of fibrinogen and osteogenic protein differentiation proteins BMP-2 and BMP-7 by staining in a rat mandible model [134].

3.7. Chitosan

Chitosan extract usually comes from the shell of crustaceans such as shrimp and crab, and chitosan is a natural polycationic compound formed by the deacetylation of chitin, which has good biocompatibility and unique film-forming properties under the action of an electric field, enabling osteoblasts to adhere and grow on the surface of the coating [135]. Chitosan is often used as a surface coating for implants, wound dressings, stents, etc., in biomaterial coatings to make better use of the biological activity of chitosan. A kind of chitosan was designed to prepare a silica–chitosan coating by the sol–gel method; the increase in chitosan content could improve the antibacterial property, and the release of silicon could promote the formation of bone [136]. The anode is made of platinum wire, the cathode is made of titanium or 316L stainless steel, the chitosan powder is dissolved in dilute hydrochloric acid, mixed with ethanol of different qualities to form an electrolyte, and a hydrophilic chitosan coating is formed by electrophoretic deposition, which has enhanced corrosion resistance with bare metal, and chitosan can be applied to the anti-corrosion of orthopedic implants [137]. Using the above cathode and anode, the electrolyte is replaced with chitosan and carbon nanotubes (MWCNTs) to prepare a hydrophilic MWCNT/chitosan coating, which can efficiently enrich calcium in the simulated body fluid, promote the formation of surface apatite layer, and can be applied to orthopedic implants to prevent corrosion [138]. Alkali heat treatment is beneficial to the combination of the negatively charged hydrate ($HTiO_3 \cdot H_2O$) on the surface of titanium and the cationic functional group (NH_3^+) of chitosan, and then the titanium is soaked in chitosan/ZnO solution for 6 h to form a composite coating; the formation of the composite coating inhibits the growth of *E. coli* and the adhesion of bacteria, and with the increase in zinc oxide content, the release of Zn^+ can combine with the negatively charged bacteria and destroy the structure, the antibacterial property is enhanced, and the compatibility with MG-63 cells is good. It can be applied to orthopedic and dental implants [139]. Khoshnood N et al. [140] used fused deposition modeling 3D-printing technology to prepare polycaprolactone/chitosan composite coating on the surface of AZ21 magnesium alloy, the corrosion resistance of the composite coating was improved, and it had antibacterial activity against both Gram-negative and -positive bacteria; the abundant functional groups on the surface of chitosan could promote the binding of cells to the scaffold surface, and the presence of chitosan on the surface of ARS staining could lead to the deposition of calcium. RT-PCR was used to detect the content of osteogenic markers (COLI [141], ALP [142], RUNX2 [143]), and it was found that the composite scaffold increased the content of osteogenic markers and promoted bone regeneration at 7 days/14 days. Therefore, 3D-printed composite scaffolds can be used for the treatment of bone defects, regeneration and repair.

Zheng Weishi et al. [144] used layer-by-layer self-assembly technology to deposit carboxymethyl chitosan, gelatin, and alginate on the surface of cotton yarn in turn to form a new type of composite dressing, and the experimental results of a mouse-liver injury model and mouse-tail docking model showed that the composite dressing had a good hemostatic effect and could promote wound healing, which was better than that of medical cotton gauze. Gao Xiang et al. [145] used a Michael addition reaction to prepare polylactic acid nanofibers with maleilated chitosan/thiolated hyaluronan composite coatings. In a diabetic mouse model, it was shown that wound healing with the composite coating is faster, and can promote angiogenesis and collagen deposition. Wang et al. [146] sequentially deposited polydopamine (PDA), ZnO nanoparticles (nZnO), and chitosan (CS)/nanocrystalline hydroxyapatite (nHA) on a titanium substrate for a bioactive coating (as shown in Figure 7). Nanocrystalline hydroxyapatite (nHA) improves the surface structure and wettability of titanium implants and inhibits the growth of bacteria. Mixed chitosan can improve cytocompatibility and promote osteoblast differentiation. Based on the antimicrobial properties, and promoting cell osteogenic differentiation, composite coatings have great potential in orthopedics and dentistry.

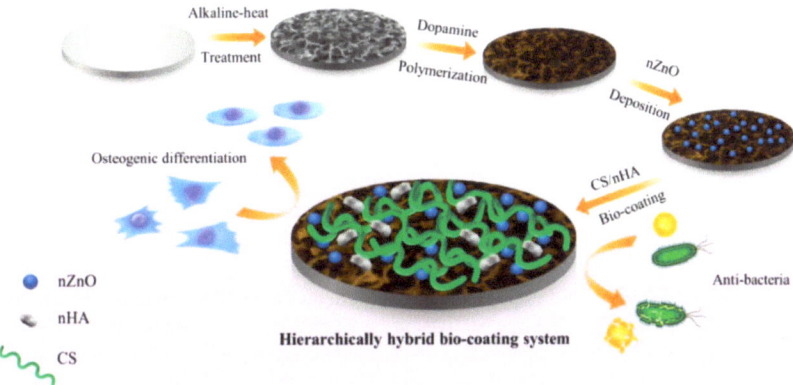

Figure 7. Hierarchically hybrid biocoatings on Ti implants are developed by gradual incorporation of polydopamine (PDA), ZnO nanoparticles (nZnO), and chitosan (CS)/nanocrystal hydroxyapatite (nHA) via oxidative self-polymerization, nanoparticle deposition, solvent casting and evaporation methods for enhancing their antibacterial activity and osteogenesis [146].

3.8. Other Applications

There are many types of natural medicines, and there are countless applications in the coating of biomaterials. The main active factor of Salvia miltiorrhizae, Tanshinone IIA(TAN), has anti-inflammatory [147], antioxidant [148], and anti-apoptotic effects [149]. TAN delivery silk-fibroin scaffolds are created by freeze-drying combined with the silk-fibroin self-crosslinking method. The prepared TAN delivery silk-fibroin scaffold can promote the activity of chondrocytes, regenerate cartilage tissue, reduce oxidative stress, and can be used to repair cartilage defects [150]. Astragalus polysaccharides in Astragalus membranaceus have anti-inflammatory [151] and angiogenic effects [152], and a nanofiber membrane composed of gelatin and Astragalus polysaccharides was prepared by electrospinning technology [153], and the release of Astragalus polysaccharides can exert anti-inflammatory effects, promote the generation of blood vessels, and can be used as wound dressings to promote wound healing. Ginkgolide is extracted from ginkgo biloba, which has antioxidant [154], anti-inflammatory [155], and proliferative properties [156], and uses hyaluronic acid and ginkgolide to form a hydrogel, which can reduce inflammation and enhance the production of blood vessels, and can be used for diabetic wound healing and promoting tissue regeneration [157]. Madian Noha G et al. [158] used chitosan mixed with different concentrations of cellulose/honey/curcumin to improve its properties, and found that the CS-Cur cross-linked film had better mechanical properties and the best antibacterial effect, which could be used as a wound dressing. Electrospun composite fibers

(PCL-COL-CUR) are prepared by incorporating curcumin into solutions of polycaprolactone (PCL) and collagen (COL), which can be used as antimicrobial dressings [159].

4. Challenges and Recommendations

With the rapid development of the biomedical field, biomaterial coatings are playing an increasingly important role in improving the performance of medical devices, promoting tissue repair, and preventing infection. Natural medicines have become a research hotspot in the field of biomaterial coatings due to their unique biological activity, low toxicity, and convenient accessibility, and common natural medicines such as berberine, allicin, chitosan, collagen, etc. can be loaded onto the surface of biomaterials by the physical adsorption method, chemical grafting method, layer-by-layer self-assembly method, the sol–gel method, and other methods to form a coating with specific functions. For example, loading the surface of orthopedic implants with natural drug coatings can promote osseointegration and reduce the risk of infection. Loading natural drugs with anticoagulant effects on cardiovascular and cerebrovascular stents can reduce the risk of thrombosis. Wound dressings are loaded with tissue-promoting drugs to accelerate wound healing and tissue regeneration.

However, the application of natural medicines to biomaterial coatings is not smooth sailing, and there are still many difficulties. The chemical structure of natural medicines is generally complex, and it is easy to be degraded or deteriorated by external factors such as light, humidity, temperature, etc., resulting in changes in biological activity. Processing conditions such as high temperature, pH, organic solvents, etc., during the preparation of coatings can cause damage to the structure and activity of natural medicines. When and how much natural medicine is released from the coating is also a major issue, as a release that is too fast may lead to side effects caused by high local concentrations, a release that is too slow may not achieve an effective effect, and bioenvironmental factors will also increase the difficulty of release. The strength of the bond between the coating and the substrate material is also critical, as if the bond is not strong, the coating will easily fall off, and if it is too strong, it will block the release of the drug load. Natural medicines have good biocompatibility, but questions remain, such as whether they will cause immune reactions or allergic reactions when they are prepared on the surface of the material and implanted in the body, whether they will disrupt certain balances in the body for a long time, and whether degradation products will cause a series of reactions in the body. The application of natural medicine involves the innovative application of medical devices, and the corresponding laws and regulations, preparation processes, and testing methods are not perfect, clinical trials may be few, and product industrialization has not yet been realized. A series of questions are raised from the negative aspects of the stability, biological activity, drug release, durability, and biocompatibility of natural medicines.

In view of the series of problems raised, future development should focus on solving these problems. With the advancement of science and technology and biotechnology, many new natural medicines have been found to have potential application value, expanding the source of new drugs. In addition, genetic engineering, synthetic biology and other methods can be used to optimize and transform natural medicines to improve biological activity and stability. The deep cross-integration of multiple disciplines promotes the new technological breakthrough of natural medicine in biomaterial coatings. For example, the application of nanotechnology can achieve precise loading and controlled release of natural medicines. The development of smart responsive materials enables coatings to automatically regulate drug release in response to changes in the in vivo environment; 3D-printing technology enables the preparation of coatings with complex structures and individualization. With the development of the concept of precision medicine, the future of biomaterial coatings will pay more attention to personalized customization. Depending on the patient's individual differences, such as the type of disease, severity of the disease, immune status, etc., the appropriate natural medicine and coating formula are selected to achieve the best treatment effect. At the same time, bioinformatics and big data analysis are used to provide patients with more accurate treatment plans. The preparation of natural medicine coatings should closely follow current development practices, focus on environmental protection and sustainability, reduce the waste of resources in the preparation process, and make more use of drug carriers of renewable resources, as well as improving the corresponding laws and regulations, preparation methods, and testing processes, increasing the number and quality of clinical trials, and slowly realizing the commercialization of products. It is of great significance to strengthen basic research and deeply explore the mechanism of action of nature. Through the study of cells and molecules, the relationship between the biological activity of natural drugs and the performance of coatings is revealed, and a theoretical basis is provided for the development of more efficient and safe coatings.

The application potential of natural medicine in biomaterial coating is huge, but it also faces some challenges. Through technological innovation, cross-integration between disciplines, the improvement of corresponding laws and regulations, strengthening the in-depth basic research, and other measures, the problems faced will be slowly solved, in order to achieve the wide application of natural medicine coating in the field of biomedicine and to contribute to the cause of human health. In the future, there will be more innovative natural medicine coatings for patients, to provide better treatment and a higher quality of life.

5. Conclusions

Natural medicines are abundant, with a wide range of sources, low toxicity, and good biological activity. The application scale of biomaterials has been popularized in all aspects of life, such as the manufacturing of artificial joints, heart valves, vascular stents in medical devices, etc. Biomaterials are used as drug carriers to achieve controlled release and targeted delivery of drugs. They are also used to construct tissue-engineering scaffolds, etc. However, a series of problems, such as the service life and biocompatibility of materials in different physiological environments, will have a certain impact. Therefore, the combination of natural medicine and biomaterials forms a protective barrier to further improve the performance and biocompatibility of medical devices. For example, curcumin and allicin on the surface of medical devices can reduce the risk of bacterial infection and inflammation. Heparin is encapsulated into a hydrogel to achieve a controlled drug release.

This review starts with the definition of natural medicine and forms coatings by using different methods according to the properties of natural medicines. The biological activity of common drugs is detailed, and many natural medicines are widely studied and used in orthopedic implants, vascular stents, wound dressings, and drug delivery. Although natural medicines have made some progress in the application of biomaterial coatings, they still face many challenges, such as the binding strength between coatings, the controllability of drug release in drug-loading systems, and biocompatibility, and a series of questions such as clinical application and whether mass production can be achieved. Constructive suggestions are made on these issues. The application of natural medicines and biomaterials will continue to be expanded in the future, and more functional coatings will be innovated upon, which will be applied to all walks of life to improve the lives, health, and quality of life of human beings.

Author Contributions: Conceptualization, J.L.; investigation, Y.W., H.D. and Z.Z.; writing—original draft preparation, Y.W.; writing—review and editing, J.L. and L.C.; supervision, J.L. and L.C.; funding acquisition, J.L. and L.C. All authors have read and agreed to the published version of the manuscript.

Funding: This research was funded by the National Natural Science Foundation of China, Grant No. 52101292/U2004164; Key Scientific and Technological Research Projects in Henan Province, Grant Nos. 232102311155 and 232102230106; Joint Founds of R&D Program of Henan Province (No. 222301420055); Zhengzhou University Major Project Cultivation Special Project, Grant No. 125-32214076.

Institutional Review Board Statement: Not applicable.

Informed Consent Statement: Not applicable.

Data Availability Statement: No new data were created or analyzed in this study.

Acknowledgments: Figure 1b reproduced from Ref. [33] with permission from Elsevier (License Number: 5878780245324); Figure 1d reproduced from Ref. [35] with permission from Elsevier (License Number: 5878790936218); Figure 1e reproduced from Ref. [36] with permission from Elsevier (License Number: 5878810372482); Figure 2 reproduced from Ref. [77] with permission from Elsevier (License Number: 5878860898599); Figure 3 was reprinted (adapted) with permission from {Han, X.; Lu, B.; Zou, D.; Luo, X.; Liu, L.; Maitz, M.F.; Yang, P.; Huang, N.; Zhao, A.S.; Chen, J. Allicin-Loaded Intelligent Hydrogel Coating Improving Vascular Implant Performance. ACS Appl. Mater. Interfaces 2023, 15, 32, 38247–38263}, copyright {2023}, the American Chemical Society; Figure 4 reproduced from Ref. [109] with permission from Elsevier (License Number: 5878871340153); Figure 5 reproduced from Ref. [129] with permission from Elsevier (License Number: 5879101122702); Figure 7 reproduced from Ref. [146] with permission from Elsevier (License Number: 5879110561069).

Conflicts of Interest: The authors declare no conflicts of interest.

References

1. Fu, Y.; Liu, T.; Wang, H.; Wang, Z.; Hou, L.; Jiang, J.; Xu, T. Applications of nanomaterial technology in biosensing. *J. Sci. Adv. Mater. Devices* **2024**, *9*, 100694. [CrossRef]
2. Vigneshvar, S.; Sudhakumari, C.C.; Senthilkumaran, B.; Prakash, H. Recent Advances in Biosensor Technology for Potential Applications—An Overview. *Front. Bioeng. Biotechnol.* **2016**, *4*, 11. [CrossRef] [PubMed]
3. Ertas, Y.N.; Mahmoodi, M.; Shahabipour, F.; Jahed, V.; Diltemiz, S.E.; Tutar, R.; Ashammakhi, N. Role of biomaterials in the diagnosis, prevention, treatment, and study of corona virus disease 2019 (COVID-19). *Emergent Mater.* **2021**, *4*, 35–55. [CrossRef] [PubMed]
4. Cai, L.; Xu, J.; Yang, Z.; Tong, R.; Dong, Z.; Wang, C.; Leong, K.W. Engineered biomaterials for cancer immunotherapy. *MedComm* **2020**, *1*, 35–46. [CrossRef]
5. Xiao, M.; Tang, Q.; Zeng, S.; Yang, Q.; Yang, X.; Tong, X.; Zhu, G.; Lei, L.; Li, S. Emerging biomaterials for tumor immunotherapy. *Biomater. Res.* **2023**, *27*, 47. [CrossRef] [PubMed]
6. Niziołek, K.; Słota, D.; Sobczak-Kupiec, A. Polysaccharide-Based Composite Systems in Bone Tissue Engineering: A Review. *Materials* **2024**, *17*, 4220. [CrossRef]
7. Gao, C.; Peng, S.; Feng, P.; Shuai, C. Bone biomaterials and interactions with stem cells. *Bone Res.* **2017**, *5*, 253–285. [CrossRef]
8. Sun, W.; Ye, B.; Chen, S.; Zeng, L.; Lu, H.; Wan, Y.; Gao, Q.; Chen, K.; Qu, Y.; Wu, B. Neuro–bone tissue engineering: Emerging mechanisms, potential strategies, and current challenges. *Bone Res.* **2023**, *11*, 65. [CrossRef]
9. Ansari, M. Bone tissue regeneration: Biology, strategies and interface studies. *Prog. Biomater.* **2019**, *8*, 223–237. [CrossRef]
10. Tang, G.; Liu, Z.; Liu, Y.; Yu, J.; Wang, X.; Tan, Z.; Ye, X. Recent Trends in the Development of Bone Regenerative Biomaterials. *Front. Cell Dev. Biol.* **2021**, *9*, 665813. [CrossRef]
11. Hou, Y.-C.; Cui, X.; Qin, Z.; Su, C.; Zhang, G.; Tang, J.-N.; Li, J.-A.; Zhang, J.-Y. Three-dimensional bioprinting of artificial blood vessel: Process, bioinks, and challenges. *Int. J. Bioprinting* **2023**, *9*, 740. [CrossRef] [PubMed]
12. Oveissi, F.; Naficy, S.; Lee, A.; Winlaw, D.; Dehghani, F. Materials and manufacturing perspectives in engineering heart valves: A review. *Mater. Today Bio* **2020**, *5*, 100038. [CrossRef] [PubMed]
13. Cordoves, E.M.; Vunjak-Novakovic, G.; Kalfa, D.M. Designing Biocompatible Tissue Engineered Heart Valves In Situ: JACC Review Topic of the Week. *J. Am. Coll. Cardiol.* **2023**, *81*, 994–1003. [CrossRef] [PubMed]
14. Shao, Z.; Tao, T.; Xu, H.; Chen, C.; Lee, I.S.; Chung, S.; Dong, Z.; Li, W.; Ma, L.; Bai, H. Recent progress in biomaterials for heart valve replacement: Structure, function, and biomimetic design. *View* **2021**, *2*, 20200142. [CrossRef]
15. Anaya-Alaminos, R.; Ibáñez-Flores, N.; Aznar-Peña, I.; González-Andrades, M. Antimicrobial biomaterials and their potential application in ophthalmology. *J. Appl. Biomater. Funct. Mater.* **2015**, *13*, 346–350. [CrossRef]
16. Bapat, R.A.; Chaubal, T.V.; Dharmadhikari, S.; Abdulla, A.M.; Bapat, P.; Alexander, A.; Dubey, S.K.; Kesharwani, P. Recent advances of gold nanoparticles as biomaterial in dentistry. *Int. J. Pharm.* **2020**, *586*, 119596. [CrossRef]
17. Robert, G.; Christian, K.; Anders, H.; Ralf, S.; Max, H.; Tessa, H. Approaches to Peripheral Nerve Repair: Generations of Biomaterial Conduits Yielding to Replacing Autologous Nerve Grafts in Craniomaxillofacial Surgery. *BioMed Res. Int.* **2016**, *2016*, 3856262.
18. Dai, W.; Yang, Y.; Yang, Y.; Liu, W. Material advancement in tissue-engineered nerve conduit. *Nanotechnol. Rev.* **2021**, *10*, 488–503. [CrossRef]
19. Carvalho, C.R.; Oliveira, J.M.; Reis, R.L. Modern Trends for Peripheral Nerve Repair and Regeneration: Beyond the Hollow Nerve Guidance Conduit. *Front. Bioeng. Biotechnol.* **2019**, *7*, 337. [CrossRef]
20. Zhang, Z.; Zhong, X.; Li, L.; Hu, J.; Peng, Z. Unmasking the delamination mechanisms of a defective coating under the co-existence of alternating stress and corrosion. *Prog. Org. Coat.* **2023**, *180*, 107560. [CrossRef]
21. Williams, G.; Kousis, C.; McMurray, N.; Keil, P. A mechanistic investigation of corrosion-driven organic coating failure on magnesium and its alloys. *npj Mater. Degrad.* **2019**, *3*, 41. [CrossRef]

22. Nazeer, A.A.; Madkour, M. Potential use of smart coatings for corrosion protection of metals and alloys: A review. *J. Mol. Liq.* **2018**, *253*, 11–22. [CrossRef]
23. Nathanael, A.J.; Oh, T.H. Biopolymer coatings for biomedical applications. *Polymers* **2020**, *12*, 3061. [CrossRef] [PubMed]
24. Feng, H.; Mohan, S. Application of process analytical technology for pharmaceutical coating: Challenges, pitfalls, and trends. *AAPS PharmSciTech* **2020**, *21*, 179. [CrossRef] [PubMed]
25. Felton, L.A.; Porter, S.C. An update on pharmaceutical film coating for drug delivery. *Expert Opin. Drug Deliv.* **2013**, *10*, 421–435. [CrossRef]
26. Zheng, T.-X.; Li, W.; Gu, Y.-Y.; Zhao, D.; Qi, M.-C. Classification and research progress of implant surface antimicrobial techniques. *J. Dent. Sci.* **2022**, *17*, 1–7. [CrossRef] [PubMed]
27. Deng, W.; Du, H.; Liu, D.; Ma, Z. The role of natural products in chronic inflammation. *Front. Pharmacol.* **2022**, *13*, 901538. [CrossRef]
28. Cao, F.; Gui, S.-Y.; Gao, X.; Zhang, W.; Fu, Z.-Y.; Tao, L.-M.; Jiang, Z.-X.; Chen, X.; Qian, H.; Wang, X. Research progress of natural product-based nanomaterials for the treatment of inflammation-related diseases. *Mater. Des.* **2022**, *218*, 110686. [CrossRef]
29. Mehta, J.; Rayalam, S.; Wang, X. Cytoprotective effects of natural compounds against oxidative stress. *Antioxidants* **2018**, *7*, 147. [CrossRef]
30. Luo, M.; Zheng, Y.; Tang, S.; Gu, L.; Zhu, Y.; Ying, R.; Liu, Y.; Ma, J.; Guo, R.; Gao, P. Radical oxygen species: An important breakthrough point for botanical drugs to regulate oxidative stress and treat the disorder of glycolipid metabolism. *Front Pharmacol.* **2023**, *14*, 1166178. [CrossRef]
31. Meier, B.P.; Lappas, C.M. The influence of safety, efficacy, and medical condition severity on natural versus synthetic drug preference. *Med. Decis. Mak.* **2016**, *36*, 1011–1019. [CrossRef]
32. Li, M.; Jiang, M.; Gao, Y.; Zheng, Y.; Liu, Z.; Zhou, C.; Huang, T.; Gu, X.; Li, A.; Fang, J. Current status and outlook of biodegradable metals in neuroscience and their potential applications as cerebral vascular stent materials. *Bioact. Mater.* **2022**, *11*, 140–153. [CrossRef]
33. Alavi, S.E.; Panah, N.; Page, F.; Gholami, M.; Dastfal, A.; Sharma, L.A.; Shahmabadi, H.E. Hydrogel-based therapeutic coatings for dental implants. *Eur. Polym. J.* **2022**, *181*, 111652. [CrossRef]
34. Elkhoury, K.; Morsink, M.; Sanchez-Gonzalez, L.; Kahn, C.; Tamayol, A.; Arab-Tehrany, E. Biofabrication of natural hydrogels for cardiac, neural, and bone Tissue engineering Applications. *Bioact. Mater.* **2021**, *6*, 3904–3923. [CrossRef]
35. Kim, J.-N.; Lee, J.; Lee, H.; Oh, I.-K. Stretchable and self-healable catechol-chitosan-diatom hydrogel for triboelectric generator and self-powered tremor sensor targeting at Parkinson disease. *Nano Energy* **2021**, *82*, 105705. [CrossRef]
36. Roy, S.; Priyadarshi, R.; Lopusiewicz, L.; Biswas, D.; Chandel, V.; Rhim, J.-W. Recent progress in pectin extraction, characterization, and pectin-based films for active food packaging applications: A review. *Int. J. Biol. Macromol.* **2023**, *239*, 124248. [CrossRef]
37. Chen, X.; Zhou, J.; Qian, Y.; Zhao, L. Antibacterial coatings on orthopedic implants. *Mater. Today Bio* **2023**, *19*, 100586. [CrossRef]
38. Cazzola, M.; Ferraris, S.; Boschetto, F.; Rondinella, A.; Marin, E.; Zhu, W.; Pezzotti, G.; Verne, E.; Spriano, S. Green Tea Polyphenols Coupled with a Bioactive Titanium Alloy Surface: In Vitro Characterization of Osteoinductive Behavior through a KUSA A1 Cell Study. *Int. J. Mol. Sci.* **2018**, *19*, 2255. [CrossRef] [PubMed]
39. Saha, K.; Ghosh, A.; Bhattacharya, T.; Ghosh, S.; Dey, S.; Chattopadhyay, D. Ameliorative effects of clindamycin-nanoceria conjugate: A ROS responsive smart drug delivery system for diabetic wound healing study. *J. Trace Elem. Med. Biol.* **2023**, *75*, 127101. [CrossRef] [PubMed]
40. Fauzi, F.; Ayu, E.S.; Hidayat, H.; Musawwa, M.M.; Swastika, P.E.; Dwandaru, W.S.B. Synthesis of polyacrylamide/graphene oxide/clove essential oil composite via physical adsorption method for potential antibacterial packaging applications. *Nano-Struct. Nano-Objects* **2022**, *32*, 100908. [CrossRef]
41. Liu, S.-S.; Chen, J.; Lin, X.-D.; Liu, H.-J.; Zeng, D.-D. Construction and Antibacterial Application of Drug-loading Chitosan/Silk Nanofiber Multilayer Film. *Acta Polym. Sin.* **2022**, *53*, 1459–1465.
42. Shahzadi, L.; Bashir, M.; Tehseen, S.; Zehra, M.; Mehmood, A.; Chaudhry, A.A.; ur Rehman, I.; Yar, M. Thyroxine impregnated chitosan-based dressings stimulate angiogenesis and support fast wounds healing in rats: Potential clinical candidates. *Int. J. Biol. Macromol.* **2020**, *160*, 296–306. [CrossRef]
43. Nijhuis, A.W.; van den Beucken, J.J.; Boerman, O.C.; Jansen, J.A.; Leeuwenburgh, S.C. 1-Step versus 2-step immobilization of alkaline phosphatase and bone morphogenetic protein-2 onto implant surfaces using polydopamine. *Tissue Eng. Part C Methods* **2013**, *19*, 610–619. [CrossRef] [PubMed]
44. Masters, K.S. Covalent Growth Factor Immobilization Strategies for Tissue Repair and Regeneration. *Macromol. Biosci.* **2011**, *11*, 1149–1163. [CrossRef] [PubMed]
45. Hu, C.; Ruan, R.; Wang, W.; Gao, A.; Xu, L. Electrochemical grafting of poly(glycidyl methacrylate) on a carbon-fibre surface. *Rsc Adv.* **2020**, *10*, 10599–10605. [CrossRef]
46. Liang, G.; Yao, F.; Qi, Y.; Gong, R.; Li, R.; Liu, B.; Zhao, Y.; Lian, C.; Li, L.; Dong, X.; et al. Improvement of Mechanical Properties and Solvent Resistance of Polyurethane Coating by Chemical Grafting of Graphene Oxide. *Polymers* **2023**, *15*, 882. [CrossRef] [PubMed]
47. Wang, H.; Yan, B.; Hussain, Z.; Wang, W.; Chang, N. Chemically graft aminated GO onto dehydro-fluorinated PVDF for preparation of homogeneous DF-PVDF/GO-NH$_2$ ultrafiltration membrane with high permeability and antifouling performance. *Surf. Interfaces* **2022**, *33*, 102255. [CrossRef]

48. Nitta, S.; Taniguchi, S.; Iwamoto, H. Preparation of hydrogel using catechin-grafted chitosan and carboxymethyl cellulose. *Macromol. Res.* **2024**, *32*, 703–715. [CrossRef]
49. Nie, C.; Shen, T.; Hu, W.; Ma, Q.; Zhang, J.; Hu, S.; Tian, H.; Wu, H.; Luo, X.; Wang, J. Characterization and antibacterial properties of epsilon-poly-L-lysine grafted multi-functional cellulose beads. *Carbohydr. Polym.* **2021**, *262*, 117902. [CrossRef]
50. Tan, X.; Liu, Y.; Wu, X.; Geng, M.; Teng, F. Layer-by-layer self-assembled liposomes prepared using sodium alginate and chitosan: Insights into vesicle characteristics and physicochemical stability. *Food Hydrocoll.* **2024**, *149*, 109606. [CrossRef]
51. Zhang, J.; Wang, D.; Jiang, X.; He, L.; Fu, L.; Zhao, Y.; Wang, Y.; Mo, H.; Shen, J. Multistructured vascular patches constructed via layer-by-layer self-assembly of heparin and chitosan for vascular tissue engineering applications. *Chem. Eng. J.* **2019**, *370*, 1057–1067. [CrossRef]
52. Viola, V.; D'Angelo, A.; Piccirillo, A.M.; Catauro, M. Si/Polymer/Natural Drug Materials Prepared by Sol–Gel Route: Study of Release and Antibacterial Activity. *Macromol. Symp.* **2024**, *413*, 2300251. [CrossRef]
53. Catauro, M.; Ciprioti, S.V. Characterization of Hybrid Materials Prepared by Sol-Gel Method for Biomedical Implementations. A Critical Review. *Materials* **2021**, *14*, 1788. [CrossRef] [PubMed]
54. Xue, J.; Wu, T.; Dai, Y.; Xia, Y. Electrospinning and Electrospun Nanofibers: Methods, Materials, and Applications. *Chem. Rev.* **2019**, *119*, 5298–5415. [CrossRef]
55. Ahmadi Bonakdar, M.; Rodrigue, D. Electrospinning: Processes, structures, and materials. *Macromol* **2024**, *4*, 58–103. [CrossRef]
56. Duygulu, N.E.; Ciftci, F.; Ustundag, C.B. Electrospun drug blended poly(lactic acid) (PLA) nanofibers and their antimicrobial activities. *J. Polym. Res.* **2020**, *27*, 232. [CrossRef]
57. Yang, Y.; Shi, Y.; Cao, X.; Liu, Q.; Wang, H.; Kong, B. Preparation and functional properties of poly(vinyl alcohol)/ethyl cellulose/tea polyphenol electrospun nanofibrous films for active packaging material. *Food Control* **2021**, *130*, 108331. [CrossRef]
58. Slowing, K.; Gomez, F.; Delgado, M.; Fernandez de la Rosa, R.; Hernandez-Martin, N.; Pozo, M.A.; Garcia-Garcia, L. PET Imaging and Neurohistochemistry Reveal that Curcumin Attenuates Brain Hypometabolism and Hippocampal Damage Induced by Status Epilepticus in Rats. *Planta Med.* **2023**, *89*, 364–376. [CrossRef]
59. Ponsiree, J.; Piyapan, S.; Hasriadi; Chawanphat, M.; Worathat, T.; Pasarapa, T.; Opa, V.; Pornchai, R. Physicochemical investigation of a novel curcumin diethyl γ-aminobutyrate, a carbamate ester prodrug of curcumin with enhanced anti-neuroinflammatory activity. *PLoS ONE* **2022**, *17*, e0265689.
60. Ak, T.; Gulcin, I. Antioxidant and radical scavenging properties of curcumin. *Chem.-Biol. Interact.* **2008**, *174*, 27–37. [CrossRef]
61. Kunnumakkara, A.B.; Bordoloi, D.; Padmavathi, G.; Monisha, J.; Roy, N.K.; Prasad, S.; Aggarwal, B.B. Curcumin, the golden nutraceutical: Multitargeting for multiple chronic diseases. *Br. J. Pharmacol.* **2017**, *174*, 1325–1348. [CrossRef]
62. Al-Thubaiti, E.H. Antibacterial and antioxidant activities of curcumin/Zn metal complex with its chemical characterization and spectroscopic studies. *Heliyon* **2023**, *9*, e17468. [CrossRef]
63. Urosevic, M.; Nikolic, L.; Gajic, I.; Nikolic, V.; Dinic, A.; Miljkovic, V. Curcumin: Biological Activities and Modern Pharmaceutical Forms. *Antibiotics* **2022**, *11*, 135. [CrossRef] [PubMed]
64. Zheng, D.; Huang, C.; Huang, H.; Zhao, Y.; Khan, M.R.U.; Zhao, H.; Huang, L. Antibacterial Mechanism of Curcumin: A Review. *Chem. Biodivers.* **2020**, *17*, e2000171. [CrossRef]
65. Dai, C.; Lin, J.; Li, H.; Shen, Z.; Wang, Y.; Velkov, T.; Shen, J. The Natural Product Curcumin as an Antibacterial Agent: Current Achievements and Problems. *Antioxidants* **2022**, *11*, 459. [CrossRef] [PubMed]
66. Kandaswamy, K.; Panda, S.P.; Subramanian, R.; Khan, H.; Shaik, M.R.; Hussain, S.A.; Guru, A.; Arockiaraj, J. Synergistic berberine chloride and Curcumin-Loaded nanofiber therapies against Methicillin-Resistant Staphylococcus aureus Infection: Augmented immune and inflammatory responses in zebrafish wound healing. *Int. Immunopharmacol.* **2024**, *140*, 112856. [CrossRef]
67. Wasim, M.; Shi, F.; Liu, J.; Zhu, K.; Liu, J.; Yan, T. Synthesis of a novel multifunctional montmorillonite/L-malic-acid/curcumin/bacterial cellulose hybrid nanofilm with excellent heat insulation, antibacterial activity and cytocompatibility. *Colloid Polym. Sci.* **2023**, *301*, 893–908. [CrossRef]
68. Sharifi, S.; Khosroshahi, A.Z.; Dizaj, S.M.; Rezaei, Y. Preparation, Physicochemical Assessment and the Antimicrobial Action of Hydroxyapatite-Gelatin/Curcumin Nanofibrous Composites as a Dental Biomaterial. *Biomimetics* **2022**, *7*, 4. [CrossRef]
69. Lee, W.H.; Rohanizadeh, R.; Loo, C.Y. In situ functionalizing calcium phosphate biomaterials with curcumin for the prevention of bacterial biofilm infections. *Colloids Surf. B Biointerfaces* **2021**, *206*, 111938. [CrossRef]
70. Hussein, Y.; Loutfy, S.A.; Kamoun, E.A.; El-Moslamy, S.H.; Radwan, E.M.; Elbehairi, S.E.I. Enhanced anti-cancer activity by localized delivery of curcumin form PVA/CNCs hydrogel membranes: Preparation and in vitro bioevaluation. *Int. J. Biol. Macromol.* **2021**, *170*, 107–122. [CrossRef]
71. Virk, R.S.; Rehman, M.A.U.; Munawar, M.A.; Schubert, D.W.; Goldmann, W.H.; Dusza, J.; Boccaccini, A.R. Curcumin-Containing Orthopedic Implant Coatings Deposited on Poly-Ether-Ether-Ketone/Bioactive Glass/Hexagonal Boron Nitride Layers by Electrophoretic Deposition. *Coatings* **2019**, *9*, 572. [CrossRef]
72. Li, P.-C.; Chen, S.-C.; Hsueh, Y.-J.; Shen, Y.-C.; Tsai, M.-Y.; Hsu, L.-W.; Yeh, C.-K.; Chen, H.-C.; Huang, C.-C. Gelatin scaffold with multifunctional curcumin-loaded lipid-PLGA hybrid microparticles for regenerating corneal endothelium. *Mater. Sci. Eng. C-Mater. Biol. Appl.* **2021**, *120*, 111753. [CrossRef]
73. Li, D.; Zhang, C.; Gao, Z.; Xia, N.; Wu, C.; Liu, C.; Tian, H.; Mei, X. Curcumin-Loaded Macrophage-Derived Exosomes Effectively Improve Wound Healing. *Mol. Pharm.* **2023**, *20*, 4453–4467. [CrossRef]

74. Fan, X.; Huang, J.; Zhang, W.; Su, Z.; Li, J.; Wu, Z.; Zhang, P. A Multifunctional, Tough, Stretchable, and Transparent Curcumin Hydrogel with Potent Antimicrobial, Antioxidative, Anti-inflammatory, and Angiogenesis Capabilities for Diabetic Wound Healing. *Acs Appl. Mater. Interfaces* **2024**, *16*, 9749–9767. [CrossRef]
75. Suresh, N.; Mauramo, M.; Waltimo, T.; Sorsa, T.; Anil, S. The Effectiveness of Curcumin Nanoparticle-Coated Titanium Surfaces in Osteogenesis: A Systematic Review. *J. Funct. Biomater.* **2024**, *15*, 247. [CrossRef]
76. Kushram, P.; Majumdar, U.; Bose, S. Hydroxyapatite coated titanium with curcumin and epigallocatechin gallate for orthopedic and dental applications. *Biomater. Adv.* **2023**, *155*, 213667. [CrossRef]
77. Gao, H.; Chen, N.; Sun, L.; Sheng, D.; Zhong, Y.; Huang, M.; Yu, C.; Yang, X.; Hao, Y.; Chen, S.; et al. Time-programmed release of curcumin and Zn^{2+} from multi-layered RSF coating modified PET graft for improvement of graft-host integration. *Int. J. Biol. Macromol.* **2024**, *272*, 132830. [CrossRef]
78. Liu, B.; Cui, L.-S.; Zhou, B.; Zhang, L.-L.; Liu, Z.-H.; Zhang, L. Monocarbonyl curcumin analog A2 potently inhibits angiogenesis by inducing ROS-dependent endothelial cell death. *Acta Pharmacol. Sin.* **2019**, *40*, 1412–1423. [CrossRef] [PubMed]
79. Xiang, L.; Zhang, Q.; Chi, C.; Wu, G.; Lin, Z.; Li, J.; Gu, Q.; Chen, G. Curcumin analog A13 alleviates oxidative stress by activating Nrf2/ARE pathway and ameliorates fibrosis in the myocardium of high-fat-diet and streptozotocin-induced diabetic rats. *Diabetol. Metab. Syndr.* **2020**, *12*, 1. [CrossRef] [PubMed]
80. Chen, M.; Qian, C.; Jin, B.; Hu, C.; Zhang, L.; Wang, M.; Zhou, B.; Zuo, W.; Huang, L.; Wang, Y. Curcumin analog WZ26 induces ROS and cell death via inhibition of STAT3 in cholangiocarcinoma. *Cancer Biol. Ther.* **2023**, *24*, 2162807. [CrossRef] [PubMed]
81. Borlinghaus, J.; Albrecht, F.; Gruhlke, M.C.H.; Nwachukwu, I.D.; Slusarenko, A.J. Allicin: Chemistry and Biological Properties. *Molecules* **2014**, *19*, 12591–12618. [CrossRef] [PubMed]
82. Sanchez-Gloria, J.L.; Arellano-Buendia, A.S.; Juarez-Rojas, J.G.; Garcia-Arroyo, F.E.; Arguello-Garcia, R.; Sanchez-Munoz, F.; Sanchez-Lozada, L.G.; Osorio-Alonso, H. Cellular Mechanisms Underlying the Cardioprotective Role of Allicin on Cardiovascular Diseases. *Int. J. Mol. Sci.* **2022**, *23*, 9082. [CrossRef] [PubMed]
83. Banerjee, S.K.; Maulik, S.K. Effect of garlic on cardiovascular disorders: A review. *Nutr. J.* **2002**, *1*, 4. [CrossRef]
84. Li, M.; Yun, W.; Wang, G.; Li, A.; Gao, J.; He, Q. Roles and mechanisms of garlic and its extracts on atherosclerosis: A review. *Front. Pharmacol.* **2022**, *13*, 954938. [CrossRef] [PubMed]
85. Wallock-Richards, D.; Doherty, C.J.; Doherty, L.; Clarke, D.J.; Place, M.; Govan, J.R.W.; Campopiano, D.C. Garlic revisited: Antimicrobial activity of allicin-containing garlic extracts against *Burkholderia cepacia* complex. *PLoS ONE* **2014**, *9*, e112726. [CrossRef] [PubMed]
86. Qidwai, W.; Ashfaq, T. Role of garlic usage in cardiovascular disease prevention: An evidence-based approach. *Evid. Based Complement. Altern. Med. ECAM* **2013**, *2013*, 125649. [CrossRef] [PubMed]
87. Pang, Y.; Zhao, M.; Xie, Y.; Wang, Y.; You, Y.; Ke, Y.; Zhang, C.; Chen, X.; Yang, Y.; Zhang, C.; et al. Multifunctional Ac@ZIF-8/AgNPs nanoplatform with pH-responsive and ROS scavenging antibacterial properties promotes infected wound healing. *Chem. Eng. J.* **2024**, *489*, 151485. [CrossRef]
88. Wang, J.; Liu, X.; Wang, Y.; An, M.; Fan, Y. Casein micelles embedded composite organohydrogel as potential wound dressing. *Int. J. Biol. Macromol.* **2022**, *211*, 678–688. [CrossRef]
89. Liu, Y.; Song, R.; Zhang, X.; Zhang, D. Enhanced antimicrobial activity and pH-responsive sustained release of chitosan/poly (vinyl alcohol)/graphene oxide nanofibrous membrane loading with allicin. *Int. J. Biol. Macromol.* **2020**, *161*, 1405–1413. [CrossRef]
90. Bose, S.; Robertson, S.F.; Vu, A.A. Garlic extract enhances bioceramic bone scaffolds through upregulating ALP & BGLAP expression in hMSC-monocyte co-culture. *Biomater. Adv.* **2023**, *154*, 213622.
91. Atef, Y.; Kinoshita, K.; Ichihara, Y.; Ushida, K.; Hirata, Y.; Kurauchi, Y.; Seki, T.; Katsuki, H. Therapeutic effect of allicin in a mouse model of intracerebral hemorrhage. *J. Pharmacol. Sci.* **2023**, *153*, 208–214. [CrossRef] [PubMed]
92. Susmita, B.; Arjak, B.; Christine, H.; Dishary, B. Allicin-Loaded Hydroxyapatite: Enhanced Release, Cytocompatibility, and Antibacterial Properties for Bone Tissue Engineering Applications. *JOM* **2022**, *74*, 3349–3356.
93. Chen, W.; Li, X.; Zeng, L.; Pan, H.; Liu, Z. Allicin-loaded chitosan/polyvinyl alcohol scaffolds as a potential wound dressing material to treat diabetic wounds: An in vitro and in vivo study. *J. Drug Deliv. Sci. Technol.* **2021**, *65*, 102734. [CrossRef]
94. Gao, X.; Zhou, Y.; Gu, J.; Liu, X.; Zhang, Z. Construction and Activity Study of a Natural Antibacterial Patch Based on Natural Active Substance-Green Porous Structures. *Molecules* **2023**, *28*, 1319. [CrossRef] [PubMed]
95. Han, X.; Lu, B.; Zou, D.; Luo, X.; Liu, L.; Maitz, M.F.; Yang, P.; Huang, N.; Zhao, A.; Chen, J. Allicin-Loaded Intelligent Hydrogel Coating Improving Vascular Implant Performance. *Acs Appl. Mater. Interfaces* **2023**, *15*, 38247–38263. [CrossRef] [PubMed]
96. Zhang, C.; Li, C.; Chen, S.; Li, Z.; Jia, X. Berberine protects against 6-OHDA-induced neurotoxicity in PC12 cells and zebrafish through hormetic mechanisms involving PI3K/AKT/Bcl-2 and Nrf2/HO-1 pathways. *Redox Biol.* **2017**, *11*, 1–11. [CrossRef]
97. Chandirasegaran, G.; Elanchezhiyan, C.; Ghosh, K.; Sethupathy, S. Berberine chloride ameliorates oxidative stress, inflammation and apoptosis in the pancreas of Streptozotocin induced diabetic rats. *Biomed. Pharmacother.* **2017**, *95*, 175–185. [CrossRef]
98. Li, X.; Wang, Q.; Shi, J.; Ju, R.; Zhu, L.; Li, J.; Guo, L.; Ye, C. Beneficial effect of berberine on atherosclerosis based on attenuating vascular inflammation and calcification. *Basic Clin. Med.* **2018**, *38*, 163–168.
99. Ai, X.; Yu, P.; Peng, L.; Luo, L.; Liu, J.; Li, S.; Lai, X.; Luan, F.; Meng, X. Berberine: A Review of its Pharmacokinetics Properties and Therapeutic Potentials in Diverse Vascular Diseases. *Front. Pharmacol.* **2021**, *12*, 762654. [CrossRef]
100. Song, D.; Hao, J.; Fan, D. Biological properties and clinical applications of berberine. *Front. Med.* **2020**, *14*, 564–582. [CrossRef]
101. Och, A.; Podgorski, R.; Nowak, R. Biological Activity of Berberine-A Summary Update. *Toxins* **2020**, *12*, 713. [CrossRef]

102. Zhong, X.J.; Liu, S.R.; Zhang, C.W.; Zhao, Y.S.; Sayed, A.; Rajoka, M.S.R.; He, Z.D.; Song, X. Natural alkaloid coptisine, isolated from *Coptis chinensis*, inhibits fungal growth by disrupting membranes and triggering apoptosis. *Pharmacol. Res.-Mod. Chin. Med.* **2024**, *10*, 100383. [CrossRef]
103. Dong, B.; Liu, X.; Li, J.; Wang, B.; Yin, J.; Zhang, H.; Liu, W. Berberine Encapsulated in Exosomes Derived from Platelet-Rich Plasma Promotes Chondrogenic Differentiation of the Bone Marrow Mesenchymal Stem Cells via the Wnt/β-Catenin Pathway. *Biol. Pharm. Bull.* **2022**, *45*, 1444–1451. [CrossRef] [PubMed]
104. Zou, Q.; Li, Y.; Zhang, L.; Zuo, Y.; Li, J.; Li, J. Antibiotic delivery system using nano-hydroxyapatite/chitosan bone cement consisting of berberine. *J. Biomed. Mater. Res. Part A* **2009**, *89*, 1108–1117. [CrossRef] [PubMed]
105. Ehterami, A.; Abbaszadeh-Goudarzi, G.; Haghi-Daredeh, S.; Niyakan, M.; Alizadeh, M.; JafariSani, M.; Atashgahi, M.; Salehi, M. Bone tissue engineering using 3-D polycaprolactone/gelatin nanofibrous scaffold containing berberine: In vivo and in vitro study. *Polym. Adv. Technol.* **2022**, *33*, 672–681. [CrossRef]
106. Maity, B.; Alam, S.; Samanta, S.; Prakash, R.G.; Govindaraju, T. Antioxidant Silk Fibroin Composite Hydrogel for Rapid Healing of Diabetic Wound. *Macromol. Biosci.* **2022**, *22*, 2200097. [CrossRef] [PubMed]
107. Wang, F.; Wang, X.; Hu, N.; Qin, G.; Ye, B.; He, J.-S. Improved Antimicrobial Ability of Dressings Containing Berberine Loaded Cellulose Acetate/Hyaluronic Acid Electrospun Fibers for Cutaneous Wound Healing. *J. Biomed. Nanotechnol.* **2022**, *18*, 77–86. [CrossRef]
108. Wang, N.; Wang, L.; Zhang, C.; Tan, H.-Y.; Zhang, Y.; Feng, Y. Berberine suppresses advanced glycation end products-associated diabetic retinopathy in hyperglycemic mice. *Clin. Transl. Med.* **2021**, *11*, e569. [CrossRef]
109. Sang, S.; Wang, S.; Yang, C.; Geng, Z.; Zhang, X. Sponge-inspired sulfonated polyetheretherketone loaded with polydopamine-protected osthole nanoparticles and berberine enhances osteogenic activity and prevents implant-related infections. *Chem. Eng. J.* **2022**, *437*, 135255. [CrossRef]
110. Tan, Y.; Sun, H.; Lan, Y.; Khan, H.M.; Zhang, H.; Zhang, L.; Zhang, F.; Cui, Y.; Zhang, L.; Huang, D.; et al. Study on 3D printed MXene-berberine-integrated scaffold for photo-activated antibacterial activity and bone regeneration. *J. Mater. Chem. B* **2024**, *12*, 2158–2179. [CrossRef]
111. Zhao, H.; Lou, Z.; Chen, Y.; Cheng, J.; Wu, Y.; Li, B.; He, P.; Tu, Y.; Liu, J. Tea polyphenols (TPP) as a promising wound healing agent: TPP exerts multiple and distinct mechanisms at different phases of wound healing in a mouse model. *Biomed. Pharmacother.* **2023**, *166*, 115437. [CrossRef]
112. Pervin, M.; Unno, K.; Takagaki, A.; Isemura, M.; Nakamura, Y. Function of Green Tea Catechins in the Brain: Epigallocatechin Gallate and its Metabolites. *Int. J. Mol. Sci.* **2019**, *20*, 3630. [CrossRef] [PubMed]
113. Zhong, W.; Tang, M.; Xie, Y.; Huang, X.; Liu, Y. Tea Polyphenols Inhibit the Activity and Toxicity of Staphylococcus aureus by Destroying Cell Membranes and Accumulating Reactive Oxygen Species. *Foodborne Pathog. Dis.* **2023**, *20*, 294–302. [CrossRef]
114. Zhang, B.; Yao, R.; Li, L.; Wang, Y.; Luo, R.; Yang, L.; Wang, Y. Green Tea Polyphenol Induced Mg^{2+}-rich Multilayer Conversion Coating: Toward Enhanced Corrosion Resistance and Promoted in Situ Endothelialization of AZ31 for Potential Cardiovascular Applications. *ACS Appl. Mater. Interfaces* **2019**, *11*, 41165–41177. [CrossRef] [PubMed]
115. Ren, Y.; Wang, F.-Y.; Chen, Z.-J.; Lan, R.-T.; Huang, R.-H.; Fu, W.-Q.; Gul, R.M.; Wang, J.; Xu, J.-Z.; Li, Z.-M. Antibacterial and anti-inflammatory ultrahigh molecular weight polyethylene/tea polyphenol blends for artificial joint applications. *J. Mater. Chem. B* **2020**, *8*, 10428–10438. [CrossRef] [PubMed]
116. Li, H.; Shen, S.; Yu, K.; Wang, H.; Fu, J. Construction of porous structure-based carboxymethyl chitosan/sodium alginate/tea polyphenols for wound dressing. *Int. J. Biol. Macromol.* **2023**, *233*, 123404. [CrossRef] [PubMed]
117. Wen, Y.; Zhao, R.; Yin, X.; Shi, Y.; Fan, H.; Zhou, Y.; Tan, L. Antibacterial and Antioxidant Composite Fiber Prepared from Polyurethane and Polyacrylonitrile Containing Tea Polyphenols. *Fibers Polym.* **2020**, *21*, 103–110. [CrossRef]
118. Wang, Z.; Lyu, T.; Xie, Q.; Zhang, Y.; Sun, H.; Wan, Y.; Tian, Y. Shape-adapted self-gelation hydrogel powder for high-performance hemostasis and wound healing. *Appl. Mater. Today* **2023**, *35*, 101948. [CrossRef]
119. Jin, H.; Zhao, Y.; Yao, Y.; Zhao, J.; Luo, R.; Fan, S.; Wei, Y.; Ouyang, S.; Peng, W.; Zhang, Y.; et al. Therapeutic effects of tea polyphenol-loaded nanoparticles coated with platelet membranes on LPS-induced lung injury. *Biomater. Sci.* **2023**, *11*, 6223–6235. [CrossRef]
120. Wick, T.V.; Roberts, T.R.; Batchinsky, A.I.; Tuttle, R.R.; Reynolds, M.M. Surface Modification of Oxygenator Fibers with a Catalytically Active Metal-Organic Framework to Generate Nitric Oxide: An Ex Vivo Pilot Study. *ACS Appl. Bio Mater.* **2023**, *6*, 1953–1959. [CrossRef]
121. Beurskens, D.M.H.; Huckriede, J.P.; Schrijver, R.; Hemker, H.C.; Reutelingsperger, C.P.; Nicolaes, G.A.F. The Anticoagulant and Nonanticoagulant Properties of Heparin. *Thromb. Haemost.* **2020**, *120*, 1371–1383. [CrossRef] [PubMed]
122. Yu, W.; Jinfa, J.; Sheng, M.; Jin, H.; Qin, Y.; Xu, W.; Wei, Z.; Liu, Z. Effects of composite Chitosan/Heparin coated membrane through a layer by layer self-assemble technique on stent thrombosis. *Acad. J. Second Mil. Med. Univ.* **2008**, *29*, 1324–1327.
123. Liu, X.; Liu, X.; Zheng, H.; Lu, K.; Chen, D.; Xiong, C.; Huang, F.; Zhang, L.; Zhang, D. Improvement of hydrophilicity and formation of heparin/chitosan coating inhibits stone formation in ureteral stents. *Colloids Surf. A-Physicochem. Eng. Asp.* **2024**, *694*, 134065. [CrossRef]
124. Awonusi, B.O.; Li, H.; Yin, Z.; Zhao, J.; Yang, K.; Li, J. Surface Modification of Zn-Cu Alloy with Heparin Nanoparticles for Urinary Implant Applications. *Acs Appl. Bio Mater.* **2024**, *7*, 1748–1762. [CrossRef]

125. Goh, M.; Min, K.; Kim, Y.H.; Tae, G. Chemically heparinized PEEK via a green method to immobilize bone morphogenetic protein-2 (BMP-2) for enhanced osteogenic activity. *RSC Adv.* **2024**, *14*, 1866–1874. [CrossRef] [PubMed]
126. Zhang, W.; Huang, Y.; Wu, H.; Dou, Y.; Li, Z.; Zhang, H. Polydopamine-heparin complex reinforced antithrombotic and antimicrobial activities of heparinized hydrogels for biomedical applications. *Compos. Part A Appl. Sci. Manuf.* **2022**, *157*, 106908. [CrossRef]
127. Jin, Y.; Zhu, Z.; Liang, L.; Lan, K.; Zheng, Q.; Wang, Y.; Guo, Y.; Zhu, K.; Mehmood, R.; Wang, B. A facile heparin/carboxymethyl chitosan coating mediated by polydopamine on implants for hemocompatibility and antibacterial properties. *Appl. Surf. Sci.* **2020**, *528*, 146539. [CrossRef]
128. Kun, W.; Ying, Y.; Wei, L.; Da, L.; Hui, L. Preparation of fully bio-based multilayers composed of heparin-like carboxymethylcellulose sodium and chitosan to functionalize poly (l-lactic acid) film for cardiovascular implant applications. *Int. J. Biol. Macromol.* **2023**, *231*, 123285.
129. Qiu, H.; Qi, P.; Liu, J.; Yang, Y.; Tan, X.; Xiao, Y.; Maitz, M.F.; Huang, N.; Yang, Z. Biomimetic engineering endothelium-like coating on cardiovascular stent through heparin and nitric oxide-generating compound synergistic modification strategy. *Biomaterials* **2019**, *207*, 10–22. [CrossRef]
130. Parolia, A.; Bapat, R.A.; Chaubal, T.; Yang, H.J.; Panda, S.; Mohan, M.; Sahebkar, A.; Kesharwani, P. Recent update on application of propolis as an adjuvant natural medication in management of gum diseases and drug delivery approaches. *Process Biochem.* **2022**, *112*, 254–268. [CrossRef]
131. Selvaraju, G.D.; Umapathy, V.R.; SumathiJones, C.; Cheema, M.S.; Jayamani, D.R.; Dharani, R.; Sneha, S.; Yamuna, M.; Gayathiri, E.; Yadav, S. Fabrication and characterization of surgical sutures with propolis silver nano particles and analysis of its antimicrobial properties. *J. King Saud Univ. Sci.* **2022**, *34*, 102082. [CrossRef]
132. Khodabakhshi, D.; Eskandarinia, A.; Kefayat, A.; Rafienia, M.; Navid, S.; Karbasi, S.; Moshtaghian, J. In vitro and in vivo performance of a propolis-coated polyurethane wound dressing with high porosity and antibacterial efficacy. *Colloids Surf. B-Biointerfaces* **2019**, *178*, 177–184. [CrossRef]
133. Isabel, M.F.A.; Sarita, M.; Sebastian, O.; Sara, L.-M.; Humberto, G.G.J.; Andres, D.-C.; Alex, L.; Carlos, P.; Alex, O.; Birgit, G.; et al. Antibacterial and osteoinductive properties of wollastonite scaffolds impregnated with propolis produced by additive manufacturing. *Heliyon* **2024**, *10*, e23955.
134. Somsanith, N.; Kim, Y.-K.; Jang, Y.-S.; Lee, Y.-H.; Yi, H.-K.; Jang, J.-H.; Kim, K.-A.; Bae, T.-S.; Lee, M.-H. Enhancing of Osseointegration with Propolis-Loaded TiO$_2$ Nanotubes in Rat Mandible for Dental Implants. *Materials* **2018**, *11*, 61. [CrossRef] [PubMed]
135. Bakshi, P.S.; Selvakumar, D.; Kadirvelu, K.; Kumar, N.S. Chitosan as an environment friendly biomaterial—A review on recent modifications and applications. *Int. J. Biol. Macromol.* **2020**, *150*, 1072–1083. [CrossRef] [PubMed]
136. Palla-Rubio, B.; Araújo-Gomes, N.; Fernández-Gutiérrez, M.; Rojo, L.; Suay, J.; Gurruchaga, M.; Goñi, I. Synthesis and characterization of silica-chitosan hybrid materials as antibacterial coatings for titanium implants. *Carbohydr. Polym.* **2018**, *203*, 331–341. [CrossRef]
137. Jian, Y.; Yang, C.; Zhang, J.; Qi, L.; Shi, X.; Deng, H.; Du, Y. One-step electrodeposition of Janus chitosan coating for metallic implants with anti-corrosion properties. *Colloids Surf. A Physicochem. Eng. Asp.* **2022**, *641*, 128498. [CrossRef]
138. Jian, Y.; Zhang, J.; Yang, C.; Qi, L.; Wang, X.; Deng, H.; Shi, X. Biological MWCNT/chitosan composite coating with outstanding anti-corrosion property for implants. *Colloids Surf. B Biointerfaces* **2023**, *225*, 113227. [CrossRef]
139. Lin, M.-H.; Wang, Y.-H.; Kuo, C.-H.; Ou, S.-F.; Huang, P.-Z.; Song, T.-Y.; Chen, Y.-C.; Chen, S.-T.; Wu, C.-H.; Hsueh, Y.-H.; et al. Hybrid ZnO/chitosan antimicrobial coatings with enhanced mechanical and bioactive properties for titanium implants. *Carbohydr. Polym.* **2021**, *257*, 117639. [CrossRef]
140. Khoshnood, N.; Frampton, J.P.; Zaree, S.R.A.; Jahanpanah, M.; Heydari, P.; Zamanian, A. The corrosion and biological behavior of 3D-printed polycaprolactone/ chitosan scaffolds as protective coating for Mg alloy implants. *Surf. Coat. Technol.* **2024**, *477*, 130368. [CrossRef]
141. Niknam, Z.; Golchin, A.; Rezaei-Tavirani, M.; Ranjbarvan, P.; Zali, H.; Omidi, M.; Mansouri, V. Osteogenic Differentiation Potential of Adipose-Derived Mesenchymal Stem Cells Cultured on Magnesium Oxide/Polycaprolactone Nanofibrous Scaffolds for Improving Bone Tissue Reconstruction. *Adv. Pharm. Bull.* **2022**, *12*, 142–154. [CrossRef] [PubMed]
142. Vimalraj, S. Alkaline phosphatase: Structure, expression and its function in bone mineralization. *Gene* **2020**, *754*, 144855. [CrossRef] [PubMed]
143. Komori, T. Regulation of Proliferation, Differentiation and Functions of Osteoblasts by Runx2. *Int. J. Mol. Sci.* **2019**, *20*, 1694. [CrossRef] [PubMed]
144. Zheng, W.; Chen, C.; Zhang, X.; Wen, X.; Xiao, Y.; Li, L.; Xu, Q.; Fu, F.; Diao, H.; Liu, X. Layer-by-layer coating of carboxymethyl chitosan-gelatin-alginate on cotton gauze for hemostasis and wound healing. *Surf. Coat. Technol.* **2021**, *406*, 126644. [CrossRef]
145. Gao, X.; Huang, R.; Jiao, Y.; Groth, T.; Yang, W.; Tu, C.; Li, H.; Gong, F.; Chu, J.; Zhao, M. Enhanced wound healing in diabetic mice by hyaluronan/chitosan multilayer-coated PLLA nanofibrous mats with sustained release of insulin. *Appl. Surf. Sci.* **2022**, *576*, 151825. [CrossRef]
146. Wang, Z.; Mei, L.; Liu, X.; Zhou, Q. Hierarchically hybrid biocoatings on Ti implants for enhanced antibacterial activity and osteogenesis. *Colloids Surf. B-Biointerfaces* **2021**, *204*, 111802. [CrossRef]

147. Gao, J.; Yang, G.; Pi, R.; Li, R.; Wang, P.; Zhang, H.; Le, K.; Chen, S.; Liu, P. Tanshinone IIA protects neonatal rat cardiomyocytes from adriamycin-induced apoptosis. *Transl. Res.* **2008**, *151*, 79–87. [CrossRef]
148. Yin, X.; Yin, Y.; Cao, F.-L.; Chen, Y.-F.; Peng, Y.; Hou, W.-G.; Sun, S.-K.; Luo, Z.-J. Tanshinone IIA Attenuates the Inflammatory Response and Apoptosis after Traumatic Injury of the Spinal Cord in Adult Rats. *PLoS ONE* **2012**, *7*, e38381. [CrossRef]
149. Fan, G.-W.; Gao, X.-M.; Wang, H.; Zhu, Y.; Zhang, J.; Hu, L.-M.; Su, Y.-F.; Kang, L.-Y.; Zhang, B.-L. The anti-inflammatory activities of Tanshinone IIA, an active component of TCM, are mediated by estrogen receptor activation and inhibition of iNOS. *J. Steroid Biochem. Mol. Biol.* **2009**, *113*, 275–280. [CrossRef] [PubMed]
150. Chen, W.; Xu, Y.; Li, H.; Dai, Y.; Zhou, G.; Zhou, Z.; Xia, H.; Liu, H. Tanshinone IIA Delivery Silk Fibroin Scaffolds Significantly Enhance Articular Cartilage Defect Repairing via Promoting Cartilage Regeneration. *Acs Appl. Mater. Interfaces* **2020**, *12*, 21470–21480. [CrossRef]
151. Liu, Q.-Y.; Yao, Y.-M.; Yu, Y.; Dong, N.; Sheng, Z.-Y. Correction: Astragalus Polysaccharides Attenuate Postburn Sepsis via Inhibiting Negative Immunoregulation of $CD4^+CD25^{high}$ T Cells. *PLoS ONE* **2018**, *6*, e19811.
152. Zheng, Y.-J.; Zhou, B.; Song, Z.-F.; Li, L.; Wu, J.; Zhang, R.-Y.; Tang, Y.-Q. Study of *Astragalus mongholicus* polysaccharides on endothelial cells permeability induced by HMGB1. *Carbohydr. Polym.* **2013**, *92*, 934–941. [CrossRef] [PubMed]
153. Wen, J.; Hu, D.; Wang, R.; Liu, K.; Zheng, Y.; He, J.; Chen, X.; Zhang, Y.; Zhao, X.; Bu, Y.; et al. Astragalus polysaccharides driven stretchable nanofibrous membrane wound dressing for joint wound healing. *Int. J. Biol. Macromol.* **2023**, *248*, 125557. [CrossRef]
154. Gao, J.; Chen, T.; Zhao, D.; Zheng, J.; Liu, Z. Ginkgolide B Exerts Cardioprotective Properties against Doxorubicin-Induced Cardiotoxicity by Regulating Reactive Oxygen Species, Akt and Calcium Signaling Pathways In Vitro and In Vivo. *PLoS ONE* **2016**, *11*, e0168219. [CrossRef] [PubMed]
155. Zheng, P.-D.; Mungur, R.; Zhou, H.-J.; Hassan, M.; Jiang, S.-N.; Zheng, J.-S. Ginkgolide B promotes the proliferation and differentiation of neural stem cells following cerebral ischemia/reperfusion injury, both in vivo and in vitro. *Neural Regen. Res.* **2018**, *13*, 1204–1211. [CrossRef] [PubMed]
156. Ge, Y.; Xu, W.; Zhang, L.; Liu, M. Ginkgolide B attenuates myocardial infarction-induced depression-like behaviors via repressing IL-1β in central nervous system. *Int. Immunopharmacol.* **2020**, *85*, 106652. [CrossRef] [PubMed]
157. Wang, L.; Xia, K.; Han, L.; Zhang, M.; Fan, J.; Song, L.; Liao, A.; Wang, W.; Guo, J. Local Administration of Ginkgolide B Using a Hyaluronan-Based Hydrogel Improves Wound Healing in Diabetic Mice. *Front. Bioeng. Biotechnol.* **2022**, *10*, 898231. [CrossRef]
158. Madian, N.G.; El-Ashmanty, B.A.; Abdel-Rahim, H.K. Improvement of Chitosan Films Properties by Blending with Cellulose, Honey and Curcumin. *Polymers* **2023**, *15*, 2587. [CrossRef]
159. San Martin-Martinez, E.; Casanas-Pimentel, R.; Almaguer-Flores, A.; Prado-Prone, G.; Garcia-Garcia, A.; Landa-Solis, C.; Hernandez-Rangel, A. Curcumin-loaded Polycaprolactone/Collagen Composite Fibers as Potential Antibacterial Wound Dressing. *Fibers Polym.* **2022**, *23*, 3002–3011. [CrossRef]

Disclaimer/Publisher's Note: The statements, opinions and data contained in all publications are solely those of the individual author(s) and contributor(s) and not of MDPI and/or the editor(s). MDPI and/or the editor(s) disclaim responsibility for any injury to people or property resulting from any ideas, methods, instructions or products referred to in the content.

Article

Friction Characteristics of Low and High Strength Steels with Galvanized and Galvannealed Zinc Coatings

Ji-Young Kim [1,2], Seung-Chae Yoon [1], Byeong-Keuk Jin [1], Jin-Hwa Jeon [1], Joo-Sik Hyun [1] and Myoung-Gyu Lee [2,*]

[1] Automotive Steel Application Engineering Team, Hyundai Steel Company, 1480 Buckbusaneop-Ro, Songak-Eup, Dangjin-Si 31719, Republic of Korea; jiyoung@hyundai-steel.com (J.-Y.K.); scyoon@hyundai-steel.com (S.-C.Y.); phentas12@hyundai-steel.com (B.-K.J.); jhjeon@hyundai-steel.com (J.-H.J.); hjs401@hyundai-steel.com (J.-S.H.)

[2] Department of Materials Science and Engineering & RIAM, Seoul National University, 1 Gwanak-Ro, Gwanak-Gu, Seoul 08826, Republic of Korea

* Correspondence: myounglee@snu.ac.kr

Abstract: As vehicle body structures become stronger and part designs more complex for lightweight, controlling frictional properties in automotive press forming has gained critical importance. Friction, a key factor in formability, is influenced by variables such as contact pressure, sliding velocity, sheet strength, and coatings. This study investigates the friction characteristics of steels with tensile strengths of 340 MPa and 980 MPa, under galvanized (GI) and galvannealed (GA) zinc coatings. Experimental results reveal that asperity flattening, a significant factor in determining friction, increases with contact pressure normalized by tensile strength, particularly for GI-coated steels. However, the relationship between friction and surface flattening deviates from conventional expectations, with the friction coefficient initially rising with increased flattening area up to ~20% before decreasing as flattening progresses. These findings suggest that traditional empirical formulas may not fully capture friction behavior under specific conditions. By understanding this inflection point, where friction reduces under high contact pressure, the study provides valuable insights for optimizing formability and improving sheet metal forming processes, especially in scenarios where precise friction control is critical for producing high-quality automotive parts.

Keywords: friction; steel sheets; zinc coating; contact pressure; sliding velocity; asperity flattening

Citation: Kim, J.-Y.; Yoon, S.-C.; Jin, B.-K.; Jeon, J.-H.; Hyun, J.-S.; Lee, M.-G. Friction Characteristics of Low and High Strength Steels with Galvanized and Galvannealed Zinc Coatings. *Materials* **2024**, *17*, 5031. https://doi.org/10.3390/ma17205031

Academic Editor: Pawel Pawlus

Received: 11 September 2024
Revised: 10 October 2024
Accepted: 13 October 2024
Published: 15 October 2024

Copyright: © 2024 by the authors. Licensee MDPI, Basel, Switzerland. This article is an open access article distributed under the terms and conditions of the Creative Commons Attribution (CC BY) license (https:// creativecommons.org/licenses/by/ 4.0/).

1. Introduction

Due to recent advances in eco-friendly vehicle technology, materials applied to automobile parts have been diversified, and the shape of the parts has become more complex. Accordingly, advanced forming technology has been applied to manufacture automotive parts, which require the optimized formability and processing parameters in the design stage [1,2].

Various process parameters in automotive manufacturing are associated with the formability of final products. For example, the formability of steel sheets highly depends on their plastic behavior along with other press-forming parameters. The typically evaluated plastic behavior of sheet metals includes isotropic or anisotropic strength and elongation, and the key forming parameters are the blank holding force and contact between the workpiece and tools. To date, besides the experimental works on the effect of process parameters on the overall formability of the sheet parts, computational approaches such as finite element (FE) modeling and simulation have also been popular for optimization. Thus, the accuracy and numerical efficiency in the computational simulation are of critical importance [3–8].

Furthermore, considerable research has focused on understanding the role of friction in sheet metal forming, as it plays a critical role in determining the quality of the final product.

Among the various processing factors, the frictional characteristics between tools and sheet metals have been extensively studied to ensure the production of high-quality sheet parts. Friction arises from complex contact conditions during forming and is influenced by factors such as sheet strength, material composition, tool materials, surface roughness, lubrication, coatings, etc. [9–13]. Thereby, studies reported that a conventional approach with constant friction coefficients in predicting the formability of automotive steel sheets could not give accurate results, while more advanced friction models, such as those based on variable friction coefficients, should be further developed [1,14–16]. For example, friction modeling with variable coefficients as a function of contact conditions during press forming increased the accuracy of formability prediction [15–17].

As the main factors influencing the friction in the sheet metal forming process, the contact pressure and sliding velocity are related to the blank holding force and forming speed [18–22]. For example, the study on the effect of forming speed on the friction coefficient of 60 K grade steel plate showed that the friction coefficient decreased as the sliding speed increased under low contact pressure. Also, friction tests for various sheet metals at a high drawing speed of 600 m/min or faster also reported that the friction coefficient decreased as the drawing speed increased. Other studies reported that the lubricating oxide can be formed on the material surface if the surface temperature increases during press forming. This resulted in reduced shear strength between tools and sheet metals. As for the effect of contact pressure on the friction coefficient, cold-rolled high-strength steel exhibited a decreased friction coefficient as the contact pressure increased at the region of low contact pressure [10,21]. However, other studies also reported that the effect of contact pressure varied and was strongly dependent on the material properties and contact conditions [10,23–26]. Particularly, friction depends on the material's hardness, surface roughness, or evolution during formation.

Bowden and Tabor [27,28] studied the effect of contact asperity on frictional behavior, and they observed that a linear relationship between contact pressure and friction coefficient changes to non-linear as the real contact area becomes large. They also found that the bulk plastic deformation changed the formation of the contact area and, eventually, the friction coefficient. Orowan et al. [29] investigated the relationship between surface contact area and stick–slip behavior, focusing on the normal load and friction coefficient.

Moreover, the type and characteristics of coatings applied to sheet metals significantly influence friction and formability [30]. Zinc coatings demonstrate this impact in two main aspects. First, depending on the coating type, surface roughness varies, affecting the contact area. Second, changes within the coating layer alter the shear yield strength of zinc-coated layers. For instance, the tensile yield strength of the coating increases with thicker Fe-Zn intermetallic phases, leading to higher shear yield strength and, consequently, an increase in the friction coefficient [31]. The formability of zinc-coated sheet metals is influenced by a complex interplay of surface friction, coating adhesion, and strain state distribution [32–34]. Surface friction affects deformation modes and can compromise sheet metal formability. Additionally, the brittleness of the coating layer leads to varying adhesion under different contact conditions. Therefore, to accurately predict the formability of galvanized steel sheets, it is essential to first understand the coefficient of friction and friction behavior based on contact conditions and the physical properties of the coating layer.

Based on the previous studies on the friction characteristics in terms of the material properties' contact conditions, the present work aims to further investigate the friction behavior of steel sheets with different strength grades, types of coating, and contact conditions. Thus, steels with significantly different tensile strengths and two zinc-coated surfaces are considered in this study. The frictional behavior is particularly related to the contact surface change between the tool and sheet metal during the press forming process. For this purpose, three sliding speeds and three normal loads are applied to the two steel sheets. These distinctive test conditions enabled us to realize various contact conditions and surface flattening of steel surfaces, which affect the characteristics of frictional behavior.

Finally, experimental results are analyzed based on the microscopic pattern of asperity flattening and compared to the classical adhesion theory.

2. Materials and Methods

2.1. Materials and Sample Preparation

Two steel sheets with different tensile strengths, 340 MPa and 980 MPa were investigated. Their coating conditions were either galvannealed (GA) or galvanized (GI). Therefore, four types of specimens were prepared in total and were labeled as TS340-GI, TS980-GI, TS340-GA, and TS980-GA, indicating both tensile strength and coating type for each specimen. The thickness of the TS340-GI and TS340-GA specimens was 0.7 mm, while the thickness of the TS980-GI and TS980-GA specimens was 1.0 mm. TS340 steel sheet is commonly used for automotive outer panels, and TS980 steel sheet is a representative grade primarily applied to automotive components requiring high stiffness. The GI coating involves applying a layer of zinc to steel sheets. GA is a type of coating applied to steel sheets through a process that involves galvanizing followed by annealing. GI and GA coatings are both commonly used in various industries, including automotive, construction, and appliance manufacturing, where corrosion resistance is crucial. Figure 1 shows the surface morphology of the four as-received specimens, and they clearly represent different initial surface conditions depending on both strength and coating. Table 1 shows the sample details. The strengths of base materials, as well as the strengths and hardness of the coatings, are listed in Table 1. The hardness of the coatings was measured using a Vickers hardness tester (Mitutoyo HV-100, Mitutoyo, Kawasaki, Japan) following ISO 6502-2 standards [35], and for comparison with substrate strength, the hardness was converted into tensile strength [36,37].

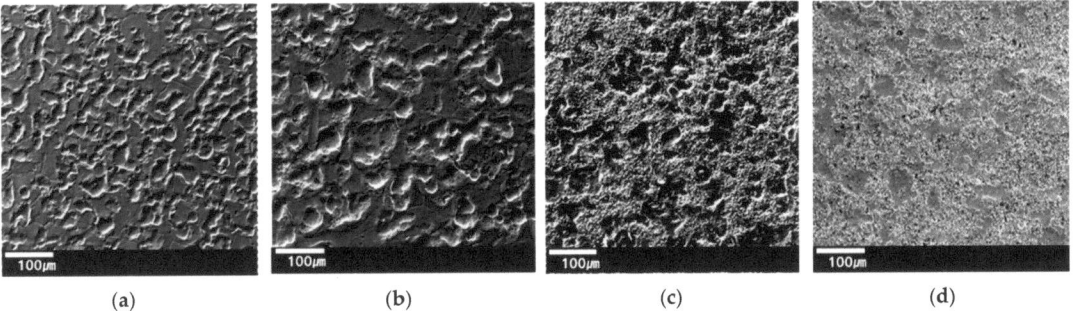

(a)   (b)   (c)   (d)

Figure 1. The top view of the four as-received specimen surfaces: (**a**) TS340-GI, (**b**) TS980-GI, (**c**) TS340-GA, and (**d**) TS980-GA.

Table 1. Specimens used in the experiments.

Specimen	Tensile Strength (TS), MPa		Hardness, HV	Roughness, μm		
	Base Metals	Coatings	Coatings	Ra	Rz	Rsk
TS 340-GI	340	34 ± 5	102 ± 16	0.92	5.32	−0.39
TS 340-GA	340	89 ± 14	266 ± 43	1.16	8.34	−0.63
TS 980-GI	980	34 ± 5	102 ± 16	0.85	6.90	−0.74
TS 980-GA	980	89 ± 14	266 ± 43	1.18	8.65	0.00

TS340-GI and TS980-GI have similar surface topography, with a lower average roughness (Ra) and maximum peak-to-valley height (Rz) compared to TS340-GA and TS980-GA, as shown in Table 1. The skewness (Rsk) values also reveal that the GI coatings exhibit more pronounced valleys on the surface, which could enhance lubricant retention. In contrast, the GA coatings show different characteristics, with TS340-GA having a moderately negative skew and TS980-GA showing a balanced surface topography, indicating an even distribution of peaks and valleys.

The original cross-sectional view of GI and GA coatings is presented in Figure 2. As shown in Figure 2, the GI coating consists solely of zinc, whereas the GA coating is formed as a zinc-iron alloy. Note that the strength and hardness of GA coating is much higher than that of the GI coating. The GI and GA coatings' thickness was approximately 13 µm and 6 µm, respectively. The rectangular-shaped specimens for the friction test were cut from the sample in the transverse direction to the roll. Steel sheets produced through a continuous rolling process typically exhibit isotropic roughness. However, slight variations in roughness and mechanical properties have been observed depending on the direction of rolling. In our study, it was recognized that the transverse direction (TD) to rolling had a higher coefficient of friction than RD. Therefore, despite these minor variations in roughness and friction coefficient across different directions, standard steel sheet product measurements are conducted in the transverse direction. This approach ensures a conservative estimate of the friction behavior in the industry. For the tests, the sample size was 220 mm × 30 mm (length × width). Before each friction test, the specimen was pre-treated following the standards ISO27831-1 [38] and ISO22462 [39]. First, any contaminants on the specimen surface were removed with ethanol solvent. Then, two drops of lubricant were applied to the specimen surface and then wiped with a clean cloth. The lubricant used in the present work was BW-80HG, and the lubrication amount applied was 2000 mg/m^2. The kinematic viscosity of the lubricant is 14.0 mm^2/s at 40 °C, and the base oil of the lubricant is composed of hydrotreated light paraffinic, hydrotreated heavy paraffinic, and sulfonic acids–petroleum–barium salt. We used the kinematic viscosity at 40 °C because, in continuous press processing, the die or punch temperature rises as the number of pressings increases.

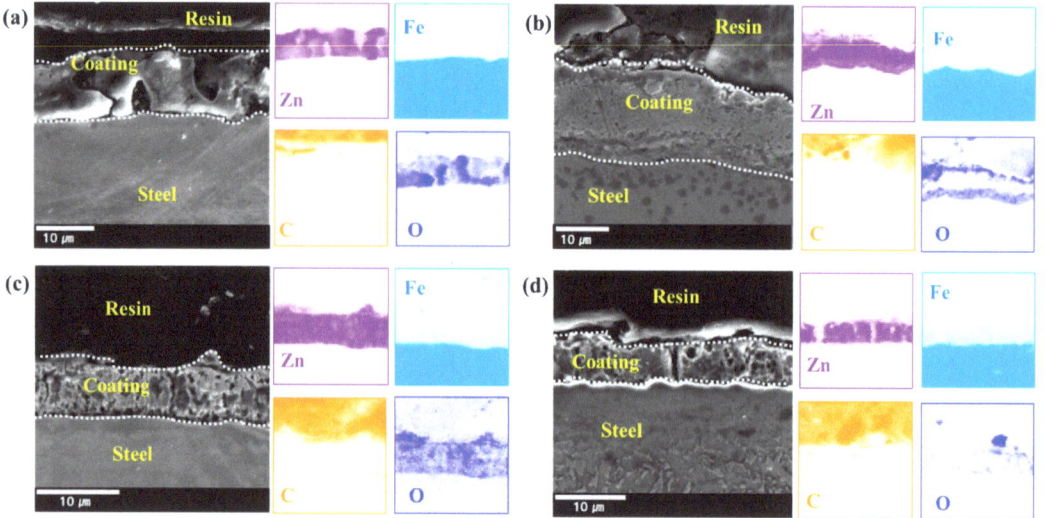

Figure 2. The cross-sectional view and EDS images of the four as-received specimen surfaces: (**a**) TS340-GI, (**b**) TS980-GI, (**c**) TS340-GA, and (**d**) TS980-GA.

2.2. Friction Test

The sliding friction test with a single-side contact was conducted. The test equipment, which is a single-sided sliding type of friction equipment, is shown in Figure 3. The test equipment is shown in Figure 3. All tests were carried out at 23 °C and 50% relative humidity (RH) according to the guideline of ISO standard [39]. Constant loads of 300, 1000, 3920, and 8000 N were applied during the friction tests. These correspond to the contact pressures of 10, 33, 130, and 267 MPa, considering the contact area of 30 mm². Before each test, the uniformity of contact pressure on the test specimen was checked with photosensitive paper. The reason for conducting coefficient of friction measurements across a range of contact pressures, from 10 MPa to over 100 MPa, in this study is to encompass the various forming conditions that can occur during the forming processes of automotive components. In actual press forming of automotive parts, most areas experience relatively low contact pressures in the range of a few MPa. However, high contact pressures may arise in certain zones of the stamping die, particularly at sharp corners with small radii or localized regions undergoing significant deformation [40]. Therefore, the range of contact pressures was selected to reflect these critical areas. Similarly, the sliding velocity conditions were determined based on forming analyses of several representative automotive outer panel components. The relative sliding velocity between the tool and specimen was set to be 0.05, 0.2, and 0.9 m/min. A constant sliding distance of 100 mm was applied to all tests. The tool material used in the present work was X153CrMoV12 steel and was ion nitrided after vacuum heat treatment with a nitride layer of about 0.07 mm. The roughness and hardness of the friction tool block were 0.1 Ra and 1000 Hv, respectively. During the friction test, the tool block was not re-polished due to the considerable difference in hardness between the tool and the material. The tool, made of SKD 11, was vacuum heat-treated and ion-nitrided, achieving a surface hardness of approximately 1000 Hv, which is substantially higher than the hardness of the Zinc coated specimens. This substantial hardness difference minimizes potential changes in the tool surface roughness over the course of the tests. The tests were conducted using the tool without re-polishing, and specifically, we conducted three repetitions for each of the nine test conditions on the same material, resulting in a total of 27 tests performed without re-polishing the tool. The applied load and contact pressure were normalized by the contact area and tensile strength (TS), respectively, using Formula (1) to exclude the material strength's effect, which is summarized in Table 2.

$$Normalised\ Contact\ Pressure(NCP) = \frac{Applied\ load(N)}{Contact\ area(mm^2) \times Tensile\ strength(MPa)}, \quad (1)$$

Figure 3. A single-sided sliding type friction test equipment.

Table 2. Contact pressure and its normalized value by the tensile strength of materials.

TS 340		TS 980	
Contact Pressure, MPa	Normalized Contact Pressure	Contact Pressure, MPa	Normalized Contact Pressure
10	0.03	33	0.03
33	0.10	100	0.10
130	0.37	267	0.27

As a result, a proportionally higher load was applied for materials with higher tensile strength, such as TS980 steel, compared to lower-strength materials like TS340 steel. This approach allows for a more accurate comparison of frictional behavior across different material grades. However, due to equipment limitations, where the applied load exceeded the operational range, a lower contact pressure was applied to the TS980 steel than to the TS340 steel under the 130 MPa condition

The friction coefficient (μ) was calculated from the normal, and friction forces were recorded during sliding. At least three tests were repeated for each test condition. The coefficient of friction is determined using the following Equation (2):

$$\mu = \frac{F_f}{F_n}, \qquad (2)$$

where F_f is the friction force, and F_n is the normal force.

2.3. Sliding Surface Analysis

After each friction test, the sliding surface of the steel sheet was observed using a scanning electron microscopy (SEM) with a Phenom Desktop SEM from Thermo Fisher Scientific (Waltham, MA, USA) and three-dimensional (3D) confocal imaging equipment using ADE's MicroXAM. The SEM specimens were cut in the direction parallel to the sliding surface, and the area of flattened asperities was evaluated by measuring the height distribution using the 3D confocal image analyzer.

The SEM analyzed the sliding surface and cross-section of the tested specimen, including the surface deformation and flattening of asperity, with varying sliding velocity and normal force. Additionally, the cross-section was analyzed for elemental composition using energy-dispersive spectroscopy (EDS). The asperity height corresponding to the local maxima is defined as the transition height, and the flattening area αh is determined as follows [41].

$$\alpha_h = \int_{h_t}^{\infty} \varphi_d(z)dz, \qquad (3)$$

where h_t and φ_d (%) are the height of the asperity corresponding to the local maximum peak and normalized height distribution of the deformed surface, respectively. A median filter with a kernel size of 3 × 3 was used for the measurement area of 2 mm × 2 mm.

3. Results

3.1. Effect of Strength and Coating Condition on Friction Coefficient

The measured friction coefficients of the four types of specimens are summarized in Figure 4 under various sliding velocities and normalized contact pressures.

For the TS340-GI specimen, the friction coefficient decreases as the sliding velocity increases. As shown in Figure 4a, under the normalized contact pressure (NCP) of 0.03, the friction coefficients decreased from 0.163 to 0.142 as the sliding velocity was increased from 0.05 to 0.9 m/min. Regarding the effect of the contact pressure, the change of friction coefficient was uni-modal. When the normalized pressure increased from 0.03 to 0.1, the change in friction coefficient was marginal. However, a significantly reduced friction coefficient was observed when the normalized contract pressure increased as high as 0.37.

For example, at the sliding velocity of 0.05, the coefficient of friction changed around 10% when the normalized contact pressure changed from 0.03 to 0.1, while it decreased over 30% for the normal contact pressure of 0.37. For the steel with higher strength and the same GI coating condition as TS980-GI, Figure 4b shows generally similar results in terms of the effect of the sliding velocity and contact pressure on friction coefficient as those of TS340-GI. The difference is that the friction coefficient is less sensitive to the sliding velocity, especially when the normalized contact pressure is as low as 0.03.

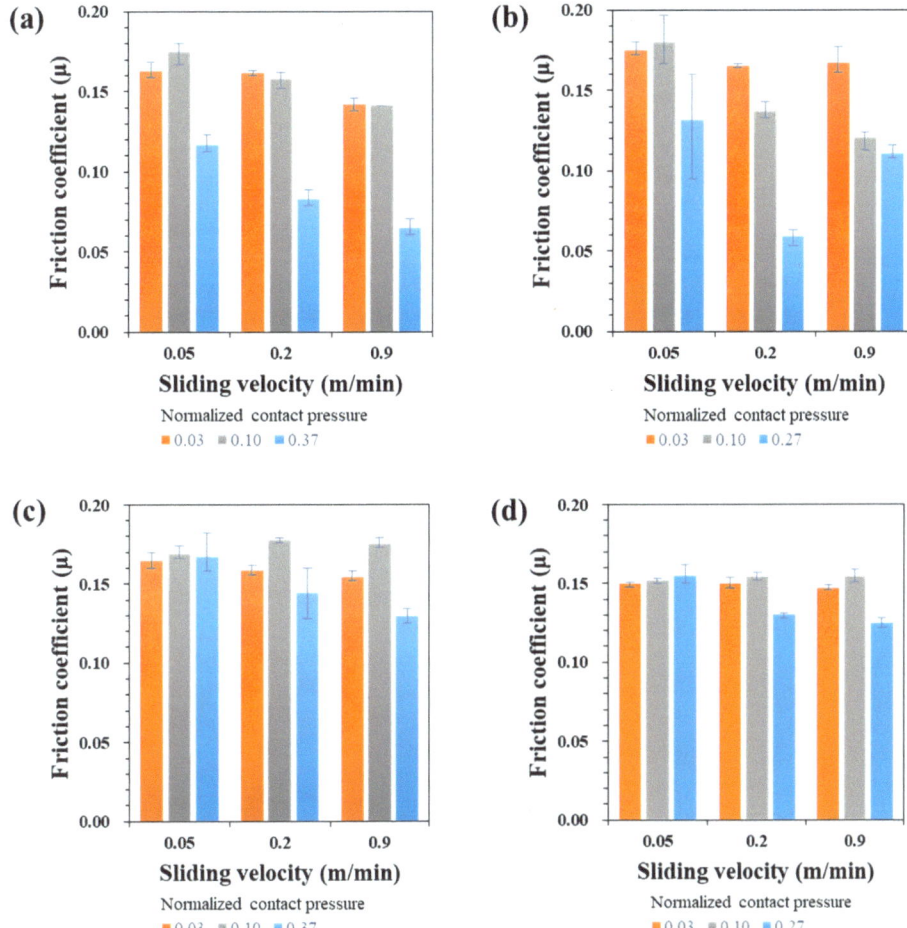

Figure 4. Measured friction coefficients with respect to the sliding velocity and normalized contact pressure: (**a**) TS 340-GI, (**b**) TS 980-GI, (**c**) TS340-GA, and (**d**) TS980-GA.

The friction coefficients of GA-coated steel sheets were less sensitive to the sliding velocity and normalized contact pressure than those of GI-coated steels, as shown in Figure 4c,d. This is particularly clear when the normalized contact pressure was an intermediate value of 0.1. That is, for the GI-coated steels, the friction coefficient exhibits an obvious decrease if the sliding velocity increases, while the friction coefficients of GA-coated steels under the contact pressure of 0.1 are virtually constant regardless of material strength. The commonly observed decrease of friction coefficient with increased sliding velocity was observed in GA-coated steels only when the normalized contact pressure was as high as

0.37 and 0.27 for TS340 and TS980 steel, respectively. Another noticeable result is that the friction coefficient variation in processing factors (sliding velocity and contact pressure) is much smaller for the GA-coated steels than that of the GI-coated steels. The GI-coated steels present friction coefficients in the range of 0.059 to 0.18, but the range becomes 0.125 to 0.177 for GA-coated steels. Moreover, the variation in the level of friction coefficient does not significantly depend on the strength of the steel.

3.2. Characteristics of Frictional Behavior

More detailed analyses of the characteristics of frictional behavior were performed for the tested specimens. Besides the friction coefficient, the friction modes and the existence of wear and surface fracture were quantified. Also, they were correlated to the sliding velocity, normalized contact pressure, material strength, and coating types. The characteristics of these frictional behavior are summarized in Table A1a–d in Appendix A. The symbols X, △, and O represent "barely observed", "occasionally observed", and "almost always observed", respectively. "Barely observed" was applied when 1 or fewer instances were found, "occasionally observed" when 2 to less than half of the instances were found, and "almost always observed" when more than half of the instances were found. As supplementary data, the raw graphs of friction coefficient change during sliding are also given in Figure A1a–d in Appendix A, which clarifies the occurrence of stick–slip behavior during the friction test. In addition, the microscopic analyses were presented using SEM, which observed the deformation and fracture of coating layers, as shown in Figure 5a–d.

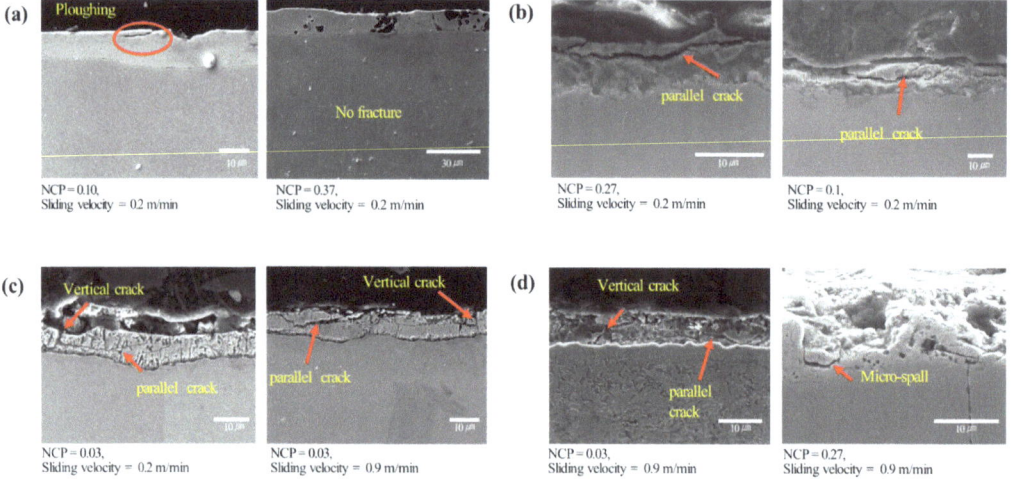

Figure 5. SEM measured cross-sectional images of the test specimens: (**a**) TS340-GI, (**b**) TS980-GI, (**c**) TS340-GA, and (**d**) TS980-GA. NCP denotes the normalized contact pressure.

The deformed surfaces of TS340-GI steel under different contact pressures are shown in Figure 5a, where flattening and plowing (or scratching) of the coating layer can be observed. As presented in Table A1a, the increased contact pressure resulted in larger asperity flattening and deformation of the coating layer. On the other hand, the effect of sliding velocity on asperity deformation was more complicated. For example, at the normalized contact pressures of 0.1 and 0.37, the increased sliding velocity led to reduced asperity flattening. The stick–slip phenomena were observed only at the normalized contact pressure of 0.03 and 0.1. The existence of stick–slip can be referred to as the serrated friction curve shown in Figure A1a. The high friction coefficient of TS340-GI under this contact pressure can be attributed to this stick–slip behavior. The stick–slip was hardly observed at higher normalized contact pressure. Besides, no crack was observed in the coating layer of TS340-GI for all tested conditions.

The friction characteristics of TS980-GI steel in terms of asperity flattening and stick–slip behavior were similar to those of TS340-GI steel (see Figures 5b and A1b). However, unlike the TS340-GI steel, evident parallel cracks in the coating layer were noticeable in TS980-GI steel in all experimental conditions (Figure 5b). The parallel cracks in TS980-GI steel during sliding may change the state of material-tool contact, which ultimately results in the reduction of stick–slip behavior compared to TS340-GI steel.

The frictional behavior of TS340-GA steel presented less asperity flattening and stick–slip than those of TS340-GI steel, as shown in Figure A1c.

From the previous references [24,25], it is known that the stick–slip behavior encountered during sliding contact can be caused by either adhesion or asperity interlocking. Enhanced chemical compatibility, increased contact area, and prolonged residence time can increase the adhesion. On the other hand, asperity interlocking occurs when a rigid asperity penetrates a softer asperity, which can be alleviated by either elevating the hardness of the softer material or reducing the applied load. The reduced stick–slipped TS340-GA-coated steel might be due to the increased hardness of the coating and decreased adhesion caused by the asperity interlocking. Also, less flattening can be related to decreased adhesion. In contrast to the GI-coated steels, the coating surface of TS340-GA steel exhibited obvious vertical cracks (Figure 5c and Table A1c). This is closely related to the strength of the coating layer because the vertical cracks are predominantly caused by tensile stress along the direction of frictional sliding [42]. Like the GI-coated steels, the GA-coated steels presented similar frictional characteristics, as shown in Table A1c. However, the GA coating on higher-strength steel, like the case of TS980-GA, showed increased vertical and parallel cracks with increased loading and sliding velocity. In addition, it was found that the coating layer's delamination (or spalling) occurred when parallel and vertical cracks occurred simultaneously, as shown in Figure 5d [43,44]. It appears that the cracks are first initiated on the surface of the coating layer and subsequently propagate in parallel, ultimately leading to the detachment of the coating layer.

Finally, it is validated that the deformation of the coating layer is intensified as the normal load increases, as explained by the evidence of plowing. While it has been reported that substrate deformation can occur as a consequence of sliding when the surface is coated [45], our present study indicates minimal signs of substrate deformation (Figure 5). Consequently, it is inferred that the increase in friction caused by substrate deformation is expected to be insignificant.

3.3. Analysis of Asperity Flattening on Sliding Surface

It has often been reported that the flattening of surface asperity is one of the major factors in determining contact characteristics and frictional force during sliding [23–25]. According to previous studies, the friction coefficient decreases as the contact normal force increases [18]. In this section, the friction behavior is further analyzed in relation to the asperity flattening for different zinc-coated steel sheets.

Figures 6–9 show the roughness images of flattened surfaces and asperity height distribution measured by the 3D confocal microscopy. The experimental results suggest that differences in surface roughness deformation appear to be influenced by the coating and substrate strength. However, since the substrate strength normalized the contact pressure, the increased flattening of surface roughness in materials with higher substrate strength may simply be a result of this normalization. Therefore, it is difficult to conclude that the substrate strength itself directly affected the roughness deformation.

Figure 6 presents the 3D confocal microscopy results for the TS340-GI specimen. Each height distribution was averaged from the data obtained at four different locations. Also, the different colors in the figure indicate the surface roughness for the sliding velocities of 0.05 m/min, 0.2 m/min, and 0.9 m/min. The surface roughness of the as-received specimen before the friction test was also included in the figure for comparison purposes. Figure 6a–c represents the surface roughness of three different contact pressures. The figures show that the change in asperity distribution noticeably depended on the contact pressure. On the other hand, sliding velocities appear to have less impact on the asperity distribution except under the high contact pressure. It was confirmed that the asperity deformation decreased as the sliding speed increased under high contact pressure. For example, the distributions of surface asperity were virtually similar for all sliding velocities when the normalized contact pressure was 0.03 (Figure 6a), but the surface roughness became much flattened with a very narrow distribution when the normalized contract pressure increased over 0.1 (Figure 6b,c). The asperity height distributions of TS340-GA for various contact conditions are depicted in Figure 7. A similar trend was noticed in the case of the TS340-GA specimen. In contrast to the findings from the TS 340-GI in Figure 6, the flattened surfaces on TS340-GA were less pronounced.

For the TS980 steel, the flattening of asperity was much more significant for GI-coated specimens (Figure 8) than for the GA-coated specimens. The GI-coated surface formed on higher-strength steel showed considerable surface deformation even under the normalized contact pressure of 0.03. However, a much higher normalized contact pressure of 0.27 was necessary to flatten the asperity for the GA-coated specimens.

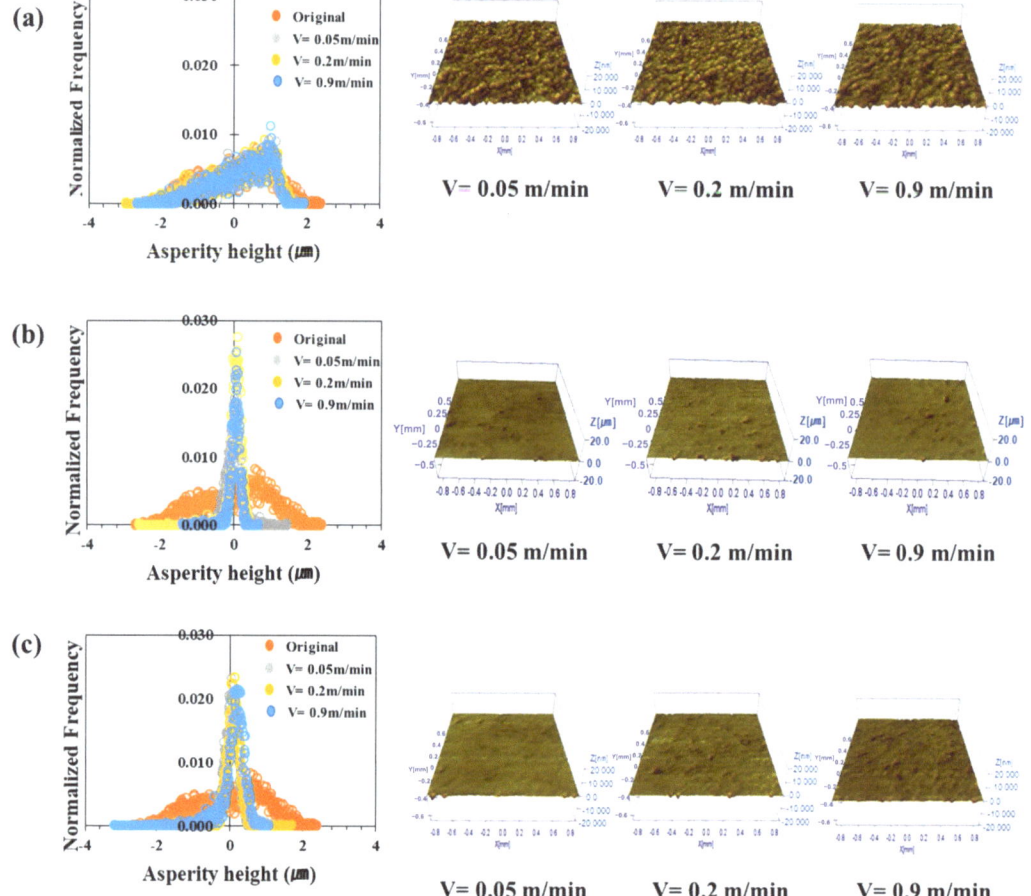

Figure 6. Asperity height distribution and 3D surface roughness images after friction tests on TS340-GI under different normalized contact pressures: (**a**) 0.03, (**b**) 0.1, and (**c**) 0.37.

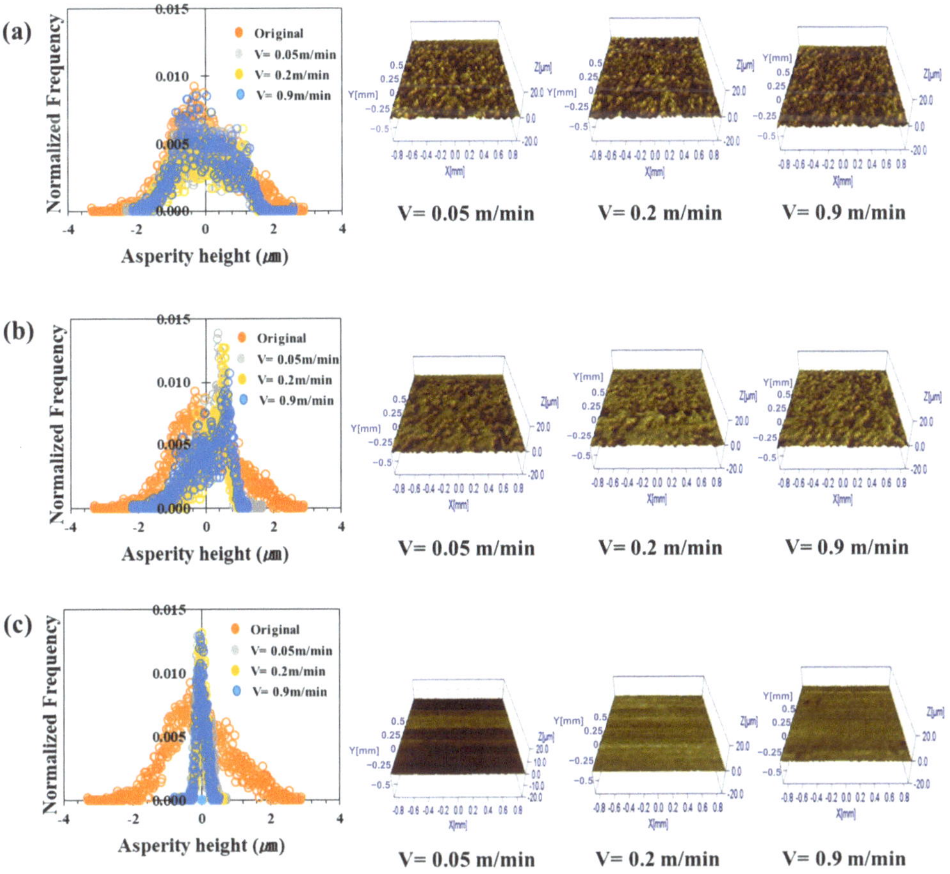

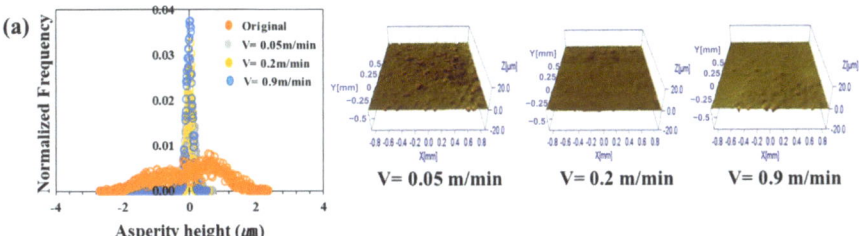

Figure 7. Asperity height distribution and 3D surface roughness images after friction tests on TS340-GA under different normalized contact pressures: (**a**) 0.03, (**b**) 0.1, and (**c**) 0.37.

Figure 8. *Cont.*

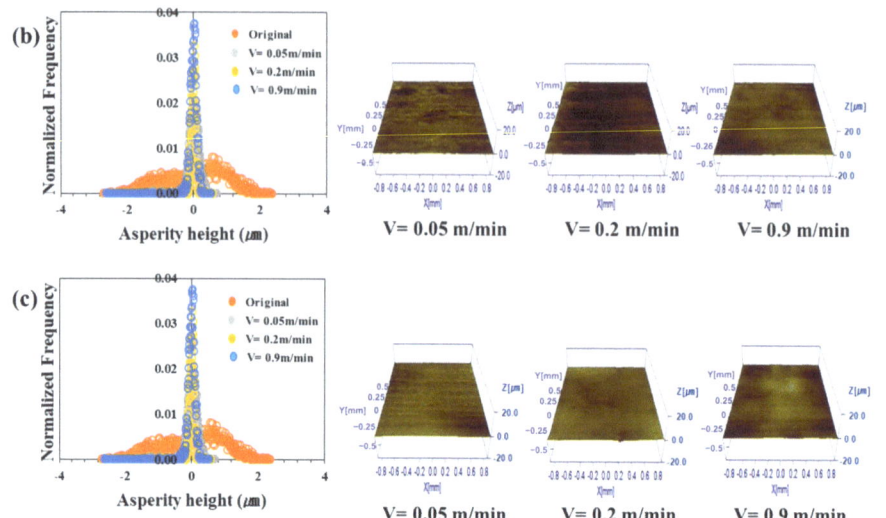

Figure 8. Asperity height distribution and 3D surface roughness images after friction tests on TS980-GI under different normalized contact pressures: (**a**) 0.03, (**b**) 0.1, and (**c**) 0.27.

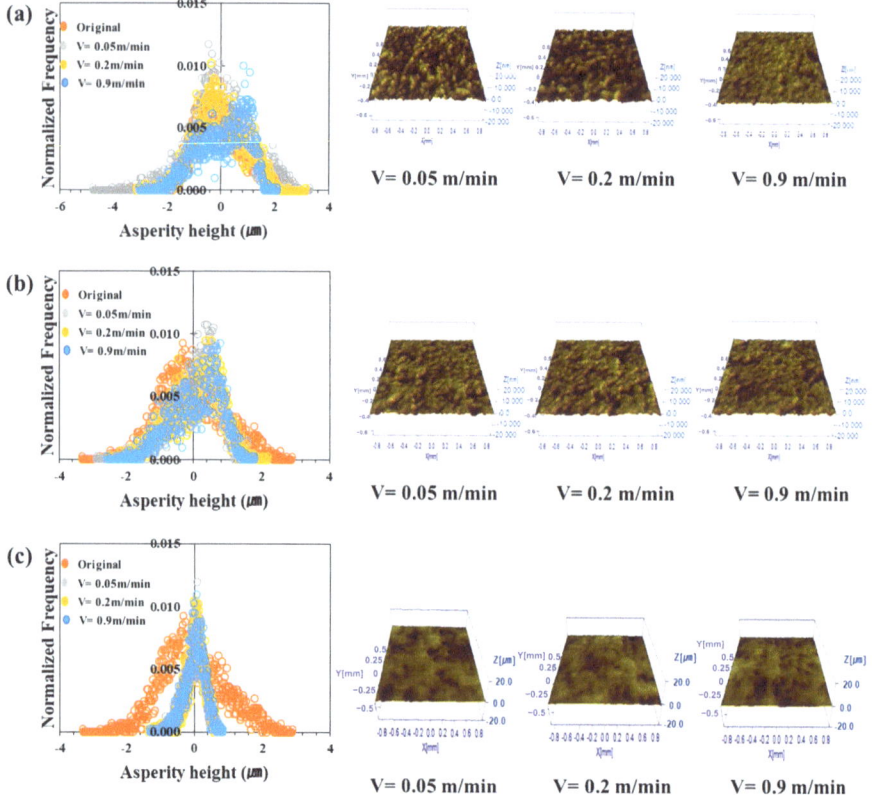

Figure 9. Asperity height distribution and 3D surface roughness images after friction tests on TS980-GA under different normalized contact pressures: (**a**) 0.03, (**b**) 0.1, and (**c**) 0.27.

4. Discussion

The primary aim of this study is to examine the effect of sliding contact conditions on the friction behavior of steels with different strengths and coatings and to identify the factors that contribute to the change of friction coefficient. For these purposes, in Section 4.1, we first analyzed the relationship between the coating type and friction behavior of the two steels with different tensile strengths. Here, we tried to confirm the validity of the conventional adhesion theory with an analysis of the stick–slip behavior at various contact pressures and sliding velocities. In Sections 4.2 and 4.3, the friction characteristics were discussed based on the observed asperity flattening as a result of contacts with different sliding conditions, which varied the friction coefficient.

4.1. Effect of Coating and Base Metal Properties on Friction Behavior

In terms of the effect of coating on friction observed in Figure 5, cracks propagated through the substrate surface of GA-coated steels regardless of the strength of the steel. Therefore, this may indicate that the strength of the coating layer has a larger effect on the formation of cracks than the other factors [46,47]. Tensile traction applied horizontally during sliding contact has been widely recognized as one of the primary factors for crack initiation and propagation. Under heavily loaded conditions, friction induces shear and plowing, which result in tensile stress within the coating layer, while compressive stress ahead of the sliding tool. The severity of this process escalates with an increase in both the hardness of the coating layer and the applied vertical load. If the coating layer becomes more brittle with higher hardness, it is more prone to fracture under tensile stress through coating thickness [48].

Meanwhile, the parallel cracks are known to be caused by stress accumulated through the sliding contact load in the coating layer. The stress can be quantified with the equivalent stress (typically defined as the von Misses stress), which increases as the applied load and sliding speed increases. The present study also validated it by showing increased parallel cracks as the contact pressure and sliding velocity increased. This is because parallel cracks are greatly affected by the shear and tensile stress simultaneously [49,50]. In our study, it was also confirmed that a larger number of fragmented cracks occurred in GA-coated steels than in GI-coated steels. This can be attributable to the higher hardness of GA coating than the GI coating, which leads to increased equivalent stress under contact surfaces [51]. Additionally, spallation by the delaminated coating layer initiated from parallel and vertical cracks was observed simultaneously [43]. Furthermore, more cracks were observed in the TS980 material (as shown in Figure 10). This is likely because the higher strength of the base metals reduces deformation under vertical loading, leading to increased stress concentration in the coating layer. Therefore, as the substrate strength increases, coating layer delamination and debris formation also appear more likely to increase.

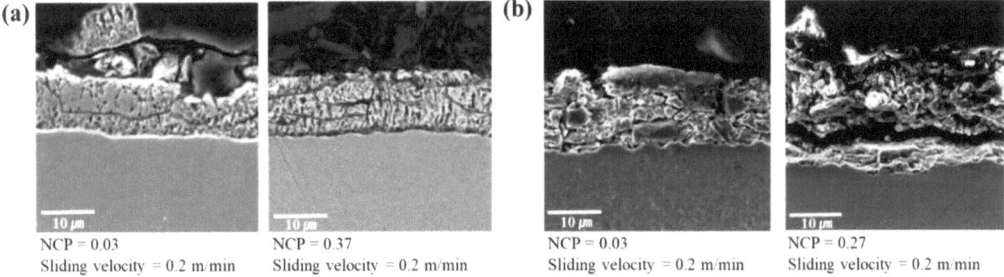

Figure 10. Comparison of coating layer cracks from friction tests: (a) TS340-GA, and (b) TS980-GA. NCP denotes the normalized contact pressure.

Adhesion can be an influencing factor for wear under sliding contact [28,52]. This study also showed adhesion behavior in the form of stick–slip response in the measured friction coefficient (See Figure A1 in Appendix A). The stick–slip is a phenomenon that occurs as evidence of adhesion. In the stick–slip motion, when force is applied to initiate sliding between the two surfaces, they initially stick together due to static friction. As the force continues to build, the static friction is overcome, and the surface suddenly slips or moves relative to each other. Compared to the GA-coated steels, the stick–slip behavior was predominantly observed in GI-coated steels. Meanwhile, the GI-coated steels showed a noticeable decrease in the friction coefficient in the region where adhesion (or stick–slip) was reduced.

As shown in Figure A1a, the TS340-GI specimen exhibited an increased adhesion (or stick–slip) as the normalized contact pressure increased from 0.03 to 0.1 for all three investigated sliding velocities. However, when the normalized contact pressure was increased to 0.37, the stick–slip decreased rapidly despite the flattening area increased. Also, in the case of the TS980-GI specimen, the stick–slip was most pronounced when the normalized contact pressure was 0.03, but it was reduced as it was increased to 0.27. This behavior was similar to the case of TS340-GI steel. These results are inconsistent with the commonly reported adhesion theory, because the real contact area and friction force increase simultaneously when the contact pressure increases, according to Orowan et al. [26,28]. However, some previous studies have suggested that under high load conditions, the deformation of asperities due to the high load can reduce the potential energy barrier for adhesion. In other words, as the vertical load lowers the roughness, the energy barrier decreases, which can lead to reduced friction [53,54].

In terms of the sliding velocity, Boden and Leben [55] reported that the relationship between adhesion and contact state (in terms of contact time and area) was proportional. Hence, the adhesion increased when the sliding speed was reduced with the increased contact time. In this study, as consistent with previous research, TS340-GI steel showed greater stick–slips at 0.05 m/min and 0.2 m/min conditions under 0.1 normalized contact pressure. However, the largest stick–slip was measured in TS980-GI steel at the sliding velocity of 0.2 m/min and normalized contact pressure of 0.03. Therefore, it seems that more complex effects, such as surface roughness and hardness, should also be considered when analyzing the stick–slip behavior [56]. In several studies, the friction coefficient decreased as the contact pressure and sliding velocity increased [9,18]. However, other studies also reported that the friction coefficient was highly dependent on the specific contact condition and material properties. For example, some researchers measured reduced friction with increasing contact pressure and lower surface roughness under lubrication. This is known to occur under the micro-EHL condition, where peaks of the most prominent asperities are removed, causing improved local hydrodynamic load-carrying capacity. The micro-EHL effect is characterized as thin-film lubrication since the height of the surface roughness is comparable to the film thickness. In the micro-EHL model, the friction coefficient tends to decrease as the contact area decreases under boundary lubrication conditions [57–61].

4.2. Summary of Sliding Velocity and Contact Pressure on the Asperity Flattening

The asperity flattening is one of the root causes of increasing friction between material and tools. To explain the change of the asperity flattening in the two investigated steel sheets, the percentage of the flattened area is plotted in Figure 11 for various contact conditions. In the figure, as the contact pressure increases, the flattening area increases. For the TS340-GI case, as the sliding velocity increased, the flattening area decreased under the normalized contact pressure over 0.1. This result appears to be related to the local side flow of the lubricant around asperities, resulting in the reduction of the lubricant film thickness with the increased sliding velocity [62]. The relationship between the sliding velocity and flattening area is still unclear in this study. The current study also showed that the increase in asperity flattening with increased contact pressure was more obvious in the GI steels. In the case of the GA-coated steels, the asperity flattening was less than that of the GI-coated steels under similar contact pressure. This is because of the higher strength of GA coating than the GI coating. The analysis can also be associated with a larger friction coefficient variation in the GI steels than that of the GA steels, regardless of the strength of the substrates (Figure 4).

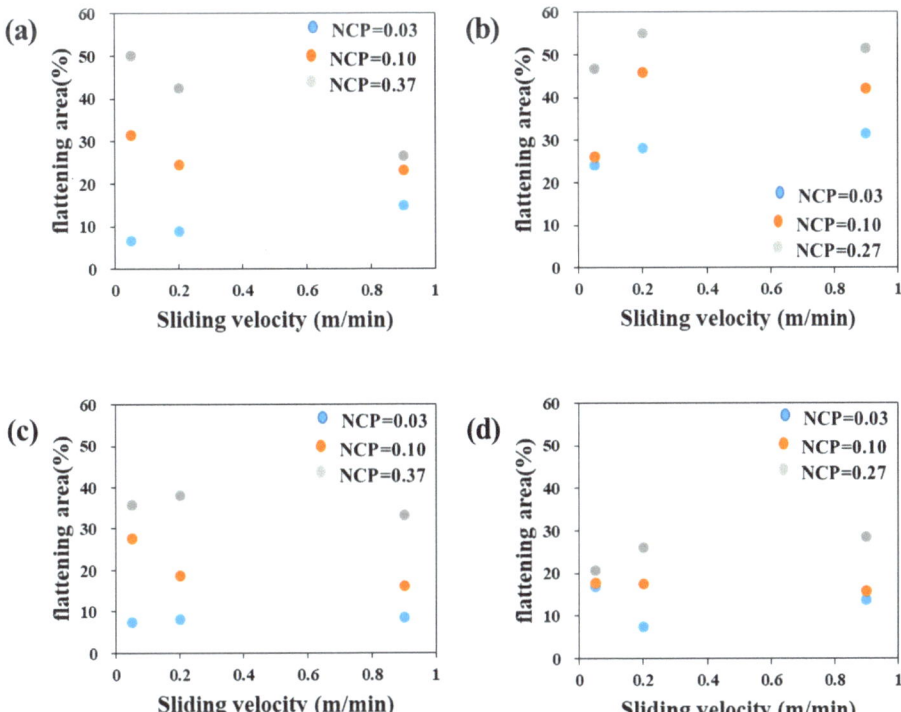

Figure 11. Measured flattening area (%) in terms of the sliding velocity and normalized contact pressure: (a) TS340-GI, (b) TS980-GI, (c) TS340-GA, and (d) TS980-GA.

4.3. Summary of Asperity Flattening on Friction Coefficient

Contrary to the classical analysis, the friction coefficient was not always linearly proportional to the asperity flattening. Figures 6–9 revealed a significant decrease in roughness with increased load. Moreover, there was a tendency for the friction coefficient to decrease as the load increased. Thus, as observed in a previous study, in a lubricated state, the flow can generate supporting force against the load, and under a larger load, it reduces direct contact between the asperities, thereby reducing the effect of asperity contact on friction and leading to lower friction coefficients [63].

The experiments showed that the asperity flattening depended highly on the applied contact pressure. The detailed results on the relationship between the friction coefficient and contact condition, coating layer, and substrate material are summarized in Figure 12. For the low normalized contact pressure of 0.03, the friction coefficient linearly decreased as the flattening area increased. However, it increased at the normalized contact pressure of 0.1 in the case of TS340-GI. When the contact pressures were maximum (0.3 and 0.27 for 340 MPa and 980 MPa steel, respectively), the overall friction coefficients were sharply decreased, especially for the GI-coated specimens. However, there was no meaningful relationship between the friction coefficient and flattening area. The most noticeable change in friction coefficient could be observed in the TS340-GI steel when the normalized contact pressure was 0.3. In this case, the friction coefficient was reduced even under 0.06 from 0.14

at the normalized contact pressure of 0.1. For the GA-coated steels, the overall effect of the contact pressure on the friction coefficient was less. These results indicate that the abrupt reduction of the friction coefficient, which cannot be simply estimated from the empirical formula in terms of contact pressure, might be attainable in the sheet-forming process if the coating and process conditions are properly controlled.

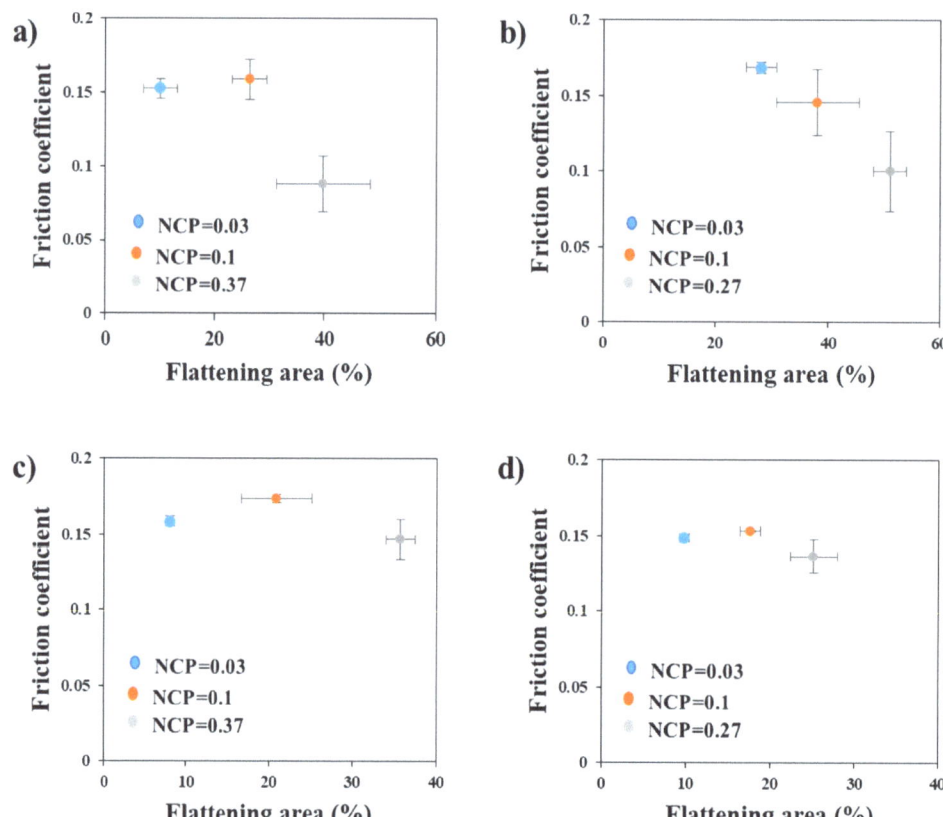

Figure 12. Friction coefficient vs. flattening area (%) under different normalized contact pressure: (**a**) TS340-GI, (**b**) TS980-GI, (**c**) TS340-GA, and (**d**) TS980-GA.

5. Conclusions

The effect of contact characteristics such as the contact pressure and sliding velocity on the frictional behavior was investigated for two types of zinc coating on two strength-grade steel sheets. The two coating conditions, GI and GA coatings, were applied to 340 MPa and 980 MPa strength steels. To analyze the effect of the strength of substrate steel, the contact pressure normalized by their tensile strength was used. More in-depth analysis of the friction was provided in relation to the surface characteristics, such as fractography analysis, coating layers, and asperity flattening behavior. The main findings of the present study can be summarized as follows.

(1) As the normalized contact pressure increased, the area of asperity flattening increased consistently regardless of the sliding velocity, substrate material, and coating condition. This is clearer and consistent in the case of GI-coated steels. At a normalized contact pressure of 0.37, the flattening area reached approximately 50% for TS340-GI and 47% for TS980-GI, confirming that higher pressure promotes surface deformation. In

(2) contrast, GA-coated steels exhibited lower asperity flattening under similar conditions, likely due to the higher strength of the coating. Meanwhile, the effect of sliding velocity on the asperity flattening was rather minor without a clear tendency.

(2) However, for the GA-coated steels, the increase of the asperity flattening area was less than that of the GI-coated steels under a similar magnitude of normalized contact pressure. The difference in the effect of contact pressure on the asperity flattening can be related to the strength of the coating layer. The greater influence of contact pressure is attributed more to the softer GI-coating layer than the GA coating.

(3) The friction coefficient can be highly related to the material strength and coating layer combination. For example, an abrupt drop in friction coefficient could be observed when the asperity flattening area exceeded a certain critical limit. The most significant reduction occurred at the highest normalized contact pressure of 0.37, where the friction coefficient dropped by over 30%. In comparison, GA-coated steels exhibited a more stable friction coefficient with less variation across different contact pressures and sliding velocities. The friction coefficient for GA-coated steels ranged between 0.125 and 0.177, while GI-coated steels showed a wider range between 0.059 and 0.18.

(4) Therefore, in the presence of a lubrication system, it is plausible that a specific regime of contact condition may exhibit a considerable reduction of friction coefficient despite the enlarged occurrence of asperity flattening caused by sliding. This result indicates that the abruptly low friction coefficient, which may not be simply estimated from the conventionally utilized empirical formula based on the contact pressure, can exist in the industrial forming process.

Author Contributions: Conceptualization, J.-Y.K., B.-K.J., J.-S.H. and M.-G.L.; methodology, J.-Y.K. and S.-C.Y.; investigation, J.-Y.K. and S.-C.Y.; validation, J.-Y.K. and J.-H.J.; writing–original draft preparation, J.-Y.K.; funding acquisitions, J.-S.H., J.-H.J. and M.-G.L.; supervision, M.-G.L.; writing–review and editing, M.-G.L. All authors have read and agreed to the published version of the manuscript.

Funding: This work was supported by Hyundai Steel Company, which is greatly appreciated. M.G.L. appreciates the partial support of the Technology Innovation Program (No. 1415185590, 20022438) funded by the MOTIE and the Institute of Engineering Research at Seoul National University.

Institutional Review Board Statement: Not applicable.

Informed Consent Statement: Not applicable.

Data Availability Statement: The original contributions presented in the study are included in the article, further inquiries can be directed to the corresponding author/s.

Conflicts of Interest: Authors Ji-Young Kim, Seung-Chae Yoon, Byeong-Keuk Jin, Jin-Hwa Jeon and Joo-Sik Hyun were employed by the company Hyundai Steel Company. The remaining authors declare that the research was conducted in the absence of any commercial or financial relationships that could be construed as a potential conflict of interest.

Abbreviation

GI galvanized
GA galvannealed
SEM scanning electron microscopy
EDS energy dispersive spectroscopy
NCP normalized contact pressure
EHL elasto-hydrodynamic lubrication

Appendix A

Detailed data for the characteristics of frictional behavior. The following figures and tables present the measured friction coefficients and frictional characteristics under various contact conditions.

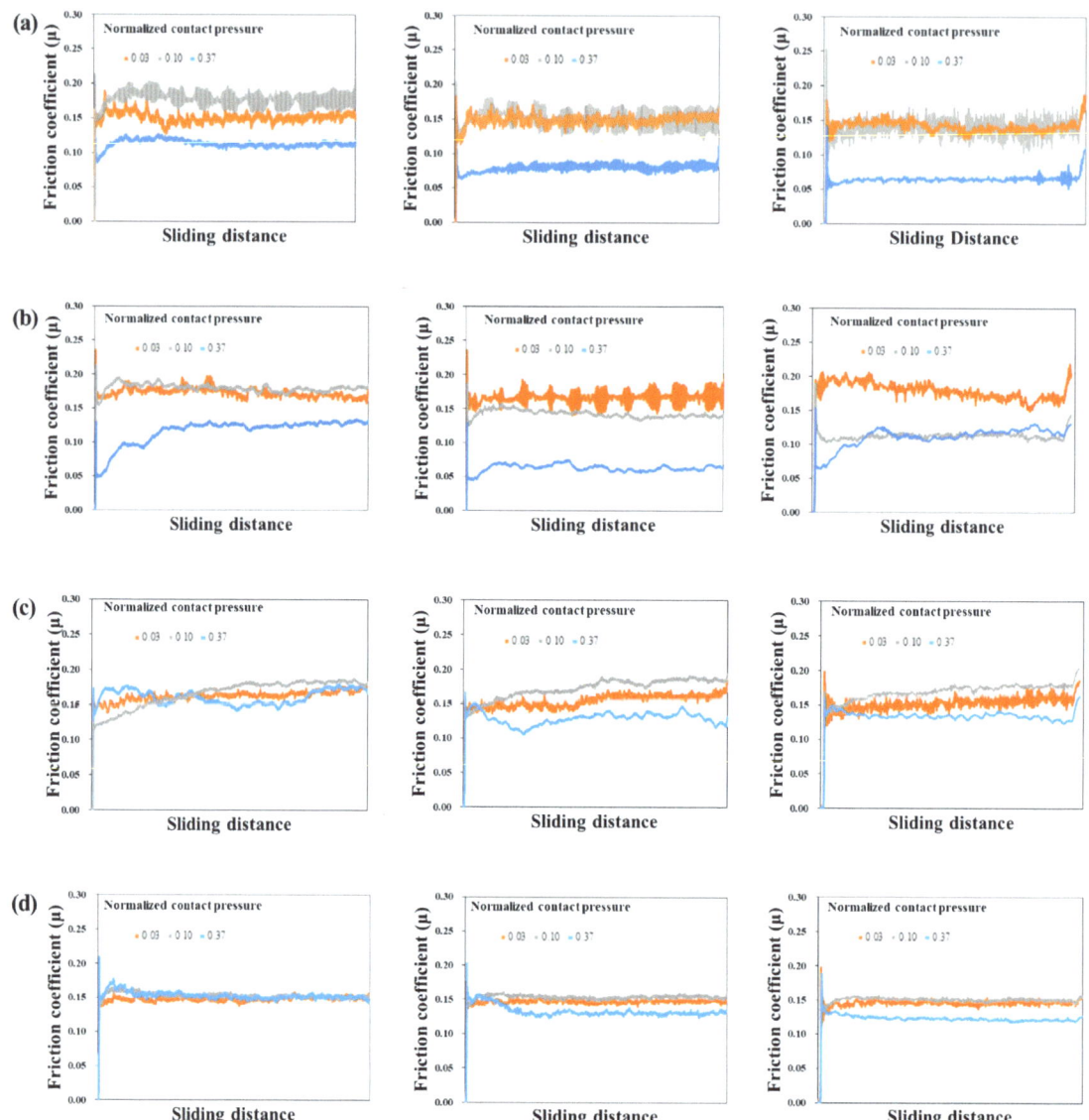

Figure A1. Comparison of Friction graphs with respect to sliding velocity and contact pressure: (**a**) TS340-GI, (**b**) TS980-GI, (**c**) TS340-GA, and (**d**) TS980-GA. For each coating condition, three sliding velocities are considered as (**left**) 0.05 m/min, (**center**) 0.2 m/min, and (**right**) 0.9 m/min.

Table A1. Frictional characteristics of tested specimens: (a) TS340-GI, (b) TS980-GI, (c) TS340-GA, and (d) TS980-GA.

Sliding Velocity	Normalized Contact Pressure	Friction Mode			Wear (Fracture)		
		Flattening (%)	Stick–Slip	Ploughing	Parallel Crack	Vertical Crack	Spalling
(a)							
0.05	0.03	6.69		medium	X	X	X
0.2	0.03	8.96		medium	X	X	X
0.9	0.03	14.97		small	X	X	X
0.05	0.1	31.52		large	X	X	X
0.2	0.1	24.57		large	X	X	X
0.9	0.1	23.28		large	X	X	X
0.05	0.37	50.20		small	X	X	X
0.2	0.37	42.41		medium	X	X	X
0.9	0.37	26.64		small	X	X	X
(b)							
0.05	0.03	24.23		medium	O	X	X
0.2	0.03	28.11		large	O	X	X
0.9	0.03	31.58		medium	O	X	X
0.05	0.1	26.12		small	O	X	X
0.2	0.1	45.86		small	O	X	X
0.9	0.1	41.98		small	O	X	X
0.05	0.27	46.72		small	O	X	X
0.2	0.27	55.08		small	O	X	X
0.9	0.27	51.39		small	O	X	X
(c)							
0.05	0.03	7.43		medium	△	△	X
0.2	0.03	8.15		medium	O	O	X
0.9	0.03	8.55		medium	O	O	O
0.05	0.1	27.65		small	O	△	O
0.2	0.1	18.82		small	O	O	O
0.9	0.1	16.24		small	O	O	O
0.05	0.37	35.69		small	△	O	O
0.2	0.37	38.06		small	O	O	O
0.9	0.37	33.14		small	O	O	O
(d)							
Sliding velocity	Normalized contact pressure	Flattening (%)	Stick–slip		Parallel crack	Vertical crack	Spalling
0.05	0.03	4.64	small		X	O	X
0.2	0.03	9.37	small		△	O	X
0.9	0.03	13.7	small		O	O	X
0.05	0.1	17.7	small		△	O	X
0.2	0.1	19.85	small		O	O	O
0.9	0.1	15.75	small		O	O	O
0.05	0.27	20.74	small		O	O	O
0.2	0.27	26.07	small		O	O	O
0.9	0.27	28.49	small		O	O	O

Note: X: barely observed, O: almost always observed, △: occasionally observed.

References

1. Lee, M.G.; Kim, C.; Pavlina, E.J.; Barlat, F. Advances in sheet forming—Materials modeling, numerical simulation, and press technologies. *J. Manuf. Sci. Eng.* **2011**, *133*, 61001. [CrossRef]
2. Yoon, M.S.; Seo, B.K.; Lim, C.S. Reverse compensation to prevent post-molding dimensional distortion of automobile parts manufactured using cf-smc. *Int. J. Adv. Manuf. Technol.* **2022**, *123*, 4181–4194. [CrossRef]

3. Barlat, F.; Brem, J.C.; Yoon, J.W.; Chung, K.; Dick, R.E.; Lege, D.J.; Pourboghrat, F.; Choi, S.H.; Chu, E. Plane stress yield function for aluminum alloy sheets—Part 1: Theory. *Int. J. Plast.* **2003**, *19*, 1297–1319. [CrossRef]
4. Song, Y.J.; Oh, I.S.; Hwang, S.H.; Choi, H.J.; Lee, M.G.; Kim, H.J. Numerically efficient sheet metal forming simulations in consideration of tool deformation. *Int. J. Autom. Technol.* **2021**, *22*, 69–79. [CrossRef]
5. Yoon, M.S.; Kang, G.H. Effect of dynamic friction and static friction in finite element analysis of carbon fiber preform. *Polym. Compos.* **2023**, *44*, 2396–2404. [CrossRef]
6. Bandyopadhyay, K.; Basak, S.; Choi, H.J.; Panda, S.K.; Lee, M.G. Influence of evolution in anisotropy during strain path change on failure limits of sheet metals. *Met. Mater. Int.* **2021**, *27*, 3225–3237. [CrossRef]
7. Lee, J.W.; Bong, H.J.; Kim, D.Y.; Lee, Y.S.; Choi, Y.M.; Lee, M.G. Mechanical properties and formability of heat-treated 7000-series high-strength aluminum alloy: Experiments and finite element modeling. *Met. Mater. Int.* **2020**, *26*, 682–694. [CrossRef]
8. Song, J.; Lan, J.; Zhu, L.; Jiang, Z.; Zhang, Z.; Han, J.; Ma, C. Finite element simulation and microstructural analysis of roll forming for dp590 high-strength dual-phase steel wheel rims. *Materials* **2024**, *17*, 3795. [CrossRef]
9. Greenwood, J.A.; Williamson, J.B.P. Contact of nominally flat surfaces. *Proc. R. Soc. A Lond.* **1966**, *295*, 300–319. [CrossRef]
10. Han, S.S. Contact pressure effect on frictional behavior of sheet steel for automotive stamping. *Trans. Mater. Process* **2011**, *20*, 99–103. [CrossRef]
11. Jang, G.H.; Cho, K.H.; Park, S.B.; Lee, W.G.; Hong, U.S.; Jang, H. Tribological properties of c/c-sic composites for brake discs. *Met. Mater. Int.* **2010**, *16*, 61–66. [CrossRef]
12. Trzepieciński, T.; Szwajka, K.; Szewczyk, M.; Barlak, M.; Zielińska-Szwajka, J. Effect of countersample coatings on the friction behaviour of dc01 steel sheets in bending-under-tension friction tests. *Materials* **2024**, *17*, 3631. [CrossRef] [PubMed]
13. Kasar, A.K.; Jose, S.A.; D'Souza, B.; Menezes, P.L. Fabrication and tribological performance of self-lubricating porous materials and composites: A review. *Materials* **2024**, *17*, 3448. [CrossRef] [PubMed]
14. Bolay, C.; Essig, P.; Kaminsky, C.; Hol, J.; Naegele, P.; Schmidt, R. Friction modelling in sheet metal forming simulations for aluminium body parts at Daimler AG. *IOP Conf. Ser. Mater. Sci. Eng.* **2019**, *651*, 012104. [CrossRef]
15. Sigvant, M.; Pilthammar, J.; Hol, J.; Wiebenga, J.H.; Chezan, T.; Carleer, B.; van den Boogaard, A.H. Friction and lubrication modeling in sheet metal forming simulations of a Volvo XC90 inner door. *IOP Conf. Ser. Mater. Sci. Eng.* **2016**, *159*, 012021. [CrossRef]
16. Lee, K.J.; Moon, C.M.; Lee, M.G. A review on friction and lubrication in automotive metal forming: Experiment and modeling. *Int. J. Autom. Technol.* **2021**, *22*, 1743–1761. [CrossRef]
17. Lee, J.Y.; Barlat, F.; Lee, M.G. Constitutive and friction modeling for accurate springback analysis of advanced high strength steel sheets. *Int. J. Plast.* **2015**, *71*, 113–135. [CrossRef]
18. Bay, N.; Wanheim, T. Real area of contact and friction stress at high pressure sliding contact. *Wear* **1976**, *38*, 201–209. [CrossRef]
19. Bang, J.H.; Park, N.S.; Song, J.H.; Kim, H.G.; Bae, G.H.; Lee, M.G. Tool wear prediction in the forming of automotive dp980 steel sheet using statistical sensitivity analysis and accelerated u-bending based wear test. *Metals* **2021**, *11*, 306. [CrossRef]
20. Kim, C.M.; Lee, J.U.; Barlat, F.; Lee, M.G. Frictional behaviors of a mild steel and a Trip780 steel under a wide range of contact stress and sliding speed. *J. Tribol.* **2014**, *136*, 21606. [CrossRef]
21. Azushima, A.; Kudo, H. Direct observation of contact behaviour to interpret the pressure dependence of the coefficient of friction in sheet metal forming. *CIRP Ann.* **1995**, *44*, 209–212. [CrossRef]
22. Wu, C.; Zhang, L.; Qu, P.; Li, S.; Jiang, Z. Multiscale interface stress characterisation in cold rolling. *Met. Mater. Int.* **2021**, *27*, 1997–2013. [CrossRef]
23. Kim, N.J.; Keum, Y.T. Experimental determination of friction characteristics for advanced high strength steel sheets. *Trans. Mater. Process* **2013**, *22*, 223–228. [CrossRef]
24. Bridgman, P.W. *The Physics of High Pressure*; Macmillan Press: New York, NY, USA, 1949.
25. Bednar, M.; Cai, B.; Kuhlmann-Wilsdorf, D. Pressure and structure dependence of solid lubricant. *Lubr. Eng.* **1993**, *49*, 741–749.
26. Suh, N.P.; Sin, H.C. The genesis of friction. *Wear* **1981**, *69*, 91–114. [CrossRef]
27. Bowden, F.P.; Tabor, D. Mechanism of metallic friction. *Nature* **1942**, *150*, 197–199. [CrossRef]
28. Bowden, F.P.; Tabor, D. *The Friction and Lubrication of Solids*; Oxford University Press: Oxford, UK, 1950; ISBN 9780198507772.
29. Orowan, E. The calculation of roll pressure in hot and cold flat rolling. *Proc. Inst. Mech. Eng.* **1943**, *150*, 140–167. [CrossRef]
30. Garza, L.G.; Van Tyne, C.J. Friction and formability of galvannealed interstitial free sheet steel. *J. Mater. Process. Technol.* **2007**, *187*, 164–168. [CrossRef]
31. Jang, Y. The Tribology and Formability of Zinc Coated Steel Sheets Subjected to Different Strain States. Bachelor's Thesis, Case Western Reserve University, Cleveland, OH, USA, 2010.
32. Ghosh, A.K. A method for determining the coefficient of friction in punch stretching of sheet metals. *Int. J. Mech. Sci.* **1977**, *19*, 457–470. [CrossRef]
33. Ghosh, A.K. The effect of lateral drawing- in on stretch formability. *Met. Eng. Quart.* **1975**, *15*, 53–64.
34. Ghosh, A.K.; Hecker, S. Failure in thin sheets stretched over rigid punches. *Met. Trans. A* **1975**, *6*, 1065–1074. [CrossRef]
35. *ISO 6502-2*; Rubber—Measurement of Vulcanization Characteristics Using Curemeters Part 2: Oscillating Disc Curemeter. ISO: Geneva, Switzerland, 2018.
36. Tabor, D. *The Hardness of Metals*; Oxford Clarendon Press: London, UK, 1951.
37. Boyer, H.E.; Gall, T.L. *Metals Handbook Desk Edition*, 2nd ed.; ASM International: Metals Park, OH, USA, 1985; pp. 1–60.

38. *ISO 27831-1*; Metallic Coatings and Other Inorganic Coatings—Cleaning and Preparation of Metal Surfaces—Part 1: Ferrous Metals and Alloys. ISO: Geneva, Switzerland, 2008.
39. *ISO 22462*; Metallic and Other Inorganic Coatings-Test Method for the Friction Coefficient Measurement of Chemical Conversion Coatings. ISO: Geneva, Switzerland, 2020.
40. Ishiwatari, A.; Urabe, M.; Inazumi, T. *Press Forming Analysis Contributing to the Expansion of High Strength Steel Sheet Applications*; JFE Technical Report No. 18; JFE: Shirur, India, April 2013.
41. Shisode, M.P.; Hazrati, J.; Mishra, T.; Rooij, M.B.; Boogaard, A.H. Semi-analytical contact model to determine the flattening behavior of coated sheets under normal load. *Tribol. Int.* **2020**, *146*, 106182. [CrossRef]
42. Moore, D.F. Chapter 13 automotive application. In *Principles and Applications of Tribology*; Oxford Pergamon Press: Dublin, Ireland, 1975. [CrossRef]
43. Belak, J.F. Nanotribology. *MRS Bull.* **1993**, *18*, 15–19. [CrossRef]
44. Halling, J. *Principles of Tribology*; Red Globe Press: London, UK, 1978. [CrossRef]
45. Holmberg, K.; Matthews, A. *Coatings Tribology Properties, Mechanisms, Techniques and Applications in Surface Engineering*; Elsevier Science: Amsterdam, The Netherlands, 2009.
46. Suh, N.P. The delamination theory of wear. *Wear* **1973**, *25*, 111–124. [CrossRef]
47. Suh, N.P. An overview of the delamination theory of wear. *Wear* **1977**, *44*, 1–16. [CrossRef]
48. Arnell, R.D. The mechanics of the tribology of thin film systems. *Surf. Coat. Technol.* **1990**, *43–44*, 674–687. [CrossRef]
49. Larsson, M.; Hedenqvist, P.; Hogmark, S. Deflection measurements as method to determine residual stress in thin hard coatings on tool materials. *Surf. Eng.* **1996**, *12*, 43–48. [CrossRef]
50. Mansour, H.A.; Arnell, R.D.; Salama, M.A.; Mustfa, A.A.F. Finite elements studies of coated surfaces under normal and tangential loads. In Proceedings of the 6th International Conference on Ion Plasma Assisted Techniques, Brighton, UK, 27–29 May 1987; pp. 179–183.
51. Cowan, R.S.; Winer, W.O. Application of the thermomechanical wear transition model to layered media. *Tribol. Ser.* **1993**, *25*, 631–639. [CrossRef]
52. Djabella, H.; Arnell, R.D. Finite element analysis of the contact stresses in elastic coating/substrate under normal and tangential load. *Thin Solid Films* **1993**, *223*, 87–97. [CrossRef]
53. Wang, A.E.; Gil, P.S.; Holonga, M.; Yavuz, Z.; Baytekin, H.T.; Sankaran, R.M.; Lacks, D.J. Dependence of triboelectric charging behavior on material microstructure. *Phys. Rev. Mater.* **2017**, *1*, 35605. [CrossRef]
54. Yu, J.X.; Qian, L.M.; Yu, B.J.; Zhou, Z.R. Nanofretting behaviors of monocrystalline silicon (100) against diamond tips in atmosphere and vacuum. *Wear* **2009**, *267*, 322–329. [CrossRef]
55. Bowden, F.P.; Leben, L. The nature of sliding and the analysis of friction. *Nature* **1938**, *141*, 691–692. [CrossRef]
56. Blau, P.J. *Friction Science and Technology*; CRC Press: Boca Raton, FL, USA, 2009; p. 148, Tables 4–7, ISBN 978-14200-5404-0.
57. Hansen, J.; Björling, M.; Larsson, R. A new film parameter for rough surface EHL contacts with anisotropic and isotropic structures. *Tribol. Lett.* **2021**, *69*, 37. [CrossRef]
58. Dowson, D. A comparative study of the performance of metallic and ceramic femoral head components in total replacement hip joints. *Wear* **1995**, *190*, 171–183. [CrossRef]
59. Spikes, H.A. Mixed lubrication—An overview. *Lubr. Sci.* **1997**, *9*, 221–253. [CrossRef]
60. Wang, Y.; Azam, A.; Zhang, G.; Dorgham, A.; Liu, Y.; Wilson, M.; Neville, A. Understanding the mechanism of load-carrying capacity between parallel rough surfaces through a deterministic mixed lubrication model. *Lubricants* **2022**, *10*, 12. [CrossRef]
61. Fowles, P.E. The application of elastohydrodynamic lubrication theory to individual asperity-asperity collisions. *ASME J. Lubr. Tech.* **1969**, *91*, 464–475. [CrossRef]
62. Chang, L.; Webster, M.N. A study of elastohydrodynamic lubrication of rough surfaces. *ASME J. Tribol.* **1991**, *113*, 110–115. [CrossRef]
63. Zheng, X.; Zhu, H.; Tieu, A.K. Roughness and lubricant effect on 3d atomic asperity contact. *Tribol. Lett.* **2014**, *53*, 215–223. [CrossRef]

Disclaimer/Publisher's Note: The statements, opinions and data contained in all publications are solely those of the individual author(s) and contributor(s) and not of MDPI and/or the editor(s). MDPI and/or the editor(s) disclaim responsibility for any injury to people or property resulting from any ideas, methods, instructions or products referred to in the content.

Article

High-Precision Instance Segmentation Detection of Micrometer-Scale Primary Carbonitrides in Nickel-Based Superalloys for Industrial Applications

Jie Zhang *, Haibin Zheng, Chengwei Zeng and Changlong Gu

College of Mechanical Engineering, Zhejiang University of Technology, Hangzhou 310014, China; 18858274591@163.com (H.Z.); 221123020460@zjut.edu.cn (C.G.)
* Correspondence: zhangjie0231@zjut.edu.cn; Tel.: +86-188-0018-5456

Abstract: In industrial production, *the* identification and characterization of micron-sized second phases, such as carbonitrides in alloys, hold significant importance for optimizing alloy compositions and processes. However, conventional methods based on threshold segmentation suffer from drawbacks, including low accuracy, inefficiency, and subjectivity. Addressing these limitations, this study introduced a carbonitride instance segmentation model tailored for various nickel-based superalloys. The model enhanced the YOLOv8n network structure by integrating the SPDConv module and the P2 small target detection layer, thereby augmenting feature fusion capability and small target detection performance. Experimental findings demonstrated notable improvements: the mAP50 (Box) value increased from 0.676 to 0.828, and the mAP50 (Mask) value from 0.471 to 0.644 for the enhanced YOLOv8n model. The proposed model for carbonitride detection surpassed traditional threshold segmentation methods, meeting requirements for precise, rapid, and batch-automated detection in industrial settings. Furthermore, to assess the carbonitride distribution homogeneity, a method for quantifying dispersion uniformity was proposed and integrated into a data processing framework for seamless automation from prediction to analysis.

Keywords: carbonitride; deep learning; instance segmentation; dispersion

Citation: Zhang, J.; Zheng, H.; Zeng, C.; Gu, C. High-Precision Instance Segmentation Detection of Micrometer-Scale Primary Carbonitrides in Nickel-Based Superalloys for Industrial Applications. *Materials* 2024, 17, 4679. https://doi.org/10.3390/ma17194679

Academic Editors: Seong-Ho Ha, Young-Ok Yoon, Young-Chul Shin and Dong-Earn Kim

Received: 23 July 2024
Revised: 4 September 2024
Accepted: 11 September 2024
Published: 24 September 2024

Copyright: © 2024 by the authors. Licensee MDPI, Basel, Switzerland. This article is an open access article distributed under the terms and conditions of the Creative Commons Attribution (CC BY) license (https://creativecommons.org/licenses/by/4.0/).

1. Introduction

Nickel-based superalloys represent a specialized alloy system designed for applications under elevated temperatures and complex stress conditions. Operating in such environments necessitates exceptional mechanical properties, including high-temperature strength, fatigue and creep resistance, fracture toughness, and microstructure stability [1]. Additionally, these alloys must exhibit surface stability to withstand environmental challenges such as high-temperature oxidation and corrosion. Due to their comprehensive performance in high-temperature operational settings, nickel-based superalloys are extensively utilized in critical hot-end components for engines and gas turbines within the aerospace industry.

The superior performance of nickel-based superalloys at elevated temperatures can be attributed to their intricate chemical composition and sophisticated manufacturing processes. Optimization of their properties predominantly involves alloying strategies, such as the introduction of specific elements like Ti and Nb into the alloys. These elements facilitate the formation of finely dispersed intermetallic compounds such as γ' phase Ni_3 (Al, Ti) or γ'' phase Ni_3 (Nb, Al, Ti), which contribute to precipitation strengthening, solid-solution strengthening, and grain-boundary strengthening. Consequently, these alloys exhibit high strength and exceptional creep resistance in high-temperature environments [2]. However, Nb and Ti also serve as carbide and nitride formers, leading to the precipitation of carbonitride phases during solidification, either in the liquid or solid two-phase region, due to microscopic segregation [3,4]. These are known as primary carbonitrides, and their sizes can be quite large, even reaching tens of micrometers. The enhancement in the mechanical properties of materials by fine precipitates in alloys is unquestionable. However, when the key alloying elements in an alloy fail to effectively form fine or nanoscale-sized precipitates and instead exist in the material as larger, hard particles, these particles are likely to have a negative impact on the mechanical properties of the steel. The detrimental effects of such large-sized hard particles on alloys are comparable to the negative impacts of large, undeformed oxide particles on the fatigue life, impact toughness, and other properties of steel [3,5]. In industrial production, the identification and characterization of micron-sized second phases, such as carbonitrides in alloys, are critical for subsequent composition adjustment and process improvement.

To measure the effect of primary carbonitrides in microstructures, researchers often performed quantitative evaluations of these carbonitrides. The traditional recognition method primarily relied on software such as Image-Pro Plus 6.0 [6]. Import the image into the software and adjust the RGB values to effectively recognize carbonitrides. However, this method mainly had the following disadvantages:

(1) It processed images singly, lacking batch processing capability and necessitating manual intervention, which was time-consuming and labor-intensive.
(2) Images with significant impurities or noise could hinder effective carbonitride recognition despite parameter adjustments, thereby impacting data processing.
(3) The method's reliance on operator expertise introduced subjectivity into results [7].

Moreover, the high demands on picture quality imposed by Image-Pro Plus 6.0 and similar software further complicate alloy sample imaging and preprocessing in industrial environments, where variability and complexity are prevalent. Consequently, such software may only recognize a subset of high-quality images.

With the development of deep learning technology, the application areas of computer vision become increasingly extensive. Many researchers have started to develop automatic identification models by combining rapid image acquisition techniques and deep learning [8] frameworks. This approach aimed to address the shortcomings of traditional methods and accommodate the unique characteristics of alloy microstructures. Brian et al. [9] investigated and discussed three feature extraction methods: BoW, the VLAD coding, and CNN networks. The SVM algorithm was used to test the performance of each feature extraction method in the task of classification of ultrahigh carbon steel microstructures. Additionally, the clustering and correlation of the microstructures were observed by reducing the high-dimensional space to a two-dimensional space using the t-SNE algorithm. Li et al. [10] utilized the U-Net architecture as a base network model and employed a deep transfer learning approach to identify the γ' phase in nickel-based high-temperature alloys at 900 °C and 1000 °C, achieving an identification accuracy of 92%. Azimi et al. [11] employed the MVFCNN network structure for the classification and segmentation of martensite, tempered martensite, bainite, and pearlite in mild steel, achieving high accuracy rates. Ghauri et al. [12] utilized the RF algorithm for the segmentation detection of carbides in HP40-Nb stainless steel, achieving notable segmentation accuracies at intergranular and grain boundaries.

Previous generations have conducted a lot of work on the combination of microstructure detection and deep learning model of nickel-based superalloy. For example, Jia et al. [13] used the Unet++ network model to detect the γ' phase in nickel-based superalloy. They obtained training data through SEM and cut the image size to 512 × 512 to facilitate model training. The Unet++ model is used to segment the γ' phase, and the mIoU (mean Intersection over Union) value reaches 0.98. That is, the Unet++ network can identify the γ' phase in nickel-based superalloys well. Senanayake et al. [14] used different methods to identify the γ' phase and γ'' phase in IN718 alloys, including digital image processing, random forest algorithm (RF), support vector machine (SVM), convolutional neural network (CNN). They used scanned electron phases of IN718 alloy taken at NASA Glenn Research Center as a training dataset. The experimental results show that the use of the convolutional neural network (CNN) can obtain the fastest recognition speed and the most accurate recognition results, and the recognition accuracy rate is 0.95. Previous studies have shown that a deep learning network is the best solution for the identification of microstructure in nickel-based superalloys.

In this study, a microstructure segmentation model was proposed for large-size primary carbonitrides in nickel-based superalloys based on an improved YOLOv8 framework. The improved model for detecting carbonitrides offered a complete alternative to the threshold segmentation method, which satisfied the requirements for high-precision, high-speed, and batch-automated detection in industrial scenarios and was not affected by the quality of the input image. Additionally, the article introduced a method for assessing carbonitride distribution homogeneity, integrating it into a data processing program for automated processing from prediction to analysis. Finally, the YOLOV8-trained model has been able to be deployed in industrial sites, directly applied to on-site production processes, and reduce the operating time of workers.

2. Improved YOLOv8n Instance Segmentation Algorithm

The YOLOv8 algorithm, released in 2023, was one of the most advanced deep-learning models available. It included the following major improvements:

(1) Compared with YOLOv5, YOLOv8 replaced the C3 module in the backbone network with the C2f module. The C2f module integrated the CSP structure and the ELAN [15] concept from YOLOv7. This integration enhanced the feature extraction capability of the YOLOv8 network and reduced computation and model complexity. The SPPF [16] module is retained at the conclusion of the backbone network, facilitating multi-scale feature fusion, thereby enhancing the detection capabilities of the model.

(2) The neck network, similar to YOLOv5, still adopted the PAN-FPN [17,18] feature fusion method. It removed the Conv module in the upsampling stage and replaced the C3 module with the C2f module. These changes maintain the advantages of the YOLOv5 network and improve the detection performance of the model in various scenarios.

(3) For the detection head, YOLOv8 used a decoupled head structure to separate the classification and detection tasks, along with the Anchor-Free [19] algorithm.

(4) The practice of using Mosaic [20] data augmentation during training and turning it off for the last 10 epochs was effective in improving model robustness [21].

Despite these advancements, YOLOv8 encountered challenges in effectively detecting small and densely clustered targets. Previous research has proposed several enhancement methods to address these limitations. For instance, Lou et al. [22] introduced the DC-YOLOv8 algorithm for the detection of small targets captured by cameras. This enhancement to the network's learning capability was achieved by revising the original downsampling module to an MDC module, swapping out the C2f module for a DC module, and refining the feature fusion process. Li et al. [23] enhanced the Neck layer of the original network by incorporating the BIFPN concept and replacing the C2f module with the Ghostblock module. They also refined the original network's CIoU with WIoU, applying these improvements to address target detection in UAV aerial imagery. The enhanced model not only reduces the complexity of the model but also the miss rate for small targets. Wang et al. [24] introduced the YOLOv8-QSD network to address the challenge of small target detection in unmanned scenarios. This network incorporated a DBB module in place of the traditional Bottleneck module within the C2f module. It also integrated a BIFPN structure to enhance the original network's Neck layer. Furthermore, the authors added a novel dynamic detection head, termed DyHead, to the network's Head layer. With these enhancements, the YOLOv8-QSD network achieved an accuracy of 64.5% on a dataset specifically designed for small target detection.

In industrial applications, the features of alloy carbonitrides were assessed through the examination of microstructures under an optical microscope. In this study, the carbonitride dataset constructed for nickel-based high-temperature alloys was obtained using a metallurgical microscope with image sizes of 2240×1524 pixels. The pixel count for individual carbonitrides ranged from approximately 50 to 5000. The MS COCO [25] dataset categorized targets as small if they were below 32 × 32 pixels in size, with the total pixel count for these small targets typically being around or under 1000. In this research, we initially quantified the size distribution of typical carbonitrides in our dataset, as illustrated in Figure 1. The findings revealed that 77.72% of the carbonitrides were classified as small targets, indicating that the majority of carbonitrides examined in this study fall into this category. For ease of calculation, it was assumed that the size of a single carbonitride is 30 × 30, and the number of pixels it occupies is 900. When the training set images were input into the YOLOv8 network, the size of the input image was reduced to 640 × 448, which was 3.5 times smaller compared to the original image, and at this time, the size of the carbonitride was about 9 × 9, which was much smaller than the size of 32 × 32 in the definition of the small target. In summary, the carbonitride dataset for nickel-based superalloys predominantly featured small targets.

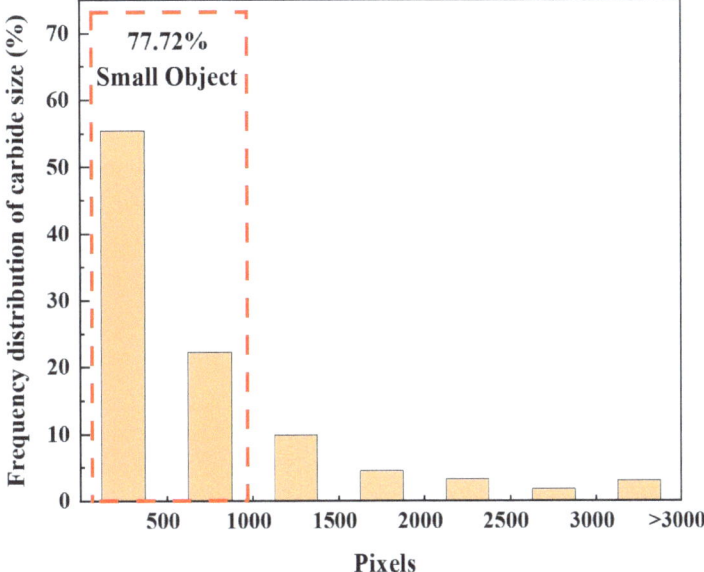

Figure 1. Carbonitride Size Distribution Map.

In order to illustrate this problem further, we have included more detailed statistics on carbonitrides in the picture. We used the thresholding segmentation method to make more detailed statistics on carbonitrides in the pictures. In order to ensure the accuracy of the statistical results, it is necessary to preprocess the pictures before statistics, such as removing noise and adjusting contrast and brightness. The specific statistical results are shown in Table 1. It can be seen from Table 1 that among the 20 carbon nitrides, only the areas (pixels) of 9#, 10#, 12#, and 18# are less than 32 × 32, which is the size definition of small targets in the MS COCO dataset. Thus, it can be clearly seen that most of the targets detected in the carbon-nitride data set are small targets.

Regarding the average area of a carbonitrides (μm^2), it can be seen that the area (μm^2) of a single carbonitrides is from a few square microns to hundreds of square microns, and the average area of these 20 carbon nitrides is about 41.7 μm^2, calculated in a square, corresponding to a side length of about 6.5 μm.

Table 1. Carbonitrides statistics.

Obj.	Type	Area (Pixels)	Area (μm^2)	Center-X	Center-Y
1#	TiN	604	34.8	737.1	463.5
2#	TiN	405	23.3	1943.0	584.0
3#	TiN	590	34.0	803.5	717.3
4#	NbC	289	16.6	314.1	895.9
5#	NbC	81	4.7	1406.2	936.9
6#	NbC	456	26.3	315.0	951.9
7#	NbC	122	7.0	714.4	969.0
8#	NbC	766	44.1	1075.6	972.7
9#	NbC	2129	122.6	1369.2	999.0
10#	NbC	2344	135.0	1162.1	1000.5
11#	NbC	625	36.0	1098.9	1007.0
12#	NbC	2043	117.7	993.8	1021.9
13#	NbC	54	3.1	967.6	1004.8
14#	NbC	858	49.4	910.4	1020.7
15#	NbC	51	2.9	952.5	1032.4
16#	NbC	122	7.0	1219.1	1041.7
17#	NbC	875	50.4	953.7	1065.4
18#	NbC	1053	60.7	1476.4	1121.9
19#	NbC	584	33.6	1246.6	1121.2
20#	NbC	416	24.0	1170.0	1127.0

To enhance the YOLOv8 network's segmentation capabilities for carbonitrides, this paper mainly improves the YOLOv8n network. We chose YOLOv8n as the basic model of this study for the following reasons: First of all, YOLOv8, as the most advanced deep learning model at present, has a good performance in detection speed and accuracy. Varghese et al. [26] conducted performance tests on YOLOv8 and its previous YOLO series models on COCO data sets. They used Average Precision Across Scales (APAS) and Frames Per Second (FPS) as the evaluation indexes of the models, among which YOLOv8 series models showed the best performance with an APAS score of 52.7 and a FPS score of 150. Compared with the YOLOv7 series model, its indexes are increased by 2.4 and 30, respectively. Secondly, according to the needs of industrial scenarios, the YOLOv8n model is the most suitable choice, considering the training time and calculation amount of the model, as well as the detection accuracy and speed of the model. Finally, the YOLOv8n model is more convenient for model improvement. During the research, we tried to use a larger model and improve it, but its training speed and detection speed were too slow to meet the needs of industrial sites. In addition, the YOLOv8n model also makes it easier to make more adjustments based on subsequent field conditions.

This research proposed several key refinements depicted in Figure 2 [27], and the refinements included:

a.   Add Space-to-Deep convolution (SPDConv) to the backbone layer.
b.   Adding a small target detection layer makes the network more focused on small target detection.

2.1. Space-to-Deep Convolution (SPDConv)

Space-to-Deep Convolution (SPDConv) was proposed by Sunkara et al. [28], and its purpose was mainly to solve the problem of target detection in the case of small targets and low-resolution images. In general detection scenarios, images possess high resolution, and the targets are of moderate size, meaning that much of the images contain redundant information. These excess data could be effectively filtered through convolutional operations, residual connections, and pooling layers, allowing the model to discern and learn the essential features of the targets. However, in low-resolution images with small targets, the presence of redundant information was minimal. Continuous downsampling by the model could result in the loss of critical feature information for these targets, potentially rendering them undetectable.

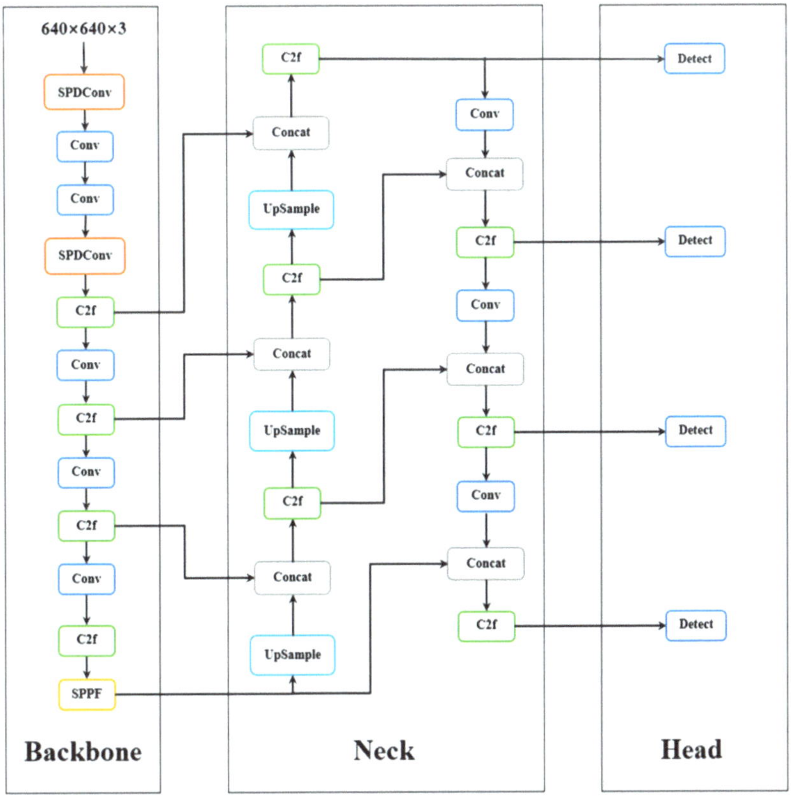

Figure 2. Improved Network Structure [28].

SPDConv primarily comprised a space-to-depth layer and a non-strided convolutional layer. It began by taking an input feature map of size (S, S, C1). The space-to-depth layer then rearranged the input X into multiple sub-feature maps, with the transformation calculated as follows:

$$f_{(0,0)} = X[0:S:scale, 0:S:scale], f_{(1,0)} = X[1:S:scale, 0:S:scale], \vdots \qquad (1)$$

$$f_{(scale-1,0)} = X[scale-1:S:scale, scale-1:S:scale]; \qquad (2)$$

$$f_{(0,1)} = X[0:S:scale, 1:S:scale], f_{(1,1)} = X[1:S:scale, 1:S:scale], \vdots \qquad (3)$$

$$f_{(scale-1,1)} = X[scale-1:S:scale, scale-1:S:scale]; \qquad (4)$$

$$f_{(0,scale-1)} = X[0:S:scale, scale-1:S:scale], \qquad (5)$$

$$f_{(1,scale-1)} = X[1:S:scale, scale-1:S:scale], \vdots \qquad (6)$$

$$f_{(scale-1,scale-1)} = X[scale-1:S:scale, scale-1:S:scale] \qquad (7)$$

Then, several sub-feature maps were concatenated along the channel direction to obtain X' with dimensions (s/scale, s/scale, scale²C1). If the scale was taken as 2, the calculation process was illustrated in Figure 3. In Figure 3, the plus sign indicates that all sub-feature graphs are spliced according to the channel direction, and the five-pointed star indicates that the convolution calculation with step size 1 is performed on the spliced feature graphs. The dimensions of X' would be (s/2, s/2, 4×C1). Subsequently, a non-strided convolution was applied to transform X' into a new feature map X″ with dimensions (s/2, s/2, C2), where C2 < scale²C1. The use of non-strided convolution here primarily aimed to retain all relevant information. While transformations from X to X″ were possible with strides greater than 1, they might result in the loss of some features.

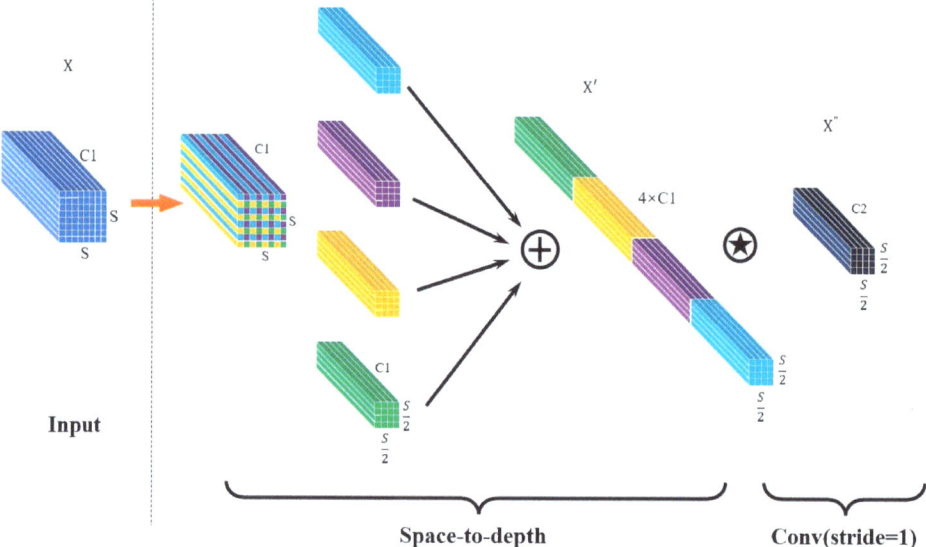

Figure 3. SPDConv Structure Diagram.

In order to further illustrate the role of the SPDConv module for small target detection, we use a concrete example. Suppose the input dimension is (640, 640, 3); after passing through the space-to-depth layer, its dimension becomes (320, 320, 12), and then the number of channels is modified by the convolution module with step 1. After such operation, the feature information of small and medium-sized objects in the picture can be well preserved. That is, the SPDConv module rearranges the spatial dimension information to the depth dimension, thereby avoiding the information loss caused by traditional step volume. If the convolution check with the size of 3 × 3 is used for the convolution operation of the input dimension, although the feature graph size of 320 × 320 can be obtained, the feature information of the small target will be lost during the convolution process, resulting in the failure of the model to recognize the small target, and the subsequent data processing will be seriously affected.

2.2. Small Target Detection Layer

The original YOLOv8n network primarily operated on feature maps sized at 80 × 80, 40 × 40, and 20 × 20. However, during downsampling, features representing carbonitrides might diminish to a few pixels or vanish entirely. To address this issue, Zhai et al. [29] investigated and enhanced the Neck and Head layers of the YOLOv8n network. The

160 × 160 scale feature maps were introduced in the Neck layer to enhance feature fusion, emphasizing that this larger scale better-preserved feature information for small targets and enhanced overall detection accuracy. Concurrently, a corresponding detection head in the Head layer was integrated, utilizing four detection heads collectively to optimize model performance.

In order to further explain the role of the P2 small target layer, we use the structure of the P2 small target layer to illustrate. Its structure diagram is shown in Figure 4. As can be seen from the structure diagram, a P2 small target layer is added to the network structure, and four detection heads are used for multi-scale detection. Among them, the feature map size of the P2 small target layer is 160 × 160, which has a higher resolution and can give a finer representation of small targets.

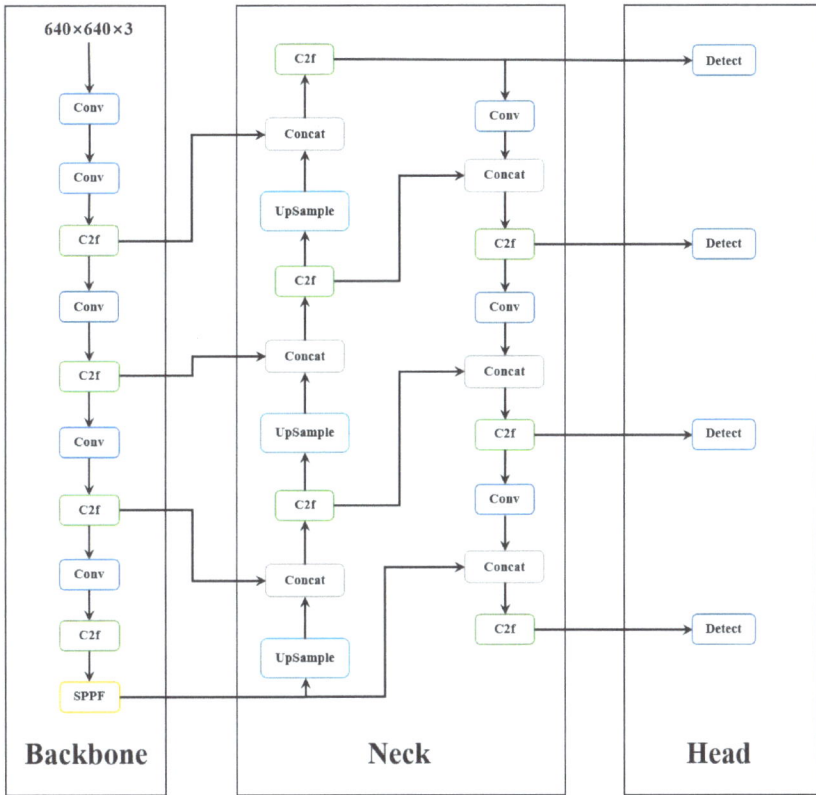

Figure 4. Small target detection layer.

3. Experiments

3.1. Image Acquisition

In this experiment, metallographic specimens sampled from nickel-based superalloy bars of different dimensions were selected. The specimens were prepared by grinding and polishing without etching. Images were captured using an optical microscope (OM, Olympus, Tokyo, Japan) at 200× magnification, resulting in a dataset where each photograph had a resolution of 2240 × 1524 pixels. This dataset included two types of compounds: TiN and NbC.

3.2. Data Annotation

All images within the dataset underwent preprocessing and were subsequently annotated for carbonitrides using the Labelme v1.8.1 software. During annotation, the delineation precisely matched the contour of the carbonitride. The detailed annotation view and software interface are shown in Figure 5, where the green contour indicates NbC and the red contour indicates TiN. After completion of the annotation process, a JSON file was generated. This file primarily contained the positional information of the carbonitride contours and the category names of the carbonitrides. Subsequently, the JSON file was converted into a TXT file to enable the improved YOLOv8 network to recognize the annotation information, ensuring the normal progression of subsequent training.

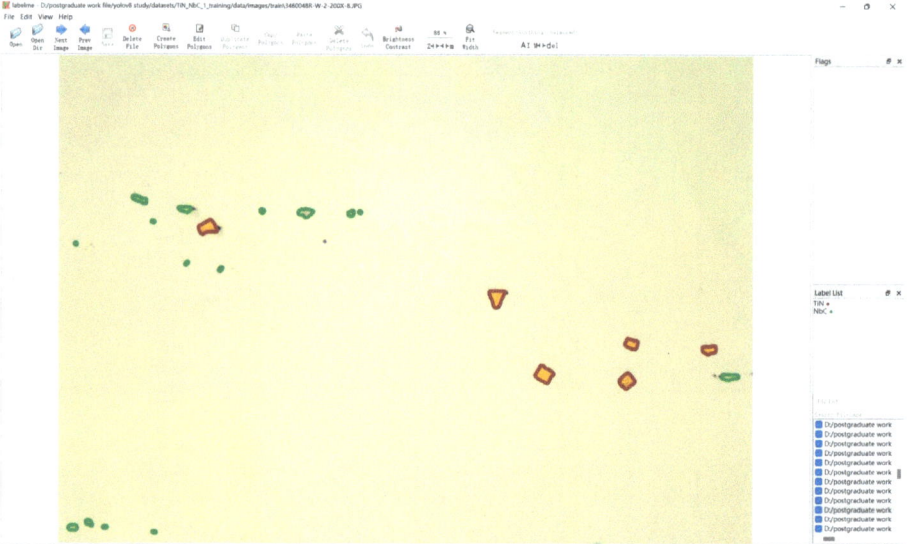

Figure 5. Labelme Labeling Software Interface.

3.3. Data Amplification

To ensure the effectiveness and enhance the generalization of the model, the original dataset was augmented. Image enhancement techniques were employed, including mirroring, Gaussian noise addition, brightness adjustment, and random point overlay. The augmented dataset comprised 1530 images, evenly split into training, validation, and test sets at a ratio of 8:1:1.

3.4. Model Training

The hyperparameters used in the experiment are shown in Table 2. Upon completion of training, a model file named 'best.pt' was generated, which was utilized for subsequent carbonitride prediction tasks.

Table 2. Hyperparameter Settings.

Parameters	Meaning	Value
imgsz	Input Image Size	640
epochs	The Rounds of Training	300
batch	Batch of Training Data	1
optimizer	Training Optimizer	SGD
lr0	Initial Learning Rate	0.01
momentum	Momentum Factor	0.937
workers	Worker Thread Count	8

4. Results and Discussion

4.1. Experimental Environment

For this experiment, the compiler used was Python 3.8, with PyTorch version 2.0.0 and CUDA version 11.8. On the hardware front, the CPU was an Intel (R) Xeon (R) Platinum 8474C, and the graphics card was an RTX 4090D with 24 GB of memory.

4.2. Model Performance Evaluation

To assess model performance, the following key metrics were selected for evaluation: the confusion matrix, Precision (P), Recall (R), and mean Average Precision (mAP).

The confusion matrix is a tabular form used to evaluate the performance of a classification model. It serves as a visualization tool, primarily for comparing classification outcomes with actual measured values, and it can display the accuracy of classification results within the matrix. Taking binary classification as an example, the confusion matrix is shown in Table 3. In it, TP (True Positive) indicates that the sample's actual value category is the positive class, and the model identification result is also the positive class. FN (False Negative) indicates that the sample's actual category is the positive class, but the model identification result is the negative class. FP (False Positive) indicates that the sample's actual category is the negative class, but the model identifies it as the positive class. TN (True Negative) indicates that the sample's actual value is the negative class, and the model also identifies it as the negative class. Subsequent advanced evaluation metrics are also calculated based on these four parameters of the confusion matrix.

Table 3. Confusion Matrix.

Reality	Prediction	
	Positive	Negative
Positive	True Positive (TP)	False Negative (FN)
Negative	False Positive (FP)	True Negative (TN)

Precision (P) refers to the proportion of data correctly predicted as the positive class among all data predicted as positive by the model. The calculation formula is:

$$P = \frac{TP}{(TP + FP)} \quad (8)$$

Recall (R) refers to the proportion of the actual positive instances in the sample that are correctly identified by the model. The calculation formula is:

$$R = \frac{TP}{(TP + FN)} \quad (9)$$

Mean Average Precision (mAP) is an important metric for evaluating model performance. A higher mAP value indicates better model performance. Before calculating mAP, you need to calculate the Average Precision (AP) for each class first. The calculation formula is:

$$AP = \int P(R)dR \tag{10}$$

In the formula, *P(R)* refers to the function curve of Precision (P)–Recall (R) for a single class. From this, the value of mAP can be calculated, with the calculation formula being:

$$mAP = \frac{1}{n}\sum_{i}^{n} AP_i \tag{11}$$

In the formula, n represents the total number of classes. In this experiment, $n = 2$.

4.3. Comparison of SPDConv Module Improvement Effects

To verify the adaptability of the SPDConv module to the overall model, comparative experiments were conducted by adding the SPDConv module to different positions within the model. The specific locations of addition are shown in Figure 6. SPD0 indicates that the SPDConv module is added at the 0th layer of the backbone. SPD01 indicates that the SPDConv module is added at the 0th and 3rd layers of the backbone, a method of addition that is consistent with the improvement approach in this research. SPD012 indicates that the SPDConv module is added at the 0th, 3rd, and 6th layers of the backbone. SPD0123 indicates that the SPDConv module is added at the 0th, 3rd, 6th, and 9th layers of the backbone. SPD01234 indicates that the SPDConv modules are added at the 0th, 3rd, 6th, 9th, and 12th layers of the backbone. The comparative experimental results for TiN and NbC are shown in Tables 4 and 5, respectively. The main evaluation metrics selected are Precision (P), Recall (R), and mAP50.

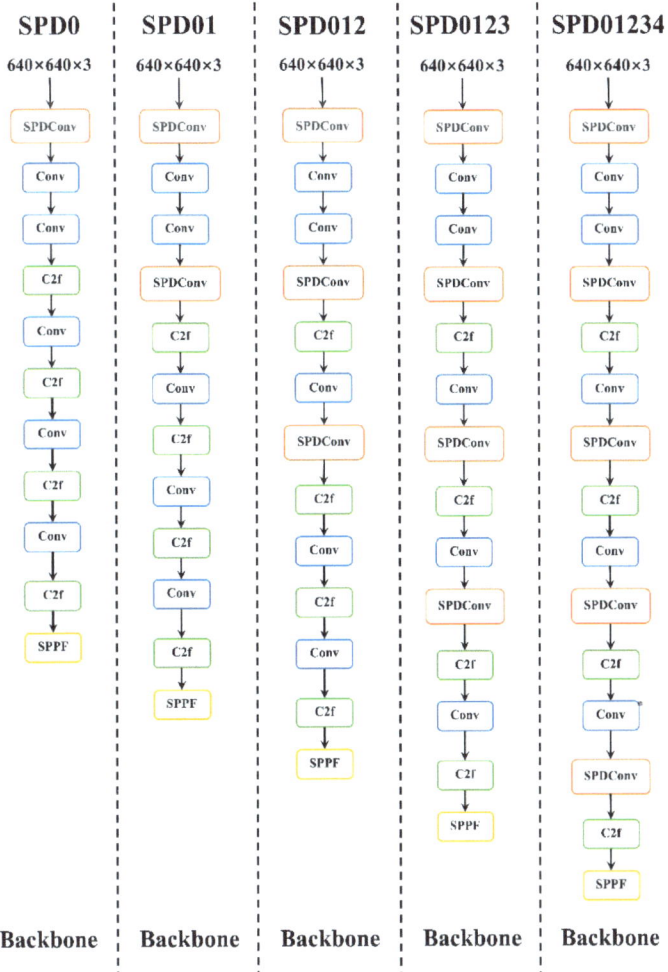

Figure 6. SPDConv Module Different Add Locations.

Table 4. TiN Test Results.

Model	Box (P)	Box (R)	mAP50	Mask (P)	Mask (R)	mAP50
YOLOv8n	0.951	0.641	0.762	0.701	0.475	0.522
SPD0	0.982	0.620	0.809	0.732	0.463	0.542
SPD01	0.975	0.649	0.808	0.749	0.500	0.576
SPD012	0.954	0.633	0.792	0.716	0.478	0.538
SPD0123	0.975	0.641	0.798	0.743	0.491	0.526
SPD01234	0.962	0.644	0.797	0.754	0.506	0.549

Table 5. NbC Test Results.

Model	Box (P)	Box (R)	mAP50	Mask (P)	Mask (R)	mAP50
YOLOv8n	0.714	0.494	0.591	0.529	0.373	0.421
SPD0	0.704	0.492	0.581	0.518	0.365	0.396
SPD01	0.737	0.514	0.622	0.557	0.396	0.439
SPD012	0.760	0.483	0.611	0.562	0.360	0.424
SPD0123	0.782	0.512	0.643	0.563	0.372	0.429
SPD01234	0.771	0.501	0.633	0.560	0.365	0.427

Data from Tables 4 and 5 show that, considering detection and segmentation accuracy alongside computational load, the SPD01 structural improvement is the most effective. For TiN, the mAP50 values for the detection box and mask are 0.808 and 0.576, increasing by 4.6% and 5.4% from the unimproved model. For NbC, these values are 0.622 and 0.439, with increases of 3.1% and 1.8%. Consequently, this research selected the SPD01 module for improvement.

4.4. Heatmap Visualization and Analysis

In order to further clarify the mechanism of the SPD01 module, this research used the GradCAM [30] method to visualize the attention region of the model by generating a heat map, and the specific effect is shown in Figure 7, where the closer the color is to red, the more attention the model pays to the region.

Figure 7 illustrates that the unimproved model attention is focused on the background and edges of the image, or it only partially captures the carbonitrides and sometimes fails to focus on them at all. After adding SPD01, the model's attention to carbonitrides was significantly improved, its attention accuracy increased substantially, and it was also able to distinguish between impurities and carbonitrides. In summary, the SPD01 module significantly improved the model recognition rate and model performance for carbonitrides by increasing the model attention to small targets.

4.5. Ablation Experiment

To verify the overall effect of the improved model, YOLOv8n was used as the baseline model, and ablation experiments were conducted on each improved module. The effectiveness of the model improvements was determined by comparing evaluation metrics such as Precision, Recall, and mAP50 values.

The ablation experiment results for TiN and NbC are detailed in Tables 6 and 7, respectively. For clarity, the small target detection layer is referred to as "small". In Tables 6 and 7, "$\sqrt{}$" indicates that the module was added to the model and "×" indicates that the module was not added to the model. As indicated in Table 6, the addition of only the SPDConv module to TiN results in a 4.6% and 5.4% increase in mAP50 values for the detection box and mask, respectively, compared to the baseline model. When the small layer is added alone, the mAP50 values for the detection box and mask rise by 9.3% and 15.9%, respectively. With both modules incorporated, the mAP50 values for the detection box and mask see an increase of 11.5% and 20.8%, respectively. Table 7 reveals that for NbC, the addition of SPDConv alone leads to a 3.1% and 1.8% increase in mAP50 values for the detection box and mask, respectively. When the "small" layer is added alone, the mAP50 values improve by 15.4% for the detection box and 13.5% for the mask. Upon adding both modules, the mAP50 values for the detection box and mask increased by 18.9% and 13.6%, respectively. These findings underscore that the enhanced network significantly bolstered model performance and enhanced the precision of carbonitride detection and segmentation.

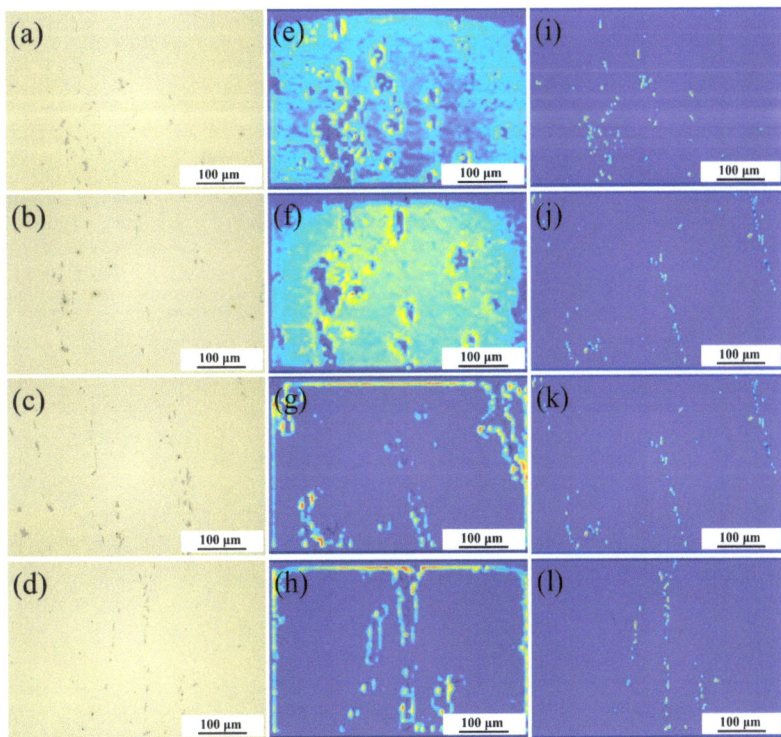

Figure 7. Visualization of Heatmap Results. (**a**–**d**): Original Image; (**e**–**h**): Unimproved model; (**i**–**l**): The Model with SPD01 Layer added.

Table 6. Results of TiN Ablation Experiments.

Model	SPD01	Small	mAP50 (Box)	mAP50 (Mask)
Model01	×	×	0.762	0.522
Model02	√	×	0.808	0.576
Model03	×	√	0.855	0.681
Model04	√	√	0.877	0.730

Table 7. Results of NbC Ablation Experiments.

Model	SPD01	Small	mAP50 (Box)	mAP50 (Mask)
Model01	×	×	0.591	0.421
Model02	√	×	0.622	0.439
Model03	×	√	0.745	0.556
Model04	√	√	0.780	0.557

4.6. Model Prediction Effect Comparison

After training the model, the batch prediction was programmed using PyCharm 2023.2.1 software. The parameters used for prediction are shown in Table 8. The prediction results before and after the improvement under the same parameter conditions are shown in Figure 8, where the blue contour represents TiN, and the red contour represents NbC. It can be observed that the unimproved model has a high false-negative rate for dense, small-sized carbonitrides, with a large number of NbCs not being correctly identified; the improved model can more easily detect them, reducing the false-negative rate, and also has good discrimination ability for impurities. Overall, the improved model was more effective for detecting carbonitrides in high-temperature alloys.

Table 8. Image prediction parameter settings.

Parameters	Implication	Value
yolo predict model	Model File	best.pt
source	Image file or path	1.JPG
save	Save the prediction results	True
save_crop	Save the cropped image with the results	True
imgsz	Image size	640
retina_masks	Use high-resolution segmentation mask	True

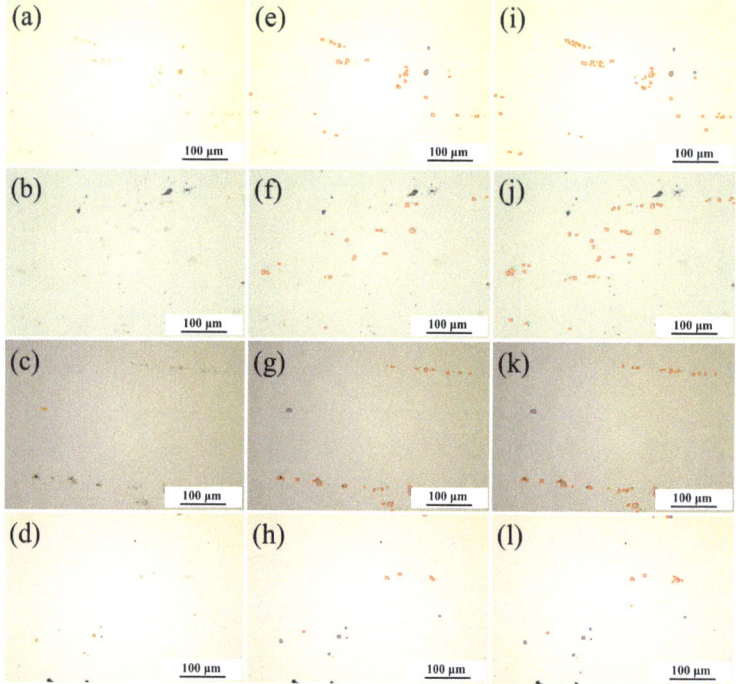

Figure 8. Comparison of The Effects Before and After Prediction. (a–d): Original Image; (e–h): Unimproved Model; (i–l): Improved Model.

4.7. Data Processing

To analyze the impact of carbonitrides on nickel-based high-temperature alloys, the data of the masked area were processed at the same time when using the model prediction. The main metrics selected for calculation include the number of carbonitrides, centroid coordinates, area of the region, area fraction, and dispersion.

4.8. Calculation of Carbonitride Dispersity

During the process improvement of nickel-based superalloys, the uniformity of the distribution of carbonitrides was found to have a critical impact on the differences in transverse and longitudinal properties of the nickel-based superalloys. It was possible to accurately quantify the average size, number, and other characteristics of carbonitride particles, but there was no suitable method for the indicator of the uniformity of the distribution of carbonitride particles. After searching through the literature, no reliable research basis could be found.

The objective of this research was to quantitatively evaluate the uniformity of the distribution of carbonitride particles. This question could be generalized into a standard computational measure: the quantification of particle dispersion. Specifically, that is calculating the standard deviation of the distances between each mass point and other mass points, as shown in Figure 9. Ideally, if the distribution of mass points was completely uniform, the standard deviation of the distance between each mass point and its nearest neighboring mass point should be zero. At this point, the dispersion of carbonitride particles was zero, indicating that the distribution of carbonitride particles was absolutely uniform.

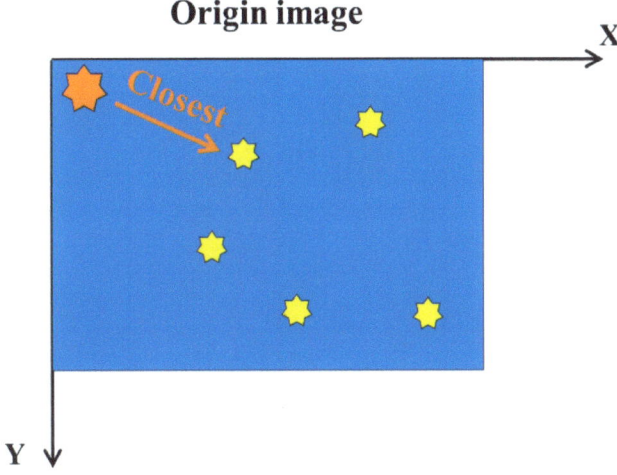

Figure 9. Schematic of Carbonitride Dispersion Statistics.

Considering the limited area that could be sampled in each image, the nearest particles of carbonitrides near the edge of the image might not be present within the same image. This could lead to an overestimation of the measured dispersion. To address this issue, this study adopted the Moore neighborhood type from cellular automata, duplicating the original image eight times and translating it to form eight adjacent neighborhoods around the original image (1 × 1), creating a new image array (3 × 3). By calculating the distance from each carbonitride particle in the central image (1 × 1) to the nearest carbonitride particle in the entire new image (3 × 3), we could mitigate the bias in the dispersion measurement compared to the actual value. This approach is illustrated in Figure 10.

In Figures 9 and 10, both the yellow and orange heptagon stars represent carbonitrides, the blue background represents the original image, and the white background represents the reproduced image.

However, simply calculating the standard deviation of the distance between each particle and the nearest particle could still lead to misjudgment. When the distribution of particles was extremely uneven, such as when large-sized primary carbonitrides were just slightly broken and clustered together, it was observed that the calculated dispersion was also relatively low. However, under actual conditions, the distribution of carbonitrides in this situation was not uniform. This was an inevitable problem encountered when calculating the nearest neighbor distance because it only considered the standard deviation of the shortest distance between particles, ignoring the distance between particles and edges or corners. Considering that the number of actual carbonitride particles was not sufficiently large, the array composed of the shortest distances was supplemented with the nearest distance from the particle group to the four corners of the rectangle. At this time, a new dispersion could be calculated.

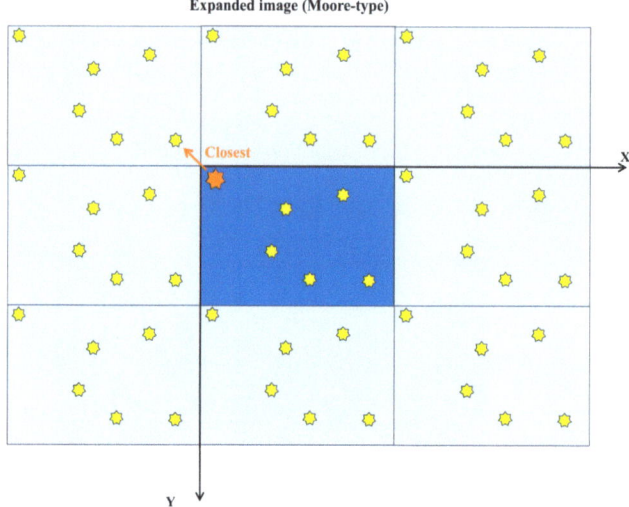

Figure 10. Schematic of the expanded image (Moore-type neighborhood).

The dispersion calculation program is embedded into the carbonitride recognition program, and the final data processing results are shown in Table 9. The corresponding image effect diagrams are shown in Figure 11.

Table 9. Data processing results.

Image	Quantities	Area Fraction	Dispersion	Dispersion-X	Dispersion-Y
1-1#	48	0.72%	2959.4372	1936.8789	2306.4717
1-2#	18	0.21%	1059.4677	836.2208	762.9323
1-3#	60	1.04%	1743.0129	1722.0014	731.2126
1-4#	38	1.08%	1935.6972	808.1959	1863.9304
1-5#	11	0.32%	815.4331	455.6615	773.5230
1-6#	7	0.27%	657.2003	364.0994	622.1773

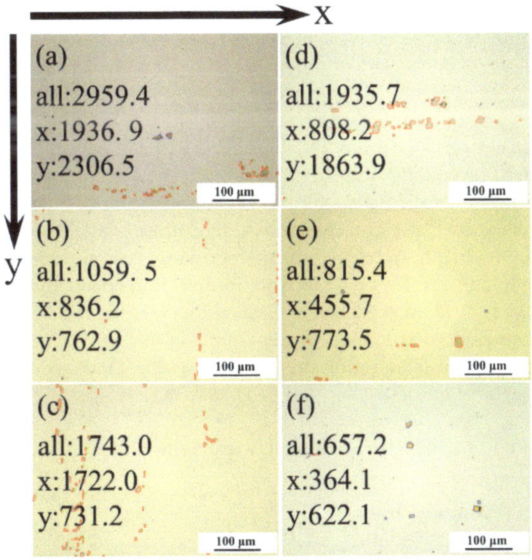

(a)
all:2959.4
x:1936.9
y:2306.5
100 μm

(d)
all:1935.7
x:808.2
y:1863.9
100 μm

(b)
all:1059.5
x:836.2
y:762.9
100 μm

(e)
all:815.4
x:455.7
y:773.5
100 μm

(c)
all:1743.0
x:1722.0
y:731.2
100 μm

(f)
all:657.2
x:364.1
y:622.1
100 μm

Figure 11. Dispersion calculation results (**a**): 1-1# (**b**) 1-2# (**c**) 1-3# (**d**) 1-4# (**e**) 1-5# (**f**) 1-6#.

4.9. Advantages in Industrial Scenarios

The complexity and uniqueness of engineering problems result in a lack of common data sets in our field, making direct comparisons with experimental results from other authors difficult. At the same time, previous methods used in this field are based on threshold segmentation, such as Image-Pro Plus 6.0 and other software for segmentation and statistics. For example, Chen et al. [31] analyzed the effects of dual melting (VIM + VAR) and triple melting (VIM + ESR + VAR) technologies on the properties of superalloy GH4738 by means of XRD, SEM, EDS, and Image-Pro Plus software. The Image-Pro Plus software is mainly used to count the number and size of inclusions in the alloy. Yang et al. [32] used characterization methods such as SEM and Image-Pro Plus software to observe and quantify the microstructure in dual-phase Ti-6AI-4V alloy to analyze its mechanical properties. The Image-Pro Plus software is mainly used to calculate the number and size of grains in the alloy. Specific statistical accuracy is not given in the literature.

In order to verify the validity and universality of our method, we will conduct a comparative analysis from two aspects:

(1) More extensive literature research. Related research in other fields has shown that approaches based on computer vision and deep learning have significant advantages over traditional approaches based on threshold segmentation. For example, Wang et al. [33] used ResNet network and Image-Pro Plus 6.0 software to count the number of nuclei in the process of cell proliferation, in which the average accuracy of Image-Pro Plus 6.0 software was 67.2%, and the average accuracy of ResNet network recognition was 90%. Zhu et al. [34] used the improved FCN network and Image-Pro Plus software to segment the eutectic silicon in Al-Si alloy, calculate the parameters of its microstructure characteristics, and compare with the theoretical results. The results obtained by the improved FCN network segmentation were closer to the theoretical values.

(2) Data statistics. In the actual industrial application environment, the improved model in this study had significant advantages in both accuracy and speed. To illustrate this, this research compared the identification results of Image-Pro Plus 6.0 software based on threshold segmentation with the instance segmentation results of the improved model and used manual statistical data as a benchmark to measure the accuracy of the two detection methods by counting the number of carbonitrides. In the Image-Pro Plus 6.0

software, the set values for R, G, and B are 200, 190, and 185, respectively, and the minimum segmentation recognition area was set to 10 to filter out the noise in the image. The specific comparison results are shown in Table 10, from which it can be seen that the number of identifications by the improved model is closer to the manual statistical results, with the recognition error between zero and five. Although the minimum recognition area was set in the Image-Pro Plus 6.0 software, due to the impact of image quality, there were still many noise points and impurities recognized. The identification results of the Image-Pro Plus 6.0 software were about three-to-seven times the manual statistical results, some of which were affected by noise and impurities, significantly affecting the subsequent data processing results. In summary, the improved model had superior performance and far exceeded the Image-Pro Plus 6.0 software in detection speed and accuracy. At the same time, the model could be used as a pre-trained weight file for automatic annotation software after processing for automatic segmentation of micron-level carbonitrides in other steel grades, significantly accelerating the detection and instance segmentation tasks of carbonitrides in other steel grades.

Table 10. Comparison results of different methods.

Image	Statistical Method (Number of Carbonitrides)		
	Manual Statistics	Image-Pro Plus	Modified Model
2-1#	47	101	46
2-2#	52	169	50
2-3#	24	296	25
2-4#	31	222	27
2-5#	57	206	54
2-6#	35	225	38
2-7#	37	203	35
2-8#	22	150	22
2-9#	67	204	62
2-10#	42	159	43

In order to further illustrate the effectiveness of the improved model, based on the actual data of the industrial field, a detailed statistical analysis is carried out. With the manual statistical data as a reference, the recognition effect of deep learning and threshold segmentation method is compared by calculating Precision and Recall. The calculation results of the two methods are shown in Table 11. From Table 11, it is obvious that the average value of the improved YOLOv8 model in terms of Precision and Recall exceeds 90%. Although the Recall of the recognition results of Image-Pro Plus 6.0 is high due to the impact of the shooting environment of industrial site pictures and the lack of preprocessing of pictures, it can recognize many impurities—that is, the Precision is low. In summary, the recognition accuracy of the improved YOLOv8 model is far higher than that of the recognition based on the threshold segmentation method.

Table 11. Comparison of calculation results.

Obj.	P (Model)	R (Model)	P (IPP)	R (IPP)
2-1#	0.978	1.000	0.446	0.957
2-2#	0.960	0.923	0.249	0.808
2-3#	0.960	1.000	0.077	0.958
2-4#	1.000	0.935	0.131	0.935
2-5#	0.981	0.929	0.262	0.947
2-6#	0.922	1.000	0.142	0.914
2-7#	1.000	0.946	0.163	0.892
2-8#	0.857	0.857	0.133	0.909
2-9#	0.984	0.910	0.328	0.910
2-10#	0.953	0.976	0.226	0.857

4.10. Practical Challenges and Industrial Relevance of Model Implementation

The ultimate goal of this research is to build a system in which the AI model can iterate itself, which also brings challenges to our work. Encouragingly, the model has significant versatility and growth. After industrial mass production has shown the characteristics of accuracy, speed, standard, and low labor cost, the model has been rapidly applied to the carbonitrides quantitative statistics of other grades of the industrial production process. In the future, we need to obtain more microstructure photos of steel varieties to continuously improve the performance of the model and build the above system.

In the subsequent research, more improvement methods will be applied, such as adding CBAM and EMA attention mechanisms, trying more loss functions, changing the feature extraction strategy of the backbone network, modifying the feature fusion mechanism of the neck network, etc. Through these methods, the performance of the model is again improved to meet the needs of the industrial field. Beyond that, existing improved models can be leveraged to process new data and added to datasets that can be used to train more general models. Implement the iterative update of the AI model.

5. Conclusions

(1) The structural improvement of SPD01 mapped spatial dimension information to depth dimension, effectively addressing the problem of feature information loss. Heatmap visualization indicated that the SPD01 module enhanced model attention on carbonitrides, with substantial accuracy gains. The mAP50 values for TiN and NbC masks improved by 5.4% and 1.8%, respectively.

(2) By incorporating a specialized P2 layer, the model was endowed with a more refined recognition capability for small targets, thereby significantly enhancing the segmentation accuracy for these targets. The mAP50 values for TiN and NbC masks improved by 15.9% and 13.5%, respectively.

(3) Compared to the original network, integrating SPD01 and the P2 layer further enhanced model performance, increasing precision in carbonitride detection and segmentation while reducing missed detection rates. The mAP50 values for TiN and NbC masks improved by 20.8% and 13.6%, respectively.

(4) A method and program for calculating the dispersity of carbonitrides developed and embedded into the carbonitride recognition program, enabling batch detection and data processing. The current model was capable of performing high-precision instance segmentation detection tasks for primary carbonitrides in industrial scenarios.

Author Contributions: Conceptualization, J.Z.; Methodology, C.G.; Software, H.Z.; Validation, H.Z.; Formal analysis, H.Z.; Investigation, J.Z.; Resources, J.Z.; Data curation, C.Z.; Writing—original draft, H.Z.; Writing—review & editing, C.Z. and C.G.; Visualization, C.Z.; Supervision, J.Z. All authors have read and agreed to the published version of the manuscript.

Funding: This research was funded by [Major Scientific and Technological Innovation Project of CITIC Group] grant number [2022ZXKYA06100].

Institutional Review Board Statement: Not applicable.

Informed Consent Statement: Not applicable.

Data Availability Statement: The original contributions presented in the study are included in the article, further inquiries can be directed to the corresponding author.

Conflicts of Interest: The authors declare no potential conflicts of interest with respect to the research, authorship, and/or publication of this article.

References

1. Reed, R.C. *The Superalloys: Fundamentals and Applications*; Cambridge University Press: Cambridge, UK, 2008.
2. Durand-Charre, M. *The Microstructure of Superalloys*; Routledge: Abingdon-on-Thames, UK, 2017.
3. Cieslak, M.J.; Knorovsky, G.A.; Headley, T.J.; Romig, A.D., Jr. *The Solidification Metallurgy of Alloy 718 and Other Nb-Containing Superalloys*; Sandia National Lab.: Albuquerque, NM, USA, 1989; Volume 20, pp. 2149–2158. [CrossRef]

4. Leonardo, I.M.; da Hora, C.S.; dos Reis Silva, M.B.; Sernik, K. Production of Nitride-Free 718 by the VIM-VAR Processing Route. In *Proceedings of the 9th International Symposium on Superalloy 718 & Derivatives: Energy, Aerospace, and Industrial Applications*; Springer: Cham, Switzerland, 2018; pp. 303–315. [CrossRef]
5. Xie, Y.; Cheng, G.G.; Chen, L.; Zhang, Y.D.; Yan, Q.Z. Characteristics and generating mechanism of large precipitates in Nb–Ti-microalloyed H13 tool steel. *ISIJ Int.* **2016**, *56*, 995–1002. [CrossRef]
6. Chen, Q.L.; LI, W.; Chen, Z. Analysis of microstructure characteristics of high sulfur steel based on computer image processing technology. *Results Phys.* **2019**, *12*, 392–397. [CrossRef]
7. Wang, Y.; Huang, X.X.; Xie, G.L.; Zhang, N.P. A high-precision automatic recognition method based on target detection for nanometer scaled precipitates or carbides in different alloys. *J. Mater. Res. Technol.* **2023**, *26*, 7767–7774. [CrossRef]
8. LeCun, Y.; Bengio, Y.; Hinton, G. Deep learning. *Nature* **2015**, *521*, 436–444. [CrossRef]
9. Brian, L.D.; Francis, T.; Holm, E.A. Exploring the microstructure manifold: Image texture representations applied to ultrahigh carbon steel microstructures. *Acta Mater.* **2017**, *133*, 30–40. [CrossRef]
10. Li, W.; Li, W.; Qin, Z.; Tan, L.; Huang, L.; Liu, F.; Xiao, C. Deep Transfer Learning for Ni-Based Superalloys Microstructure Recognition on γ' Phase. *Materials* **2022**, *15*, 4251. [CrossRef]
11. Azimi, S.M.; Britz, D.; Engstler, M.; Fritz, M.; Mucklich, F. Advanced Steel Microstructural Classification by Deep Learning Methods. *Sci Rep.* **2018**, *8*, 2128. [CrossRef]
12. Ghauri, H.; Tafreshi, R.; Mansoor, B. Toward automated microstructure characterization of stainless steels through machine learning-based analysis of replication micrographs. *J. Mater. Sci. Mater. Eng.* **2024**, *4*, 19. [CrossRef]
13. Jia, K.; Li, W.F.; Wang, Z.L.; Qin, Z.J. Accelerating Microstructure Recognition of Nickel-Based Superalloy Data by UNet++. *Int. Symp. Intell. Autom. Soft Comput. (IASC)* **2021**, *80*, 863–870. [CrossRef]
14. Senanayake, N.M.; Carter, J.L.W. Computer Vision Approaches for Segmentation of Nanoscale Precipitates in Nickel-Based Superalloy IN718. *Integr. Mater. Manuf. Innov.* **2020**, *9*, 446–458. [CrossRef]
15. Wang, C.Y.; Bochkovskiy, A.; Liao, H.Y.M. YOLOv7: Trainable bag-of-freebies sets new state-of-the-art for real-time object detectors. In Proceedings of the IEEE/CVF Conference on Computer Vision and Pattern Recognition (CVPR), Vancouver, BC, Canada, 17–24 June 2023; pp. 7467–7475. [CrossRef]
16. He, K.M.; Zhang, X.Y.; Ren, S.Q.; Sun, J. Spatial Pyramid Pooling in Deep Convolutional Networks for Visual Recognition. *IEEE Trans. Pattern Anal. Mach. Intell.* **2015**, *37*, 1904–1916. [CrossRef] [PubMed]
17. Lin, T.Y.; Dollár, P.; Girshick, R.; He, K.M.; Hariharan, B.; Belongie, S. Feature Pyramid Networks for Object Detection. In Proceedings of the IEEE Conference on Computer Vision and Pattern Recognition (CVPR), Honolulu, HI, USA, 21–26 July 2017; pp. 936–944. [CrossRef]
18. Li, H.C.; Xiong, P.F.; An, J.; Wang, L.X. Pyramid attention network for semantic segmentation. *arXiv* **2018**. [CrossRef]
19. Zhu, C.C.; He, Y.H.; Savvides, M. Feature Selective Anchor-Free Module for Single-Shot Object Detection. In Proceedings of the IEEE/CVF Conference on Computer Vision and Pattern Recognition (CVPR), Long Beach, CA, USA, 16–20 June 2019; pp. 840–849. [CrossRef]
20. Bochkovskiy, A.; Wang, C.Y.; Liao, H.Y.M. Yolov4: Optimal speed and accuracy of object detection. *arXiv* **2020**. [CrossRef]
21. Ge, Z.; Liu, S.T.; Wang, F.; Li, Z.M.; Sun, J. Yolox: Exceeding yolo series in 2021. *arXiv* **2021**. [CrossRef]
22. Lou, H.T.; Duan, X.H.; Guo, J.M.; Liu, H.Y.; Gu, J.; Bi, L.Y.; Chen, H.N. DC-YOLOv8: Small-Size Object Detection Algorithm Based on Camera Sensor. *Electronics* **2023**, *12*, 2323. [CrossRef]
23. Li, Y.T.; Fan, Q.S.; Huang, H.S.; Han, Z.G.; Gu, Q. A Modified YOLOv8 Detection Network for UAV Aerial Image Recognition. *Drones* **2023**, *7*, 304. [CrossRef]
24. Wang, H.; Liu, C.Y.; Cai, Y.F.; Chen, L.; Li, Y.C. YOLOv8-QSD: An Improved Small Object Detection Algorithm for Autonomous Vehicles Based on YOLOv8. *IEEE Trans. Instrum. Meas.* **2024**, *73*, 1–16. [CrossRef]
25. Lin, T.Y.; Maire, M.; Belongie, S.; Hays, J.; Perona, P.; Ramanan, D.; Dollár, P.; Zitnick, C.L. Microsoft COCO: Common Objects in Context. In Proceedings of the European Conference on Computer Vision (ECCV), Zurich, Switzerland, 6–12 September 2014; Volume 8693, pp. 740–755. [CrossRef]
26. Varghese, R.; Sambath, M. YOLOv8: A Novel Object Detection Algorithm with Enhanced Performance and Robustness. In Proceedings of the 2024 International Conference on Advances in Data Engineering and Intelligent Computing Systems (ADICS), Chennai, India, 18–19 April 2024; pp. 1–6. [CrossRef]
27. Zhong, R.; Peng, E.D.; Li, Z.Q.; Ai, Q.; Han, T.; Tang, Y. SPD-YOLOv8: An small-size object detection model of UAV imagery in complex scene. *J. Supercomput.* **2024**, *80*, 17021–17041. [CrossRef]
28. Sunkara, R.; Luo, T. No More Strided Convolutions or Pooling: A New CNN Building Block for Low-Resolution Images and Small Objects. In Proceedings of the European Conference on Machine Learning and Principles and Practice of Knowledge Discovery in Databases (ECML PKDD), Grenoble, France, 19–23 September 2022; pp. 443–459. [CrossRef]
29. Zhai, X.X.; Huang, Z.H.; Li, T.; Liu, H.Z.; Wang, S.Y. YOLO-Drone: An Optimized YOLOv8 Network for Tiny UAV Object Detection. *Electronics* **2023**, *12*, 3664. [CrossRef]
30. Selvaraju, R.R.; Cogswell, M.; Das, A.; Vedantam, R.; Parikh, D.; Batra, D. Grad-CAM: Visual Explanations from Deep Networks via Gradient-Based Localization. In Proceedings of the IEEE International Conference on Computer Vision (ICCV), Venice, Italy, 22–29 October 2017; pp. 618–626. [CrossRef]

31. Chen, Z.Y.; Yang, S.F.; Qu, J.L.; Li, J.S.; Dong, A.P.; Gu, Y. Effects of Different Melting Technologies on the Purity of Superalloy GH4738. *Materials* **2018**, *11*, 1838. [CrossRef]
32. Yang, D.; Liu, Z.Q. Quantification of Microstructural Features and Prediction of Mechanical Properties of a Dual-Phase Ti-6Al-4V Alloy. *Materials* **2016**, *9*, 628. [CrossRef] [PubMed]
33. Wang, H.; Lv, X.Y.; Wu, G.H.; Lv, G.D.; Zheng, X.X. Cell proliferation detection based on deep learning. In Proceedings of the 2020 2nd International Conference on Information Technology and Computer Application (ITCA), Guangzhou, China, 18–20 December 2020; pp. 208–212. [CrossRef]
34. Zhu, L.F.; Luo, Q.; Chen, C.H.; Zhang, Y.; Zhang, L.J.; Hu, B.; Han, Y.X.; Li, Q. Prediction of ultimate tensile strength of Al-Si alloys based on multimodal fusion learning. *MGE Adv.* **2024**, *2*, 1. [CrossRef]

Disclaimer/Publisher's Note: The statements, opinions and data contained in all publications are solely those of the individual author(s) and contributor(s) and not of MDPI and/or the editor(s). MDPI and/or the editor(s) disclaim responsibility for any injury to people or property resulting from any ideas, methods, instructions or products referred to in the content.

Article

Ultra-High Strength in FCC+BCC High-Entropy Alloy via Different Gradual Morphology

Ziheng Ding [1,2,3], Chaogang Ding [1,2,3,*], Zhiqin Yang [2,3,4], Hao Zhang [3], Fanghui Wang [1,2,3], Hushan Li [1,2,3], Jie Xu [1,2,3,*], Debin Shan [1,2,3] and Bin Guo [1,2,3]

1. Key Laboratory of Micro-Systems and Micro-Structures Manufacturing of Ministry of Education, Harbin Institute of Technology, Harbin 150001, China; 17805602723@163.com (Z.D.); wangfanghui1014@163.com (F.W.); hushan1124@163.com (H.L.); shandb@hit.edu.cn (D.S.); bguo@hit.edu.cn (B.G.)
2. National Key Laboratory for Precision Hot Processing of Metals, Harbin Institute of Technology, Harbin 150001, China; zhiqinyang@foxmail.com
3. School of Materials Science and Engineering, Harbin Institute of Technology, Harbin 150001, China; haozhang_1838@163.com
4. Department of Materials Science and Engineering, Pohang University of Science and Technology, Pohang 37673, Republic of Korea
* Correspondence: dingcg@hit.edu.cn (C.D.); xjhit@hit.edu.cn (J.X.)

Abstract: In this study, high-pressure torsion (HPT) processing is applied to the as-cast $Al_{0.5}CoCrFeNi$ high-entropy alloy (HEA) for 1, 3, and 5 turns. Microstructural observations reveal a significant refinement of the second phase after HPT processing. This refinement effect is influenced by the number of processing turns and the distance of the processing position from the center. As the number of processing turns or the distance of the processing position from the center increases, the fragmentation effect on the second phase becomes more pronounced. The hardness of the alloy is greatly enhanced after HPT processing, but there is an upper limit to this enhancement. After increasing the number of processing turns to 5, the increase in hardness at the edge becomes less significant, while the overall hardness becomes more uniform. Additionally, the strength of the processed alloy is significantly enhanced, while its ductility undergoes a noticeable decrease. With an increase in the number of processing turns, the second phase is further refined, resulting in improvement of strength and ductility.

Keywords: dual-phase high-entropy alloys; high-pressure torsion; microstructure; mechanical properties

1. Introduction

High-entropy alloys (HEAs) are those that contain at least five principal elements [1]. The highly uniform distribution of chemical composition endows HEAs with outstanding properties, making them a popular research topic in recent years [2–5]. As research on HEAs progresses, the relationship between mechanical properties and microstructure of HEAs has attracted increasing attention [6–9]. Compared to traditional single-phase HEAs, dual-phase HEAs simultaneously contain two different crystal phases: a face-centered cubic (FCC) phase and a body-centered cubic (BCC) phase. A complex reticular structure is formed by these different crystal phases in the alloy, resulting in excellent mechanical properties [10–12]. Dual-phase HEAs may offer a solution to overcome the strength–ductility trade-off by deliberately exploiting their heterophase nature to achieve superior mechanical properties. The presence of heterointerfaces in dual-phase HEA microstructures forms barriers for slip, resulting in higher strength than conventional single-phase HEAs. For instance, as-cast $Al_{0.5}CoCrFeNi$ alloy with a duplex FCC+BCC microstructure exhibits exceptional strain hardening rates of 6 GPa at high strains (>30%) at room and cryogenic temperatures, which was attributed to the formation of deformation twinning in the FCC phase [13]. The demand for improved damage tolerance has pushed scientific research towards the development of new materials possessing both high strength and high ductility. Therefore, dual-phase high-entropy alloys that can meet these demands are highly worthy of in-depth study.

The AlxCoCrFeNi system is currently one of the most studied HEA systems [14–17]. When $x \leq 0.4$, the alloy is a single-FCC-phase alloy. When x reaches 0.5, the BCC phase begins to appear in the alloy, resulting in differences in the mechanical properties of the alloy. J. Joseph [18] investigated the influence of Al content on the microstructure and mechanical properties of the alloy, finding that as the Al content increased, the BCC phase content of the alloy gradually increased, resulting in increased strength but decreased ductility. Generally, the FCC phase exhibits high ductility and low strength, while the BCC phase exhibits the opposite, being brittle and hard. The composition and structure of these two phases have a significant impact on the alloy's mechanical properties. Alloys with lower BCC phase content can have higher strength while maintaining a certain degree of plasticity. $Al_{0.5}CoCrFeNi$ is composed of certain BCC and FCC phases. It utilizes the high strength of the BCC phase and the good plasticity of the FCC phase, with a good balance of mechanical properties, and is widely used in aerospace, automotive and other fields. Aizenshtein et al. [19] investigated the microstructure, kinetics, and thermodynamics of $Al_{0.5}CoCrFeNi$ at $T \geq 800$ °C. They not only proved that the content of the two phases in the alloy is related to temperature but also found that deformation before heat treatment can change the BCC phase morphology, thereby effectively reducing the yield stress. Niu et al. [20] successfully fabricated nanostructured alloys through heat treatment approaches. By controlling the content of the BCC phase and the size of the precipitates, the yield strength and elongation, respectively, reached 834 MPa and 25%. The microstructure of the alloy, including the second-phase size, also greatly influences its mechanical properties [21–23]. Therefore, studying how to adjust the microstructure of HEAs to achieve balanced and excellent mechanical properties is an important research direction.

In recent years, several plastic deformation processes such as equal-channel angular pressing (ECAP) and high-pressure torsion (HPT) have received considerable attention [24–29]. Compared to several other plastic deformation processes, HPT has the advantages of simple operation and the ability to introduce controlled plastic strain. In terms of microstructure control, HPT can achieve second-phase fragmentation and refinement by applying shear strain to the material, thereby improving its mechanical properties [30–32]. Zheng et al. [33] conducted HPT on FeNiCoCu high-entropy alloy (HEA) and $(FeNiCoCu)_{86}Ti_7Al_7$ HEA with up to 10 turns. Microstructural observations and mechanical property tests showed significant second-phase refinement and strength enhancement in both alloys after HPT treatment. Hyogeon Kim [34] studied the mechanical properties and microstructural evolution of 7075 aluminum alloy processed by high-pressure torsion, indicating significant enhancement in the mechanical properties of nanocrystalline materials due to microstructural changes in high-temperature plastic deformation metal alloys.

Based on recent research of dual-phase HEAs and high-pressure torsion, this study subjected $Al_{0.5}CoCrFeNi$ cast alloy to high-pressure torsion processing with different numbers of turns, investigating the influence of HPT processing turns on the microstructure and mechanical properties of the alloy and providing a detailed analysis of the strengthening effect of high-pressure torsion on the alloy. Due to the uneven shear strain introduced by HPT processing, the microstructure exhibited varying degrees of refinement, presenting a gradient structure. Mechanical performance testing showed that the strength and hardness of the alloy after HPT processing had been greatly improved, with a tensile strength increase of about 4.6 times and a hardness increase of about 1.5 times compared to the cast alloy. Meanwhile, the processed alloy still retained a certain degree of plasticity. The ultra-high strength brought about by HPT processing provides a theoretical basis and practical basis for the widespread engineering application of $Al_{0.5}CoCrFeNi$ EHEAs. This article first introduces the materials and experimental methods, then it analyzes the experimental data results, and finally, it obtains three main conclusions based on the analysis.

2. Experimental Details

A nominal composition $Al_{0.5}CoCrFeNi$ (in molar radio) high-entropy alloy (HEA) was produced by arc melting in an argon atmosphere using high-purity metals (more than 99.99 wt.%). The chemical composition of the $Al_{0.5}CoCrFeNi$ EHEAs is detailed in Table 1. To improve chemical homogeneity, the ingot was re-melted at least five times. The molten alloy was cast into a mold to obtain a cylindrical rod having dimensions of 100 mm in length and 20 mm in diameter. Samples having a diameter of 10 mm and a thickness of 1.5 mm were fabricated and then deformed though high-pressure torsion (HPT) at room temperature using a constrained HPT apparatus. The high-pressure torsion apparatus consists of a press and a torsion mold. While the press applies pressure in the height direction of the deformed body, it also applies a torque on its cross-section through active friction, causing the deformed body to undergo plastic deformation of axial compression and tangential shear deformation. The torsion mold is shown in Figure 1a. The pressures and the rotation speeds were construed at 6 GPa and 1 rpm, respectively. The HPT turn was set as 1, 3, and 5, and five HPT samples was repeated for each turn.

Table 1. Chemical composition of the experimental HEAs.

Element	Al	Co	Cr	Fe	Ni
at.%	11.21	22.19	22.24	22.23	22.13

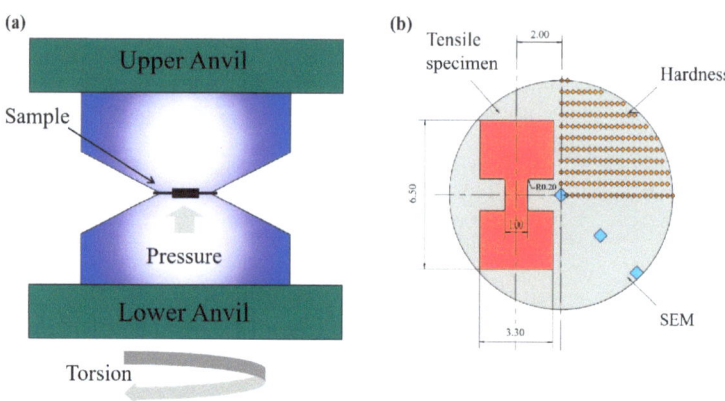

Figure 1. Schematic illustration of (**a**) HPT method and (**b**) sample and procedure used for different characterization methods.

In order to accurately assess the influence of different HPT processing turns on hardness of surface, Vickers hardness testing was carried out via a semi-automatic Vickers hardness tester with 200 g loading force and 10 s holding. A quarter-circle area was selected with a transverse point spacing and rows spacing of 0.25 mm and 0.5 mm, which means that 200 points in total were measured.

The tensile properties of the specimens for different HPT processes were performed at room temperature with a strain rate of 0.001 s^{-1} using an AG-X100kN universal testing machine (Shimadzu Corporation, Kyoto, Japan). The dog-bone-shaped specimens were designed with a gauge length of 2 mm and width of 1 mm [29]. In order to avoid the influence of the processing process on the structure or composition of the specimen, the tensile specimens were cut using low-speed WEDM. In order to obtain more accurate measurement results, digital image correlation (DIC) technology was employed for strain measurement. Before tensile testing, the surface of the tensile specimens was polished and the surface speckles were created using ink powder. Each tensile experiment was replicated four times to mitigate experimental variability, thereby enhancing the scientific validity and precision of the experiments.

X-ray diffraction (XRD) was carried out on the obtained HPT disk-shaped specimens to determine the phase structure by comparing the diffraction pattern of the sample with the diffraction pattern in the standard database and finding the pattern with the highest matching degree. The operating parameters were 40 kV and 30 mA in the 2θ range of 20° to 100°, with a scanning step of 0.01° and a scanning speed of 2°/min. Microstructural characterization was performed employing scanning electron microscopy (SEM). The equipment used was a Zeiss Gemini-560 field emission scanning electron microscope (Carl Zeiss AG, Oberkochen, Germany) with an acceleration voltage of 25 kV. According to the different deformations in each turn, the specimens were divided into the central region, half-radius region, and edge region. SEM was used to capture the microstructure of the fracture morphology in order to better characterize the tensile properties of the material after the tensile tests.

3. Results and Discussion

3.1. Microstructure Evolution of $Al_{0.5}CoCrFeNi$ High-Entropy Alloy

3.1.1. XRD Analysis

The XRD scan results in Figure 2a reveal that the initial cast alloy exhibited the structure of FCC and BCC phases, consistent with the SEM image results. With an increase in the number of HPT processing cycles, the intensity of the XRD diffraction peaks underwent a two-stage variation. In the first stage, applying high-pressure torsion deformation led to a decrease in the diffraction peak intensity, but the decline was limited to only 4.3%. In the second stage, as the number of processing cycles increased, the diffraction peak intensity rose significantly. Compared to when the number of processing turns was 3, the strength increased by 70.9% when the number of processing turns was 5. Similar phenomena were observed by P. F. Yu et al. [35] when studying the effects of high-pressure torsion on $Al_{0.1}CoCrFeNi$ high-entropy alloys. As their research has shown, equivalent strain, ε_{eq}, during HPT is calculated using the following equation:

$$\varepsilon_{eq} = \frac{2\pi r N}{\sqrt{3}h} \tag{1}$$

where N represents the number of rotations in HPT, r denotes the distance from the center of the circular sample, and h stands for the sample thickness. As indicated by Formula (1), significant shear strain is applied to the material during HPT processing, in a manner proportional to the number of torsion cycles. In the first stage, as the number of processing cycles is small, the shear strain generated by processing leads to an increase in lattice strain, fragmentation, and rearrangement of grains, resulting in the loss of crystalline perfection and a decrease in X-ray scattering, which causes the reduction in diffraction peak intensity. The variation of XRD peak intensities is similar to that caused by the addition of multi-principal elements with different atomic sizes [36]. In the second stage, as the number of processing cycles increases, the shear strain acting on the alloy becomes large enough to cause the grains to be fully broken and form a more uniform microstructure, thereby reducing the scattering effect of differently oriented grains on X-rays and increasing the intensity of XRD diffraction peaks [37]. During the high-pressure torsion processing, the peak width of the XRD gradually broadens, especially for the (111) plane. The FWHM values at different cycles of the diffraction peaks of the (111) plane were measured using Origin software (https://www.originlab.com/ (accessed on 12 September 2024)), and they were, respectively, 0.534, 0.579, and 0.602. It was demonstrated that as the number of processing turns increased, the FWHM values continued to increase and the peak width continued to widen. Figure 2b shows the changes in grain size and second-phase size under different processing turns. As the number of processing turns increased, the grain size decreased from 118 nm to 92 nm. The decrease in grain size will lead to a broadening of diffraction peak width, which was consistent with the XRD pattern. It is noteworthy that no distinct BCC phase peak was observed in the XRD results of the HPT-processed specimens. Xue et al. [38] found that the size of the phase was greatly decreased, due to the phenomenon of second-phase fragmentation refinement introduced by HPT processing. Additionally, it may also be due to grain boundary slip, causing the small grains of the second phase to fracture, leading to thin layers on the grain boundaries of the primary phase [39]. Consequently, the characteristic peaks of the second phase are undetectable once grain size reaches a certain threshold. Zhang et al. [40] also found, in the CoCrNiMo high-entropy alloy, that although the second phase can be observed via SEM, it cannot be detected by XRD due to the small content of the second phase. As described in the subsequent scanning images, the alloy remains as FCC+BCC phases after HPT processing, indicating that high-pressure torsion processing did not alter the alloy phase structure.

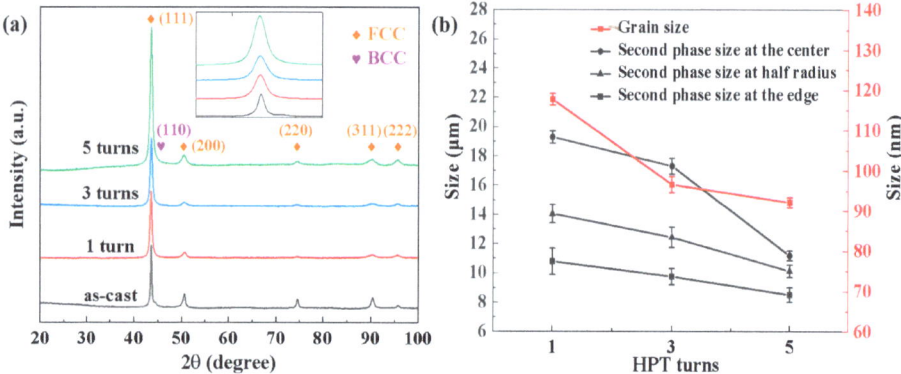

Figure 2. (a) XRD analysis for the as-cast sample and after HPT through 1, 3, and 5 turns; (b) grain size and second-phase size under different processing cycles.

3.1.2. SEM-BSE Images of $Al_{0.5}CoCrFeNi$ High-Entropy Alloy

Figure 3 shows the microstructure of the $Al_{0.5}CoCrFeNi$ as-cast specimen. The initial cast alloy exhibited a typical dendritic structure, with the matrix phase being the face-centered cubic (FCC) phase and the inter-dendritic second phase being the body-centered cubic (BCC) phase. The element distribution map and the results of the lines scanning pointed by the arrows in Figure 3 also indicates an enrichment of the Cr and Fe elements in the matrix phase and a lack of the Al and Ni elements. Conversely, the inter-dendritic region exhibits the opposite elemental distribution. The Co elements are evenly distributed in both phases. This result is in accord with other research findings on $Al_{0.5}CoCrFeNi$ high-entropy alloys [41,42].

Figure 4 displays the microstructure of different samples processed by 1, 3, and 5 torsion cycles. In comparison to the initial as-cast sample, the second phase was significantly fragmented after HPT processing. The initial inter-dendritic structure was disrupted, showing a trend of dispersed distribution. High shear strain was introduced during the HPT processing, resulting in the fragmentation and refinement of the second phase [30–32]. Different regions of the same HPT-processed sample also exhibited distinct microstructures. According to the vertical comparison shown in Figure 4, the level of refinement of the second phase was significantly higher at the edge region compared to the central. According to Formula (1), the magnitude of shear strain induced by HPT processing relates to the distance from the processing position to the center [43]. The closer to the center, the smaller the shear strain induced by HPT. The level of fragmentation and refinement of the second phase was notably higher at the edge compared to the central region. As is shown in Figure 2b, when the number of processing cycles is 1, the diameter of the second phase at the edge is 10.78 μm, and the diameter at the center is 19.28 μm, with a refinement of about 44%. Especially during the initial stages of HPT processing (e.g., at 1 cycle), some dendritic structures were still kept in the central region because of the lower shear strain. However, when the torsion turns reached 3 and 5, the difference in the level of refinement of the second phase between the edge and the 1/2 radius region was relatively small, which means that there is a limit to the refining effect of shear strain on the second phase. Due to the lower shear strain experienced, the level of refinement of the second phase in the central region was limited and gradually increased with the number of processing cycles [44]. After increasing the number of processing cycles, most of the dendritic structures were completely disrupted. Due to the strain induced by circumferential shear strain, the second phase was fragmented and tended to distribute along the rotational streamline direction.

Figure 5 shows the distribution of elements when the number of torsion cycles is 3. The application of high-pressure twisting processing did not cause a change in the chemical composition of the BCC and FCC phases but rather caused the second phase to break into smaller sizes, as reported earlier in other materials [45,46].

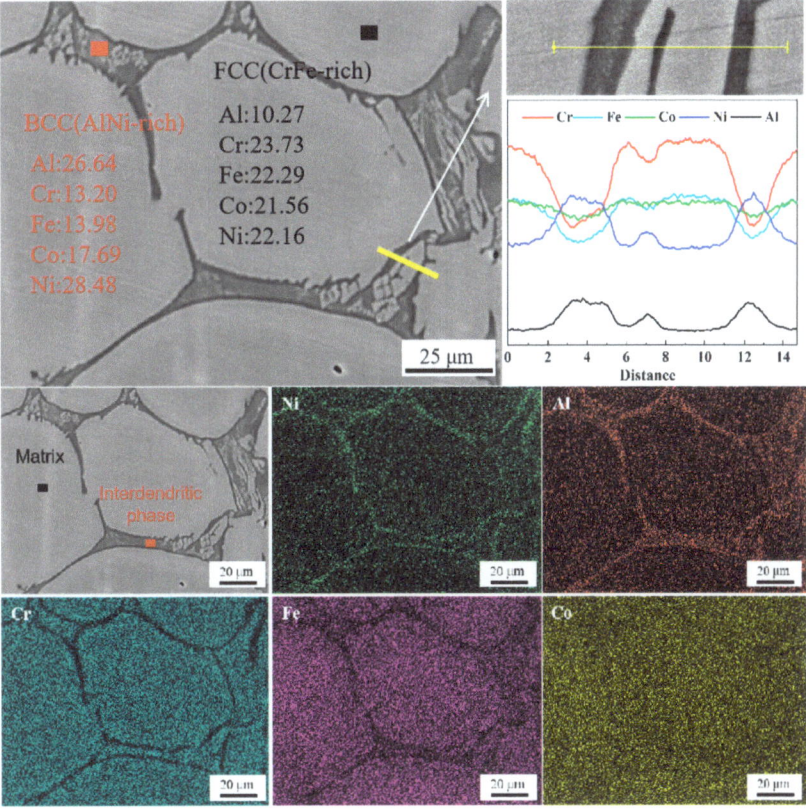

Figure 3. Microstructure and element distribution map of $Al_{0.5}CoCrFeNi$ as-cast alloy.

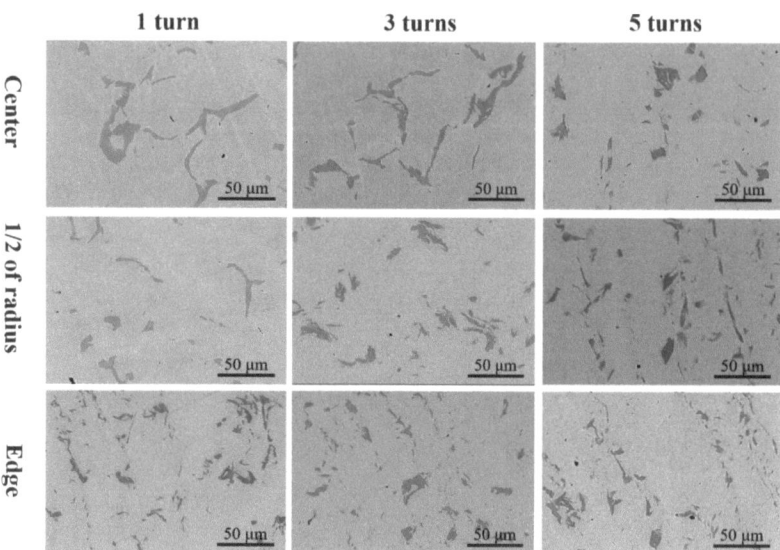

Figure 4. SEM-BSE images showing the microstructure at the center, 1/2 of radius, and edge region after HPT for 1, 3, and 5 turns.

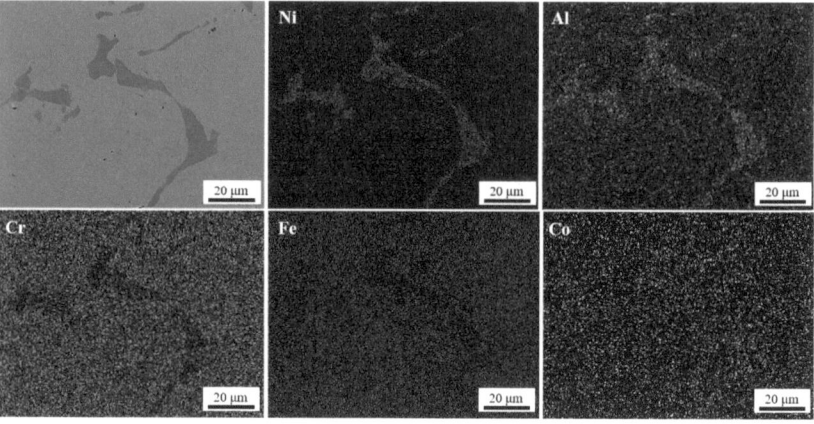

Figure 5. Element distribution map at the center after HPT for 1 turn.

3.2. Mechanical Properties

3.2.1. Hardness Evolution after HPT Processing

Figure 6a–c illustrate the variation trend of Vickers microhardness along different positions of the $Al_{0.5}CoCrFeNi$ alloy under different HPT processing cycles, while Figure 6d displays the radial variation of Vickers microhardness. Due to the uniform surface hardness distribution of the initial as-cast samples, the hardness measurement method used for the HPT samples was not adopted. Instead, five random points were selected on the surface with a 1mm spacing between points. The test results show that the surface hardness of the as-cast sample was approximately 249 HV. Observing the test results in Figure 6, it is found that compared to the as-cast sample, the surface hardness was significantly increased after HPT processing. This is mainly due to the surface work hardening induced by HPT processing. When the number of torsion cycles was 1, the lowest surface hardness was 352 HV, which was about 30% higher than that of the as-cast sample. The surface hardness gradually increased with the increase of HPT processing cycles. When the number of torsion cycles rose to 3 and 5, the hardness reached a saturation level of around 530 HV. Further increasing the number of processing cycles did not significantly increase the hardness. Different shear strain was introduced into different regions during HPT processing. This resulted in lower hardness in the center of the sample and higher hardness at the edges when the number of processing cycles was small. After reaching saturation, continuing to increase the number of processing cycles introduced larger shear strain, which help to make the surface hardness of the sample more uniform. It could be observed that when the number of processing cycles was 1 and 3, the hardness in the center was about 140 HV lower than that at the edges; when the number of processing cycles was 5, the difference in hardness between the center and the edges decreased, which meant that the hardness in different regions became more uniform. The high shear and compressive stresses introduced by HPT processing generated a significant second-phase refinement [47,48]. This was the primary reason for the increase of hardness. This is also consistent with the results shown in Figure 2b. On the one hand, as the number of torsion cycles increased and the distance from the processing position to the center kept increasing, the grain size and second-phase size decreased, leading to an increase in hardness. On the other hand, after three cycles of processing, the refinement effect brought about by HPT was significantly weakened, resulting in only a smaller increase of hardness. When the processing position was at the edge, the size of the second phase did not change significantly when the number of processing cycles was increased, resulting in little change in hardness. When the processing position was located at the center of the circle, increasing the number of processing cycles could significantly refine the second phase and greatly increase hardness. Figure 6e shows the functional relationship between the equivalent strain introduced by HPT processing and hardness, where the equivalent strain was calculated by Formula (1) and was proportional to distance and number of torsion cycles. As the number of machining cycles or distance from the center increased, the hardness increased continuously until saturation.

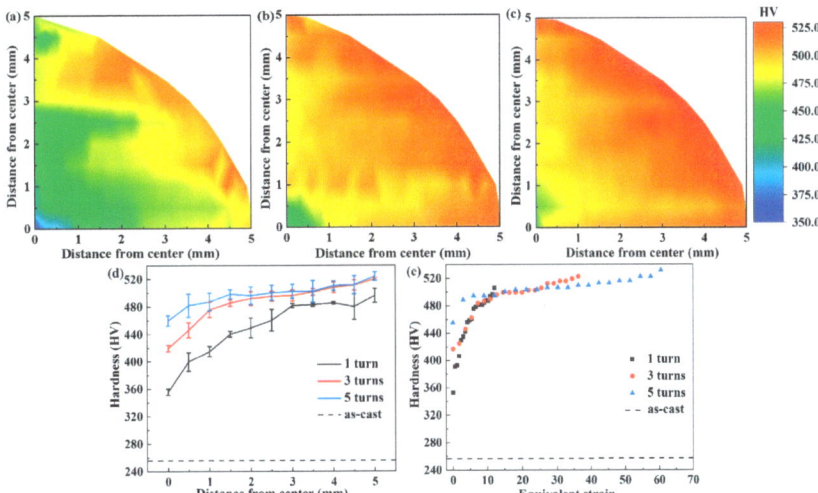

Figure 6. The Vickers microhardness plotted against (**a**) 0.5 turns, (**b**) 1 turn, (**c**) 3 turns, (**d**) distance from the disk center, and (**e**) equivalent strain.

According to previous research results [49], the hardness change after HPT processing can be classified into three types: no recovery type, with recovery type, and with softening type. The hardness changes of the $Al_{0.5}CoCrFeNi$ alloy after HPT processing conform to the no recovery type, where the Vickers microhardness increases with the increase of equivalent strain and reaches saturation with the increase of processing cycles, stabilizing thereafter. This type is typical for most high-entropy alloys [50–52].

3.2.2. Tensile Properties

To visually represent the trend of performance changes between the as-cast and HPT-processed specimens, Figure 7b compares the elongation, tensile strength, and yield strength. The as-cast $Al_{0.5}CoCrFeNi$ alloy exhibited a relatively high elongation of 36.5% and a lower strength of 323MPa. Compared to the as-cast specimens, the elongation of the HPT-processed specimens decreased significantly to around 4%, while the strength increased substantially. Consistent with the findings of most studies [53,54], on one hand, the HPT processing introduced a large number of dislocations and low-angle grain boundaries. During tensile deformation, dislocations became entangled and accumulated at grain boundaries, hindering the movement of grain boundaries and leading to the formation of numerous microcracks at grain boundaries, which resulted in lower elongation. On the other hand, the HPT processing promoted a more uniform distribution of the BCC and FCC phases and refined the BCC phase. These actions could suppress dislocation motion, leading to an increase in the alloy's strength.

The engineering stress–strain curves of the HPT-processed specimens are depicted in Figure 7a. As the number of processing revolutions increased, both the strength and elongation of the specimens improved. Consistent with the results shown in Figure 2b, with the increase in the number of processing cycles, the shear strain on the sample increased, and the size of the second phase was further refined. As the size of the second phase gradually decreased, the strengthening effect of the second phase on the tensile strength of the alloy was further improved, resulting in a continuous increase in the tensile strength of the alloy with an increase in the number of machining cycles, which is consistent with previous research findings [55,56]. At the same time, as the number of processing revolutions increased, the second phase tended to be more thoroughly fragmented, leading to an increase in plasticity. Zheng et al. [33] conducted HPT on FeNiCoCu high-entropy alloy (HEA) and (FeNiCoCu)86Ti7Al7 HEA with up to 10 turns. They found that HPT processing can cause complex changes in the microstructure of alloys. Although the shear stress introduced by HPT processing generates some dislocations, it also leads to a decrease in the size of the second phase, thereby reducing the stress concentration level around it. Therefore, during the process of tensile deformation, severe stress concentration points will be delayed, thereby improving a certain degree of ductility. Meanwhile, it was observed that as the number of processing revolutions increases, the elongation gradually increases. This is contrary to the typical work hardening effect [57]. According to the study by Hyogeon Kim [34] on the influence of high-pressure torsion on Al7075 alloy, it was found that HPT processing not only provides large shear strain but also provides a large amount of frictional heat, which is conducive to generating complex microstructural changes. Large shear strain can lead to grain refinement, while the large amount of frictional heat generated by processing can cause the alloy to undergo certain dynamic recovery. The former is beneficial for improving the strength of the alloy, while the latter can lead to the disappearance of dislocations, resulting in a decrease in the work hardening rate of the alloy. Therefore, the samples having 3 processing turns exhibited higher working hardening rates than those having 5 processing turns during the tension process of the current study.

Figure 7d lists the elongation and tensile strength comparison of the $Al_{0.5}CoCrFeNi$ alloy under HPT processing and other processing methods [58–60]. It can be clearly seen that the alloy after HPT processing is located at the top left corner of the graph, demonstrating a significant increase in strength after HPT processing, far exceeding that of the alloy under other processing methods. Conversely, the alloy processed by HPT exhibited shortcomings in plasticity. We considered that there exists a clear trade-off relationship between strength and plasticity after HPT processing. However, according to the other literature, it is possible to perform heat treatment on the alloy after HPT processing to improve plasticity to a certain extent while sacrificing a certain level of strength, thus achieving a better combination of strength and plasticity [31]. These conjectures need further experimental investigation for verification.

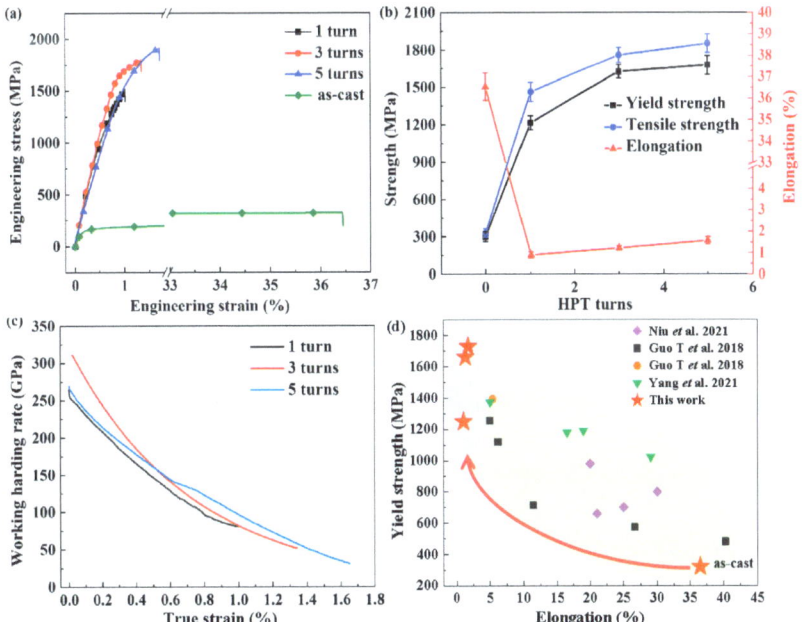

Figure 7. Tensile properties of Al$_{0.5}$CoCrFeNi HEAs processed by HPT: (**a**) engineering stress–strain curves; (**b**) changing tendencies of yield strength, tensile strength, and elongation; (**c**) work hardening rate plotted against true strain; (**d**) comparison with the data collected from other studies [58–60].

3.2.3. Fracture Analysis

Figure 8 shows the microstructure of the alloy's tensile specimen after HPT processing. The fracture mode of the Al$_{0.5}$CoCrFeNi alloy mainly involved tearing of the soft FCC phase, while the harder BCC phase particles remained in the matrix, which could be intuitively observed in the fracture morphology [61,62]. Near the BCC phase particles, a brittle fracture was observed, and the fracture surface appeared stepped, indicating the role of the second phase in hindering dislocation motion and reducing material plasticity. Consistent with the results shown in Figure 7b, when the number of processing revolutions was 1, the specimen exhibited low plasticity, with a brittle fracture and almost no observable dimples on the fracture surface. After increasing the number of processing revolutions to 3, the plasticity of the specimen improved, and the fracture surface showed a combination of dimples and river-like cleavage patterns. As the number of processing revolutions increases, a broken second phase can be observed in Figure 8. The second phase gradually refined, and its quantity increased. The hinderance effect of refined second phase on dislocations decreased, leading to an increase in the material's ductility [30–32].

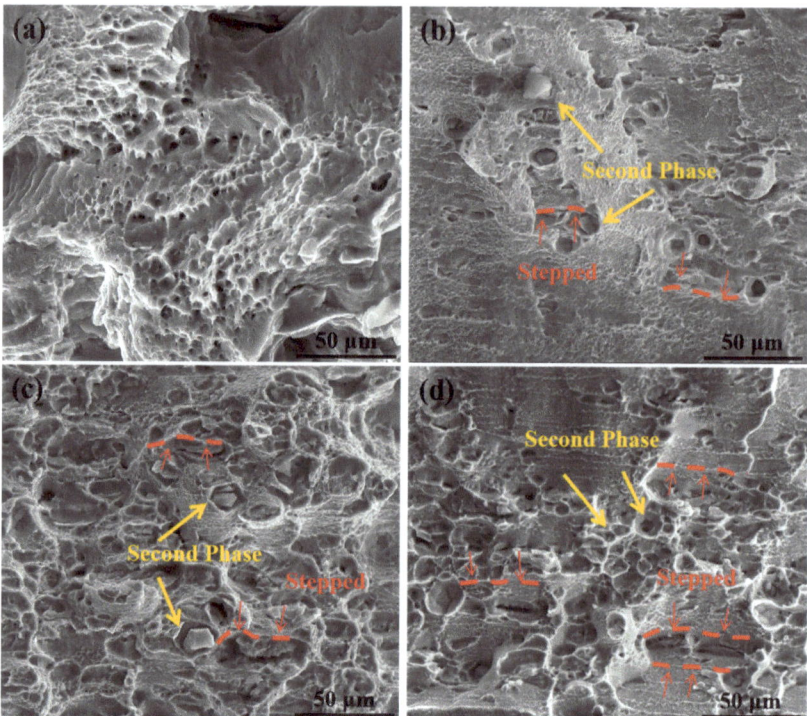

Figure 8. The microscopic morphology of fracture surface: (**a**) as-cast; (**b**) 1 turn; (**c**) 3 turns; (**d**) 5 turns.

4. Conclusions

In this study, high-pressure torsion processing was applied to the as-cast $Al_{0.5}CoCrFeNi$ high-entropy alloy for 1, 3, and 5 turns, investigating its effects on the microstructure and mechanical properties of the alloy. The following conclusions were drawn:

1. The shear strain introduced by high-pressure torsion led to the refinement of the second phase, grain fragmentation, and rearrangement effects, without altering the alloy's phase composition.
2. Significant enhancement in hardness was observed after high-pressure torsion processing of the $Al_{0.5}CoCrFeNi$ high-entropy alloy. The hardness gradually increased with the number of processing turns, and the hardness of various regions on the sample surface became progressively uniform. The hardness enhancement reached saturation, with a saturation hardness of approximately 530 HV, representing an increase of about 113% compared to the initial as-cast state.
3. The strength of the alloy was greatly improved after high-pressure torsion processing, increasing from 323 MPa to over 1500 MPa, but there was a decrease in ductility. The second phase was fragmented and refined by the introduction of shear strain, resulting in an increase in the ductility with an increasing number of processing turns.

Overall, high-pressure torsion significantly affected the properties of the as-cast $Al_{0.5}CoCrFeNi$ high-entropy alloy, enhancing its hardness and strength while sacrificing some ductility.

Author Contributions: Conceptualization, methodology, writing—original draft preparation, and writing—review and editing, Z.D.; methodology, F.W. and Z.Y.; data curation, methodology, and writing—review and editing, H.L. and H.Z.; conceptualization, methodology, writing—review and editing, and funding acquisition, C.D.; writing—review and editing and project administration, J.X., D.S. and B.G. All authors have read and agreed to the published version of the manuscript.

Funding: This research was funded by Heilongjiang Touyan Team grant number HITTY-20190036, Heilongjiang Provincial Natural Science Foundation of China grant number LH2023E033, and CGN-HIT Advanced Nuclear and New Energy Research Institute grant number CGN-HIT202305.

Institutional Review Board Statement: Not applicable.

Informed Consent Statement: Not applicable.

Data Availability Statement: The original contributions presented in the study are included in the article; further inquiries can be directed to the corresponding authors.

Conflicts of Interest: The authors declare no conflicts of interest.

References

1. George, E.P.; Raabe, D.; Ritchie, R.O. High-Entropy Alloys. *Nat. Rev. Mater.* **2019**, *4*, 515–534. [CrossRef]
2. George, E.P.; Curtin, W.A.; Tasan, C.C. High Entropy Alloys: A Focused Review of Mechanical Properties and Deformation Mechanisms. *Acta Mater.* **2020**, *188*, 435–474. [CrossRef]
3. Ye, Y.F.; Wang, Q.; Lu, J.; Liu, C.T.; Yang, Y. High-Entropy Alloy: Challenges and Prospects. *Mater. Today* **2016**, *19*, 349–362. [CrossRef]
4. Zhu, Z.; Wang, M.; He, T.; Li, T.; Di, Y.; Yan, H.; Zhang, Y.; Lu, Y. Ultrastrong High-Ductility $Ni_{35}Co_{35}Fe_{10}Al_{10}Ti_8B_2$ High-Entropy Alloy Strengthened with Super-High Concentration L12 Precipitates. *Adv. Eng. Mater.* **2023**, *25*, 2300689. [CrossRef]
5. Shi, Z.; Fang, Q.; Liaw, P.K.; Li, J. Corrosion-Resistant Biomedical High-Entropy Alloys: A Review. *Adv. Eng. Mater.* **2023**, *25*, 2300968. [CrossRef]
6. Zheng, Q.; Lu, W.; Han, D.; Li, T.; Guo, H.; Qiu, K.; Yang, B.; Wang, J. Effect of Al Content On Microstructure, Mechanical, and Corrosion Properties of $(Fe_{33}Cr_{36}Ni_{15}Co_{15}Ti_1)_{100-X}Al_X$ High-Entropy Alloys. *Adv. Eng. Mater.* **2023**, *25*, 2301013. [CrossRef]
7. Liu, J.; Jiang, Z.; Chen, W.; Fan, H.; Fu, Z. Microstructural Evolution and Mechanical Behavior of Co-Free $(Fe_{40}Ni_{30}Cr_{20}Al_{10})_{100-X}Ti_X$ High-Entropy Alloys. *Adv. Eng. Mater.* **2023**, *25*, 2300946. [CrossRef]
8. Abbasi, A.; Zarei-Hanzaki, A.; Charkhchian, J.; Moshiri, A.; Malayeri, N.A.; Lee, J.; Park, N.; Abedi, H.R. Revealing the Effects of Friction Stir Processing On the Microstructural Evolutions and Mechanical Properties of as-Cast Interstitial Femncocrn High-Entropy Alloy. *Adv. Eng. Mater.* **2024**, *26*, 2301908. [CrossRef]
9. Zhang, K.B.; Fu, Z.Y.; Zhang, J.Y.; Wang, W.M.; Wang, H.; Wang, Y.C.; Zhang, Q.J.; Shi, J. Microstructure and Mechanical Properties of $CoCrFeNiTiAl_X$ High-Entropy Alloys. *Mater. Sci. Eng. A* **2009**, *508*, 214–219. [CrossRef]
10. Shaysultanov, D.G.; Salishchev, G.A.; Ivanisenko, Y.V.; Zherebtsov, S.V.; Tikhonovsky, M.A.; Stepanov, N.D. Novel $Fe_{36}Mn_{21}Cr_{18}Ni_{15}Al_{10}$ High Entropy Alloy with Bcc/B2 Dual-Phase Structure. *J. Alloys Compd.* **2017**, *705*, 756–763. [CrossRef]
11. Zhang, T.; Zhao, R.D.; Wu, F.F.; Lin, S.B.; Jiang, S.S.; Huang, Y.J.; Chen, S.H.; Eckert, J. Transformation-Enhanced Strength and Ductility in a Fecocrnimn Dual Phase High-Entropy Alloy. *Mater. Sci. Eng. A* **2020**, *780*, 139182. [CrossRef]
12. Wang, Q.; Zeng, L.; Gao, T.; Du, H.; Liu, X. On the Room-Temperature Tensile Deformation Behavior of a Cast Dual-Phase High-Entropy Alloy $CrFeCoNiAl_{0.7}$. *J. Mater. Sci. Technol.* **2021**, *87*, 29–38. [CrossRef]
13. Bönisch, M.; Wu, Y.; Sehitoglu, H. Twinning-Induced Strain Hardening in Dual-Phase $FeCoCrNiAl_{0.5}$ at Room and Cryogenic Temperature. *Sci. Rep.* **2018**, *8*, 10663. [CrossRef] [PubMed]
14. Wang, Y.P.; Li, B.S.; Ren, M.X.; Yang, C.; Fu, H.Z. Microstructure and Compressive Properties of Alcrfeconi High Entropy Alloy. *Mater. Sci. Eng. A* **2008**, *491*, 154–158. [CrossRef]
15. Gwalani, B.; Soni, V.; Choudhuri, D.; Lee, M.; Hwang, J.Y.; Nam, S.J.; Ryu, H.; Hong, S.H.; Banerjee, R. Stability of Ordered L12 and B2 Precipitates in Face Centered Cubic Based High Entropy Alloys-$Al_{0.3}CoFeCrNi$ and $Al_{0.3}CuFeCrNi_2$. *Scr. Mater.* **2016**, *123*, 130–134. [CrossRef]
16. Huang, L.; Sun, Y.; Amar, A.; Wu, C.; Liu, X.; Le, G.; Wang, X.; Wu, J.; Li, K.; Jiang, C.; et al. Microstructure Evolution and Mechanical Properties of $Al_xCoCrFeNi$ High-Entropy Alloys by Laser Melting Deposition. *Vacuum* **2021**, *183*, 109875. [CrossRef]
17. Zhiqin, Y.; Jianxing, B.; Chaogang, D.; Sujung, S.; Zhiliang, N.; Jie, X.; Debin, S.; Bin, G.; Seop, K.H. Electroplasticity in the Al0.6Cocrfenimn High Entropy Alloy Subjected to Electrically-Assisted Uniaxial Tension. *J. Mater. Sci. Technol.* **2023**, *148*, 209–221.
18. Joseph, J.; Stanford, N.; Hodgson, P.; Fabijanic, D.M. Understanding the Mechanical Behaviour and the Large Strength/Ductility Differences Between Fcc and Bcc Alxcocrfeni High Entropy Alloys. *J. Alloys Compd.* **2017**, *726*, 885–895. [CrossRef]
19. Aizenshtein, M.; Strumza, E.; Brosh, E.; Hayun, S. Microstructure, Kinetics and Thermodynamics of Hea $Al_{0.5}CoCrFeNi$ at $T \geq 800$ °C. *Mater. Charact.* **2021**, *171*, 110738. [CrossRef]
20. Niu, S.; Kou, H.; Guo, T.; Zhang, Y.; Wang, J.; Li, J. Strengthening of Nanoprecipitations in an Annealed $Al_{0.5}CoCrFeNi$ High Entropy Alloy. *Mater. Sci. Eng. A* **2016**, *671*, 82–86. [CrossRef]
21. Fang, Q.; Chen, Y.; Li, J.; Jiang, C.; Liu, B.; Liu, Y.; Liaw, P.K. Probing the Phase Transformation and Dislocation Evolution in Dual-Phase High-Entropy Alloys. *Int. J. Plast.* **2019**, *114*, 161–173. [CrossRef]

22. Li, Z.; Tasan, C.C.; Pradeep, K.G.; Raabe, D. A Trip-Assisted Dual-Phase High-Entropy Alloy: Grain Size and Phase Fraction Effects On Deformation Behavior. *Acta Mater.* **2017**, *131*, 323–335. [CrossRef]
23. Gludovatz, B.; George, E.P.; Ritchie, R.O. Processing, Microstructure and Mechanical Properties of the Crmnfeconi High-Entropy Alloy. *Jom* **2015**, *67*, 2262–2270. [CrossRef]
24. Arivu, M.; Hoffman, A.; Duan, J.; Poplawsky, J.; Zhang, X.; Liou, F.; Islamgaliev, R.; Valiev, R.; Wen, H. Comparison of the Thermal Stability in Equal-Channel-Angular-Pressed and High-Pressure-Torsion-Processed Fe–21Cr–5Al Alloy. *Adv. Eng. Mater.* **2023**, *25*, 2300756. [CrossRef]
25. Hu, H.; Zhang, D.; Yang, M.; Ming, D. Grain Refinement in Az31 Magnesium Alloy Rod Fabricated by Extrusion-Shearing Severe Plastic Deformation Process. *Trans. Nonferrous Met. Soc. China* **2011**, *21*, 243–249. [CrossRef]
26. Maziarz, W.; Greger, M.; Długosz, P.; Dutkiewicz, J.; Wójcik, A.; Rogal, A.; Stan-Głowińska, K.; Hilser, O.; Pastrnak, M.; Cizek, L. Effect of Severe Plastic Deformation Process On Microstructure and Mechanical Properties Ofalsi/Sic Composite. *J. Mater. Res. Technol.* **2022**, *17*, 948–960. [CrossRef]
27. Lowe, T.C.; Valiev, R.Z. *Investigations and Applications of Severe Plastic Deformation*; Springer Science & Business Media: Dordrecht, The Netherlands, 2000; Volume 80.
28. Gutkin, M.Y.; Ovid'Ko, I.; Pande, C.S. Theoretical Models of Plastic Deformation Processes in Nanocrystalline Materials. *Rev. Adv. Mater. Sci.* **2001**, *2*, 80–102.
29. Chen, W.; Xu, J.; Liu, D.; Bao, J.; Sabbaghianrad, S.; Shan, D.; Guo, B.; Langdon, T.G. Microstructural Evolution and Microhardness Variations in Pure Titanium Processed by High-Pressure Torsion. *Adv. Eng. Mater.* **2020**, *22*, 1901462. [CrossRef]
30. Mohamed, I.F.; Masuda, T.; Lee, S.; Edalati, K.; Horita, Z.; Hirosawa, S.; Matsuda, K.; Terada, D.; Omar, M.Z. Strengthening of a2024 Alloy by High-Pressure Torsion and Subsequent Aging. *Mater. Sci. Eng. A* **2017**, *704*, 112–118. [CrossRef]
31. Zhilyaev, A.P.; Langdon, T.G. Using High-Pressure Torsion for Metal Processing: Fundamentals and Applications. *Prog. Mater. Sci.* **2008**, *53*, 893–979. [CrossRef]
32. Zhilyaev, A.P.; Nurislamova, G.V.; Kim, B.; Baró, M.D.; Szpunar, J.A.; Langdon, T.G. Experimental Parameters Influencing Grain Refinement and Microstructural Evolution During High-Pressure Torsion. *Acta Mater.* **2003**, *51*, 753–765. [CrossRef]
33. Zheng, R.; Chen, J.; Xiao, W.; Ma, C. Microstructure and Tensile Properties of Nanocrystalline (FeNiCoCu)$_{1-x}$Ti$_x$Al$_x$ High Entropy Alloys Processed by High Pressure Torsion. *Intermetallics* **2016**, *74*, 38–45. [CrossRef]
34. Kim, H.; Ha, H.; Lee, J.; Son, S.; Kim, H.S.; Sung, H.; Seol, J.B.; Kim, J.G. Outstanding Mechanical Properties of Ultrafine-Grained Al7075 Alloys by High-Pressure Torsion. *Mater. Sci. Eng. A* **2021**, *810*, 141020. [CrossRef]
35. Yu, P.F.; Cheng, H.; Zhang, L.J.; Zhang, H.; Jing, Q.; Ma, M.Z.; Liaw, P.K.; Li, G.; Liu, R.P. Effects of High Pressure Torsion On Microstructures and Properties of an Al$_{0.1}$CoCrFeNi High-Entropy Alloy. *Mater. Sci. Eng. A* **2016**, *655*, 283–291. [CrossRef]
36. Yeh, J.; Chang, S.; Hong, Y.; Chen, S.; Lin, S. Anomalous Decrease in X-Ray Diffraction Intensities of Cu–Ni–Al–Co–Cr–Fe–Si Alloy Systems with Multi-Principal Elements. *Mater. Chem. Phys.* **2007**, *103*, 41–46. [CrossRef]
37. Zhilyaev, A.P.; McNelley, T.R.; Langdon, T.G. Evolution of Microstructure and Microtexture in Fcc Metals During High-Pressure Torsion. *J. Mater. Sci.* **2007**, *42*, 1517–1528. [CrossRef]
38. Xue, K.; Luo, Z.; Xia, S.; Dong, J.; Li, P. Study of Microstructural Evolution, Mechanical Properties and Plastic Deformation Behavior of Mg-Gd-Y-Zn-Zr Alloy Prepared by High-Pressure Torsion. *Mater. Sci. Eng. A* **2024**, *891*, 145953. [CrossRef]
39. Fátay, D.; Bastarash, E.; Nyilas, K.; Dobatkin, S.; Gubicza, J.; Ungár, T. X-Ray Diffraction Study On the Microstructure of an Al–Mg–Sc–Zr Alloy Deformed by High-Pressure Torsion. *Int. J. Mater. Res.* **2022**, *94*, 842–847. [CrossRef]
40. Tong, Y.; Zhang, H.; Huang, H.; Yang, L.; Hu, Y.; Liang, X.; Hua, M.; Zhang, J. Strengthening Mechanism of Cocrnimox High Entropy Alloys by High-Throughput Nanoindentation Mapping Technique. *Intermetallics* **2021**, *135*, 107209. [CrossRef]
41. Ghaderi, A.; Moghanni, H.; Dehghani, K. Microstructural Evolution and Mechanical Properties of Al$_{0.5}$CoCrFeNi High-Entropy Alloy After Cold Rolling and Annealing Treatments. *J. Mater. Eng. Perform.* **2021**, *30*, 7817–7825. [CrossRef]
42. Zhuang, Y.; Zhang, X.; Gu, X. Effect of Annealing On Microstructure and Mechanical Properties of Al$_{0.5}$CoCrFeMo$_x$Ni High-Entropy Alloys. *Entropy* **2018**, *20*, 812. [CrossRef]
43. Liu, X.; Ding, H.; Huang, Y.; Bai, X.; Zhang, Q.; Zhang, H.; Langdon, T.G.; Cui, J. Evidence for a Phase Transition in an AlCrFe$_2$Ni$_2$ High Entropy Alloy Processed by High-Pressure Torsion. *J. Alloys Compd.* **2021**, *867*, 159063. [CrossRef]
44. Yang, J.; Wang, G.; Park, J.M.; Kim, H.S. Microstructural Behavior and Mechanical Properties of Nanocrystalline Ti-22Al-25Nb Alloy Processed by High-Pressure Torsion. *Mater. Charact.* **2019**, *151*, 129–136. [CrossRef]
45. Renk, O.; Pippan, R. Saturation of Grain Refinement During Severe Plastic Deformation of Single Phase Materials: Reconsiderations, Current Status and Open Questions. *Mater. Trans.* **2019**, *60*, 1270–1282. [CrossRef]
46. Edalati, P.; Mohammadi, A.; Ketabchi, M.; Edalati, K. Microstructure and Microhardness of Dual-Phase High-Entropy Alloy by High-Pressure Torsion: Twins and Stacking Faults in Fcc and Dislocations in Bcc. *J. Alloys Compd.* **2022**, *894*, 162413. [CrossRef]
47. Kawasaki, M.; Ahn, B.; Langdon, T.G. Microstructural Evolution in a Two-Phase Alloy Processed by High-Pressure Torsion. *Acta Mater.* **2010**, *58*, 919–930. [CrossRef]
48. Sabbaghianrad, S.; Kawasaki, M.; Langdon, T.G. Microstructural Evolution and the Mechanical Properties of an Aluminum Alloy Processed by High-Pressure Torsion. *J. Mater. Sci.* **2012**, *47*, 7789–7795. [CrossRef]
49. Kawasaki, M. Different Models of Hardness Evolution in Ultrafine-Grained Materials Processed by High-Pressure Torsion. *J. Mater. Sci.* **2014**, *49*, 18–34. [CrossRef]

50. Wei, D.; Koizumi, Y.; Yamanaka, A.; Yoshino, M.; Li, Y.; Chiba, A. Control of Γ Lamella Precipitation in Ti–39at.% Al Single Crystals by Nanogroove-Induced Dislocation Bands. *Acta Mater.* **2015**, *96*, 352–365. [CrossRef]
51. Shahmir, H.; Langdon, T.G. Using Heat Treatments, High-Pressure Torsion and Post-Deformation Annealing to Optimize the Properties of Ti-6Al-4V Alloys. *Acta Mater.* **2017**, *141*, 419–426. [CrossRef]
52. Heczel, A.; Kawasaki, M.; Lábár, J.L.; Jang, J.; Langdon, T.G.; Gubicza, J. Defect Structure and Hardness in Nanocrystalline Cocrfemnni High-Entropy Alloy Processed by High-Pressure Torsion. *J. Alloys Compd.* **2017**, *711*, 143–154. [CrossRef]
53. Volokitin, A.; Naizabekov, A.; Volokitina, I.; Kolesnikov, A. Changes in Microstructure and Properties of Austenitic Steel Aisi 316 During High-Pressure Torsion. *J. Chem. Technol. Metall.* **2022**, *57*, 809–815.
54. Skrotzki, W.; Pukenas, A.; Odor, E.; Joni, B.; Ungar, T.; Völker, B.; Hohenwarter, A.; Pippan, R.; George, E.P. Microstructure, Texture, and Strength Development During High-Pressure Torsion of Crmnfeconi High-Entropy Alloy. *Crystals* **2020**, *10*, 336. [CrossRef]
55. Zhang, W.; Ma, Z.; Zhao, H.; Ren, L. Refinement Strengthening, Second Phase Strengthening and Spinodal Microstructure-Induced Strength-Ductility Trade-Off in a High-Entropy Alloy. *Mater. Sci. Eng. A* **2022**, *847*, 143343. [CrossRef]
56. Shi, Z.; Li, C.; Li, M.; Li, X.; Wang, L. Second Phase Refining Induced Optimization of Fe Alloying in Zn: Significantly Enhanced Strengthening Effect and Corrosion Uniformity. *Int. J. Miner. Metall. Mater.* **2022**, *29*, 796–806. [CrossRef]
57. Srinivasarao, B.; Zhilyaev, A.P.; Langdon, T.G.; Perez-Prado, M.T. On the Relation Between the Microstructure and the Mechanical Behavior of Pure Zn Processed by High Pressure Torsion. *Mater. Sci. Eng. A* **2013**, *562*, 196–202. [CrossRef]
58. Niu, S.; Kou, H.; Wang, J.; Li, J. Improved Tensile Properties of $Al_{0.5}$CoCrFeNi High-Entropy Alloy by Tailoring Microstructures. *Rare Met.* **2021**, *40*, 1–6. [CrossRef]
59. Guo, T.; Li, J.; Wang, J.; Wang, W.Y.; Liu, Y.; Luo, X.; Kou, H.; Beaugnon, E. Microstructure and Properties of Bulk $Al_{0.5}$CoCrFeNi High-Entropy Alloy by Cold Rolling and Subsequent Annealing. *Mater. Sci. Eng. A* **2018**, *729*, 141–148. [CrossRef]
60. Yang, H.; Li, J.; Pan, X.; Wang, W.Y.; Kou, H.; Wang, J. Nanophase Precipitation and Strengthening in a Dual-Phase $Al_{0.5}$CoCrFeNi High-Entropy Alloy. *J. Mater. Sci. Technol.* **2021**, *72*, 1–7. [CrossRef]
61. Hossain, A.M.; Kumar, N. Microstructure and Mechanical Properties of a Dual Phase Transformation Induced Plasticity Fe-Mn-Co-Cr High Entropy Alloy. *J. Alloys Compd.* **2022**, *893*, 162152. [CrossRef]
62. Park, J.M.; Moon, J.; Bae, J.W.; Kim, D.H.; Jo, Y.H.; Lee, S.; Kim, H.S. Role of Bcc Phase On Tensile Behavior of Dual-Phase $Al_{0.5}$CoCrFeMnNi High-Entropy Alloy at Cryogenic Temperature. *Mater. Sci. Eng. A* **2019**, *746*, 443–447. [CrossRef]

Disclaimer/Publisher's Note: The statements, opinions and data contained in all publications are solely those of the individual author(s) and contributor(s) and not of MDPI and/or the editor(s). MDPI and/or the editor(s) disclaim responsibility for any injury to people or property resulting from any ideas, methods, instructions or products referred to in the content.

Article

Preparation of Metallized Pellets for Steelmaking by Hydrogen Cooling Reduction with Different Cooling Rates

Guanwen Luo, Zhiwei Peng *, Kangle Gao, Wanlong Fan, Ran Tian, Lingyun Yi and Mingjun Rao

School of Minerals Processing and Bioengineering, Central South University, Changsha 410083, China; 225601020@csu.edu.cn (G.L.); 8204223001@csu.edu.cn (K.G.); 225611071@csu.edu.cn (W.F.); 215601041@csu.edu.cn (R.T.); yilingyun@csu.edu.cn (L.Y.); mj.rao@csu.edu.cn (M.R.)
* Correspondence: zwpeng@csu.edu.cn; Tel.: +86-731-88877656

Abstract: To utilize the sensible heat of hot roasted iron ore pellets with no CO_2 emission in the production of metallized pellets for direct steelmaking, the pellets were reduced in H_2 during their cooling process with variable cooling rates. When the cooling rate decreased from 5.2 °C/min to 2.0 °C/min, the total iron content, reduction degree, and iron metallization degree of the pellets increased continuously from 74.0 wt%, 52%, and 31.1% to 84.9 wt%, 93.4%, and 89.2%, respectively. However, the compressive strength of the pellets increased initially from 2100 N/p to 2436 N/p and then decreased considerably to 841 N/p. As the cooling rate decreased, more Fe_2O_3 was reduced to Fe with diminishing FeO and Fe_2SiO_4. The porosity of the pellets increased from 23.9% to 54.3%, with higher distribution uniformity of pores. The morphology of metallic iron particles also transited from a layered form to a spherical form and lastly to a porous reticular form. Meanwhile, the metallic iron particles in the pellets grew evidently with more uniform distributions. When the cooling rate was 3.7 °C/min, the resulting metallized pellets had the reduction degree of 74.2%, iron metallization degree of 66.9%, and the highest compressive strength of 2436 N/p, in association with the spherical morphology and relatively large size of metallic iron particles.

Keywords: hydrogen reduction; iron ore pellets; cooling rate; metallization degree; porosity; metallic iron particles

Citation: Luo, G.; Peng, Z.; Gao, K.; Fan, W.; Tian, R.; Yi, L.; Rao, M. Preparation of Metallized Pellets for Steelmaking by Hydrogen Cooling Reduction with Different Cooling Rates. *Materials* **2024**, *17*, 4362. https://doi.org/10.3390/ma17174362

Academic Editors: Seong-Ho Ha and Hideki Hosoda

Received: 1 August 2024
Revised: 20 August 2024
Accepted: 28 August 2024
Published: 3 September 2024

Copyright: © 2024 by the authors. Licensee MDPI, Basel, Switzerland. This article is an open access article distributed under the terms and conditions of the Creative Commons Attribution (CC BY) license (https:// creativecommons.org/licenses/by/ 4.0/).

1. Introduction

With the continuous consumption of fossil fuels in the process of economic development, the global CO_2 emissions have been kept increasing in the past decades [1]. Among carbon-intensive industries, the iron and steel industry is one of the biggest CO_2 emitters [2]. China's iron and steel industry is dominated by the "blast furnace-basic oxygen furnace" process, which has a huge demand for traditional fossil fuels such as coke and coal [2,3]. Although the CO_2 emission from blast furnace ironmaking has been significantly reduced through technological innovation and the development of carbon dioxide capture technology, developing green ironmaking technology without the involvement of fossil fuels is still the main strategic direction for the industry [4,5].

The use of hydrogen for ironmaking has become essential to address the problems associated with the declining availability and quality of natural resources and to fulfill China's imminent "dual carbon" target [6]. Currently, the main hydrogen ironmaking technologies include hydrogen-rich blast furnace ironmaking [7], hydrogen direct reduction (H-DR) [8–10], hydrogen plasma smelting reduction (HPSR) [11–13], and hydrogen flash ironmaking [14]. Hydrogen-rich blast furnace ironmaking has been pioneered in industrial production in Germany and Japan [15,16]. However, because coke acts as the structural framework of the blast furnace and cannot be completely replaced, hydrogen-rich blast furnace ironmaking is expected to reduce CO_2 emission limitedly (up to 20%) [17]. H-DR was developed based on the shaft furnace direct reduction process. Midrex and HYL are two typical H-DR processes which use a mixed gas of H_2 and CO for reduction, showing a huge CO_2 reduction capacity [18,19]. HPSR and hydrogen flash ironmaking may use 100% H_2 for smelting. In theory, they can achieve zero CO_2 emission [5]. Nevertheless, they are far away from industrial applications.

For hydrogen reduction, it has a better reduction performance than carbothermic reduction at high temperatures associated with the smaller specific gravity and higher diffusivity of H_2. After reduction, there are usually obvious changes in the physicochemical properties of the feed materials, usually iron ore pellets which are prepared by drying of green pellets produced via traditional pelletizing of iron concentrate with a small amount of binder, followed by high-temperature preheating and then roasting [20,21]. It was shown that the temperature and atmosphere during the reaction process had significant effects on the pore structure and swelling characteristics of the resulting metallized pellets or direct reduced iron (DRI). The changes in pore structure also affected the diffusivity of H_2 inside the pellets and ultimately the reaction kinetics. It was found that nascent metallic iron particles would transform into abundant whiskers, which might cause serious sticking between the pellets that lowered the product quality and production efficiency [22].

To make use of sensible heat of hot iron ore pellets after high-temperature roasting, which could avoid cooling and reheating needed for traditional hydrogen reduction processes, this study explored the hydrogen reduction characteristics of the pellets during their cooling process, called the hydrogen cooling reduction (HCR) process. The effect of cooling rate on the reduction process and the properties of the resulting metallic pellets was evaluated in detail.

2. Materials and Methods

2.1. Materials

Table 1 shows the main chemical composition of iron ore pellets which were produced from iron concentrate in a conventional way after pelletizing, drying, preheating, and roasting successively. The total iron content (TFe) of the pellets was 64.00 wt% and those of SiO_2, Al_2O_3, and CaO were 8.51 wt%, 0.42 wt%, and 0.18 wt%, respectively. The contents of harmful elements, such as P and S, were relatively low. Table 2 shows the basic physical properties of the pellets. They had a relatively uniform size and high compressive strength. Figure 1 shows the phase composition of the pellets. Its main mineral phase was hematite, with small amounts of magnetite and quartz.

Table 1. Main chemical composition of iron ore pellets (wt%).

TFe	SiO_2	Al_2O_3	CaO	Na_2O	MgO	K_2O	P	S	LOI
64.00	8.51	0.42	0.18	0.18	0.15	0.10	0.045	0.0046	2.58

Table 2. Physical characteristics of iron ore pellets.

Item	Size (mm)	Compressive Strength (N/p)	True Density (g/cm^3)	Bulk Density (g/cm^3)	Porosity (%)
Value	14–16	2100	4.45	3.39	23.9

Figure 2 shows the scanning electron microscopy and energy-dispersive X-ray spectroscopy (SEM-EDS) analysis of iron ore pellets. They had a relatively compact structure. Moreover, the EDS analysis of spots A, B, and C confirmed the existence of hematite, magnetite, and quartz, agreeing with the result in Figure 1.

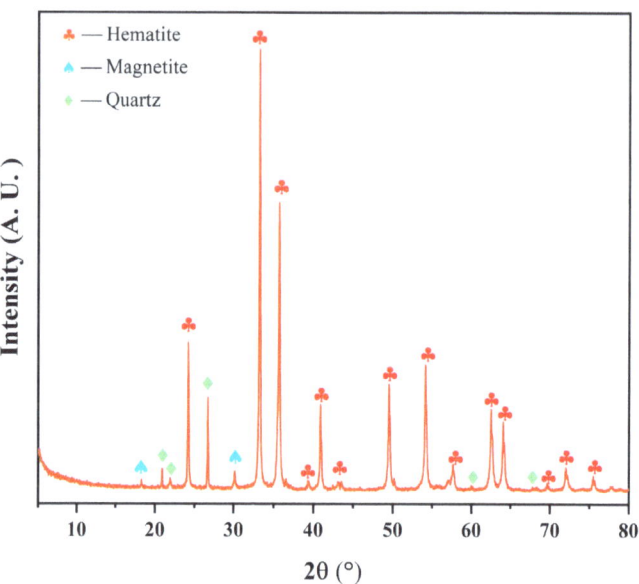

Figure 1. XRD pattern of iron ore pellets.

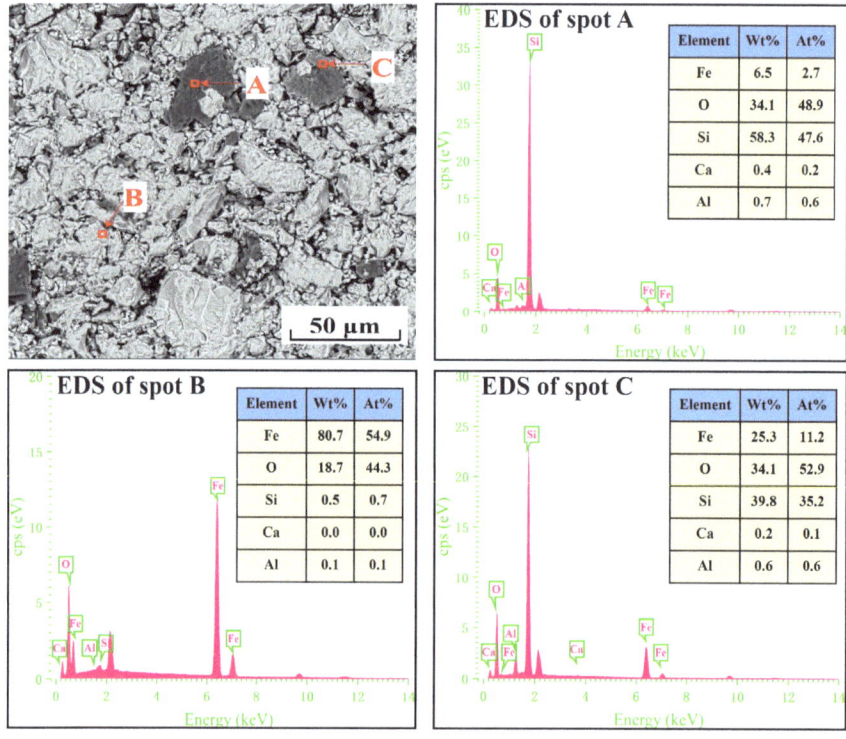

Figure 2. SEM-EDS analysis of iron ore pellets.

2.2. Methods

Before the reduction, 40 g of iron ore pellets were firstly loaded in the quartz tube (ø 40 mm × 510 mm) of a vertical tube furnace (KHGT-30, Kehui Furnace Industry Technology Co., Ltd., Changsha, Hunan, China), as shown in Figure 3. The pellets were then heated to 1150 °C with the heating rate of 10 °C/min in 0.5 L/min N_2. After keeping the pellets at this temperature for 30 min, the gas was switched to H_2 with the flow rate of 0.4 L/min to start the reduction process with different cooling rates, i.e., 5.2 °C/min, 4.5 °C/min, 3.7 °C/min, and 2.0 °C/min, respectively, which were determined by considering the variation and control of the rate in the entire HCR process. When the temperature dropped to 450 °C, the gas was switched back to 0.5 L/min N_2 for protection and cooling to room temperature. During the reduction process, an electronic balance was used to record the mass changes of the pellets.

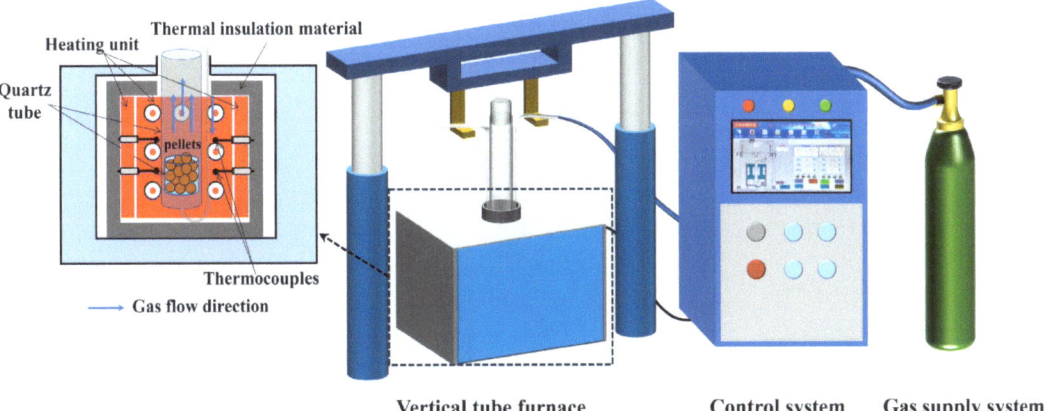

Figure 3. Schematic illustration of experimental setup for reduction.

2.3. Characterizations

The total iron contents of the pellets were determined by chemical titration according to the Chinese National Standard Test Method GB/T 6730.5-2022 [23]. The ferrous iron and metallic iron contents of the pellets after reduction were determined to calculate the reduction degree and iron metallization degree according to the Chinese National Standard Test Method GB/T 24236-2009 [24]. The reduction degree was calculated using the following equation:

$$RD = \left\{ \frac{0.111W_1}{0.430W_2} + \frac{m_0 - m_t}{m_0 \times 0.430W_2} \times 100 \right\} \times 100\% \quad (1)$$

where W_1 is the ferrous content of the pellets before reduction, wt%; W_2 is the total iron content of the pellets before reduction, wt%; m_0 is the mass of the pellets before reduction, g, and m_t is the mass of the pellets after reduction, g. The iron metallization degree (MD) was calculated using the following equation:

$$MD = \frac{MFe}{TFe} \times 100\% \quad (2)$$

where MFe is the metallic iron content of the metallized pellets, wt.%, and TFe is the total iron content of the metallized pellets, wt%. The compressive strength of the pellets was measured according to the Chinese National Standard Test Method GB/T 14201-2018 [25].

The phase compositions of the pellets were determined using an X-ray diffraction spectrometer (D8 Advance, Bruker, Karlsruhe, Germany). The morphologies/microstructures of the metallized pellets, which were polished and embedded in epoxy resin, were characterized using a scanning electron microscope equipped with an energy-dispersive X-ray spectrometer (Quanta 250FEG, FEI, Brno, Czech Republic). The porosities of different areas of the metallized pellets were determined based on the analysis of SEM images of cross-sections of the pellets using the method proposed by Otsu [26,27]. The pore distributions of the metallized pellets, which were embedded in epoxy resin and polished in advance, were determined based on the EDS line scan results of the elements including carbon, iron, and oxygen [20]. The total porosities of the pellets were characterized according to the Chinese National Standard Test Method GB/T 24586-2009 [28]. The iron particle sizes after hydrogen reduction with different cooling rates were calculated by processing the sectioned microscopic images of qualified metallized pellets using the software Image J 2.0 (National Institutes of Health, Bethesda, MD, USA). The Feret diameter, defined as the distance between two boundary lines of the projected contour of an object's particles measured along any direction, was used to describe the average iron particle size [29,30].

3. Results and Discussion

3.1. Effect of Cooling Rate on the Reduction Indexes

To investigate the effect of cooling rate on the HCR process, Figure 4 shows the variations of reduction indexes of the pellets with cooling rate. As the cooling rate declined from 5.2 °C/min to 2.0 °C/min, the total iron content of metallized pellets increased from 74.0 wt% to 84.9 wt%. Meanwhile, the reduction degree and iron metallization degree increased from 52.0% and 31.1% to 93.4% and 89.2%, respectively.

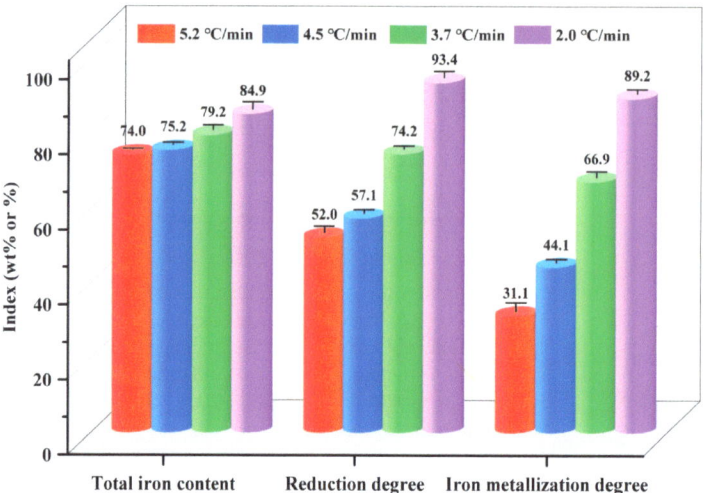

Figure 4. Total iron contents, reduction degrees, and iron metallization degrees of the metallized pellets obtained by reduction with different cooling rates.

Along with the changes in the above indexes, the compressive strength of the pellets varied evidently. Figure 5 shows the compressive strength of the metallized pellets obtained by HCR with different cooling rates. As shown in Figure 5, when the cooling rate was 5.2 °C/min, the compressive strength of metallized pellets was only 710 N/p, indicating a huge decrease in the strength compared with the iron ore pellets. This was because under the condition of rapid cooling, the internal stress in the pellets could not be released in time, which would induce "cold embrittlement" [31]. As the cooling rate declined, more iron metallization occurred and the compressive strength tended to increase. Under the influence of these two factors, the compressive strength of the pellets reached the maximum, 2436 N/p, when the cooling rate was 3.7 °C/min. However, further decreasing the cooling rate negatively affected the compressive strength. In specific, as the cooling rate was

2.0 °C/min, the compressive strength decreased to only 841 N/p. This phenomenon was mainly associated with the increases in reduction degree and iron metallization degree, which enhanced the plastic deformation capacity of the pellets [32]. It was also partially ascribed to the formation of pores in the reduction process due to various gas–solid reactions which reduced the strength of the pellets [33].

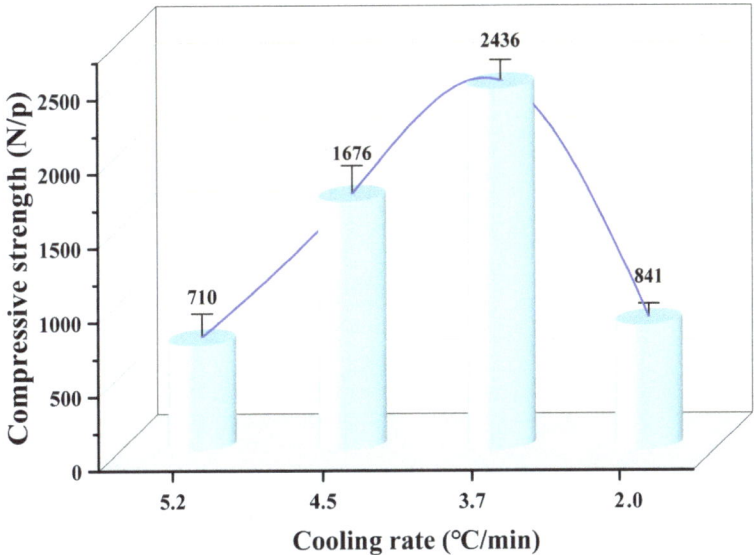

Figure 5. Compressive strength of the metallized pellets obtained by reduction with different cooling rates.

It was reported that the compressive strength of iron ore pellets needs to be higher than 2500 N/p for ironmaking in large blast furnaces [34]. When the cooling rate was 3.7 °C/min, the compressive strength of the metallized pellets basically met this requirement [35]. Therefore, the pellets could be used as a feedstock for the blast furnace ironmaking process. In addition, they could act as an iron-containing coolant in the steelmaking process [36].

During the reduction process, the phase composition of the pellets was expected to change. Figure 6 shows the XRD patterns of the pellets after HCR with different cooling rates. When the cooling rate was 5.2 °C/min, hematite and magnetite in the pellets were reduced by H_2 to metallic iron and FeO. Partial trivalent iron was converted to divalent iron which combined with silicon to form Fe_2SiO_4, indicating insufficient reduction [37]. As the cooling rate decreased, the diffraction peaks of metallic iron became stronger with declining peaks of FeO and Fe_2SiO_4, confirming more complete reduction of the pellets.

For further analysis of the differences of the metallized pellets obtained by reduction with different cooling rates, Figures 7–10 show the SEM-EDS analyses of the metallized pellets. After reduction with the cooling rate of 5.2 °C/min, the metallic iron particles were distributed in a fragmented manner, with clear gaps between them. With decreasing cooling rate, the particles tended to aggregate. Moreover, the gaps between the particles were reduced and a certain quantity of pores were generated. It explained the changing trend of compressive strength from the microscopic point of view. When the cooling rate declined to 2.0 °C/min, the metallic iron particles became more aggregated, with large pores between them, which reduced the compressive strength of the metallized pellets.

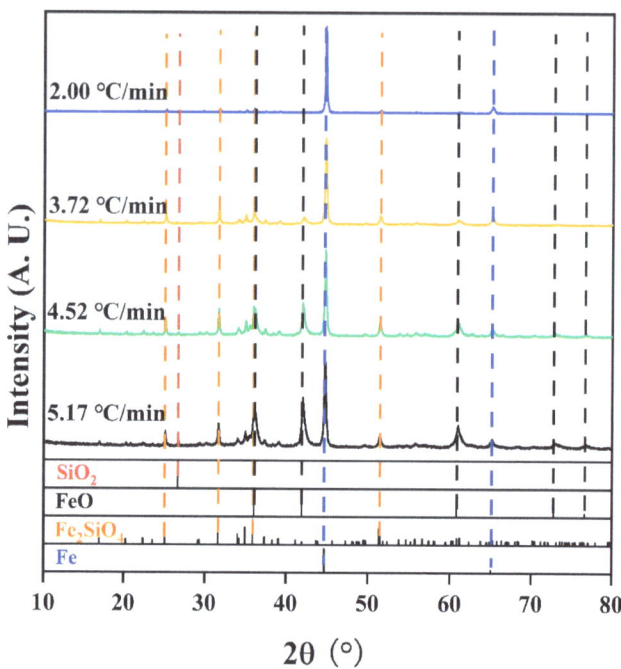

Figure 6. XRD patterns of the metallized pellets obtained by reduction with different cooling rates.

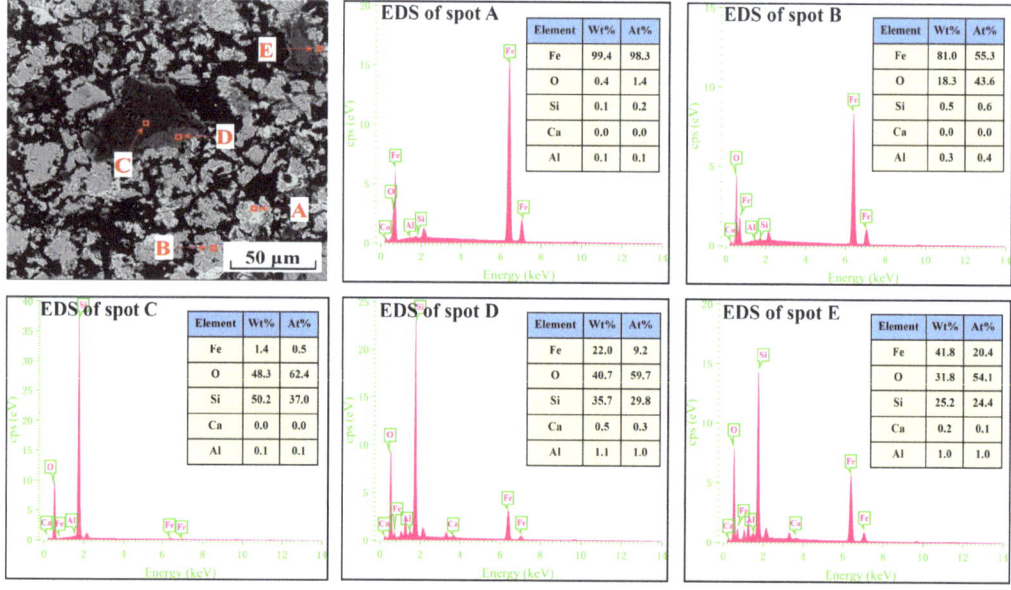

Figure 7. SEM-EDS analysis of the metallized pellets obtained by reduction with the cooling rate of 5.2 °C/min.

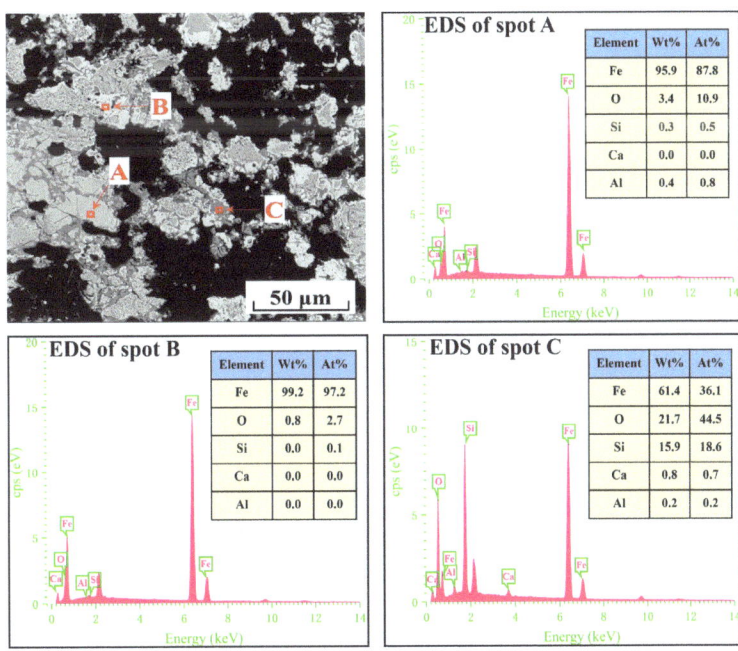

Figure 8. SEM-EDS analysis of the metallized pellets obtained by reduction with the cooling rate of 4.5 °C/min.

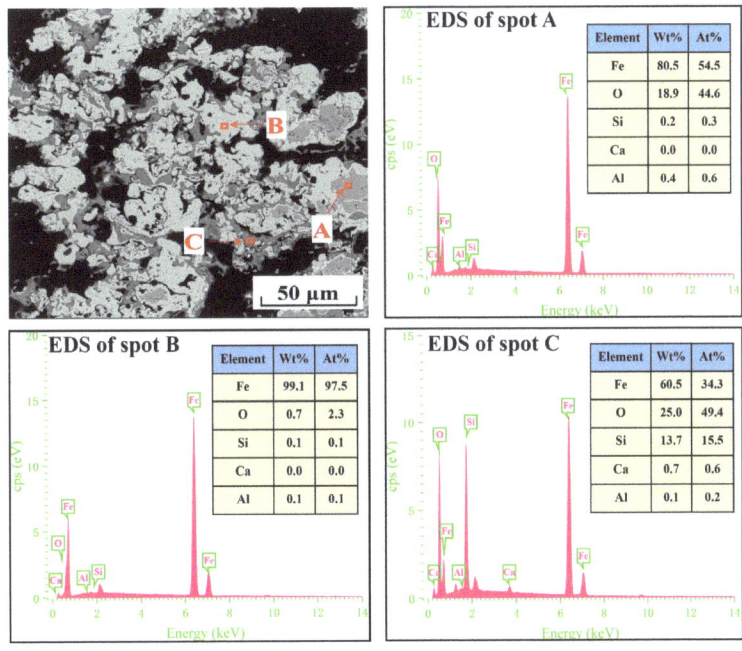

Figure 9. SEM-EDS analysis of the metallized pellets obtained by reduction with the cooling rate of 3.7 °C/min.

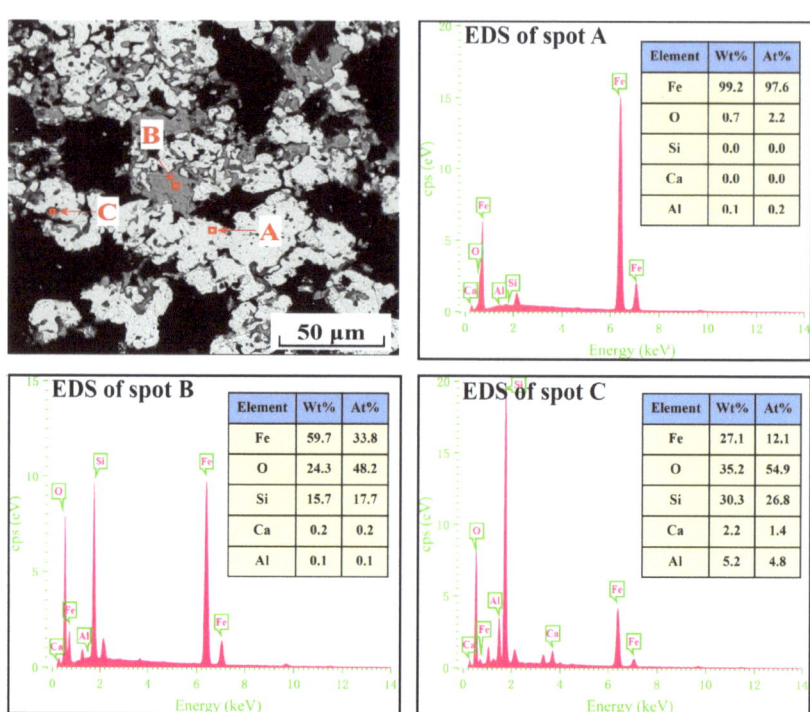

Figure 10. SEM-EDS analysis of the metallized pellets obtained by reduction with the cooling rate of 2.0 °C/min.

The EDS analysis of the spots in Figures 7 and 8 revealed that metallic iron particles were preferentially generated from the edges of FeO particles and continuously diffused inward as the cooling rate decreased. Large SiO_2 particles gradually disappeared and transformed into Fe_2SiO_4 existing between the FeO particles. As the cooling rate decreased further, according to the EDS analysis of spots A, B, and C in Figure 9, the nascent iron aggregates kept growing to encapsulate unreacted FeO and Fe_2SiO_4. When the cooling rate was 2.0 °C/min, complete reduction was identified, as confirmed by the EDS analysis in Figure 10. In specific, FeO was completely converted to Fe and Fe_2SiO_4 was partially converted to Fe and SiO_2.

To further examine the extent of reduction in different regions inside the metallized pellets obtained under different conditions, the microstructures of the metallized pellets at their edge, 1/4 diameter, and central areas were characterized, as shown in Figure 11. The metallization degree of the metallized pellets increased from the center to the edge after reduction with different cooling rates. When the cooling rate was 5.2 °C/min, the metallized pellets had obvious metallization at the edges. Only a few metallic iron particles were formed at the 1/4 diameter and at the center. It verified that with a high cooling rate, the temperature of the system decreased rapidly and could not provide sufficient thermal energy for reduction reactions. Decreasing cooling rate significantly promoted the reactions at both the edge and 1/4 diameter area, especially for reduction with the cooling rate of 3.7 °C/min. As the cooling rate was 2.0 °C/min, the difference between these regions almost disappeared.

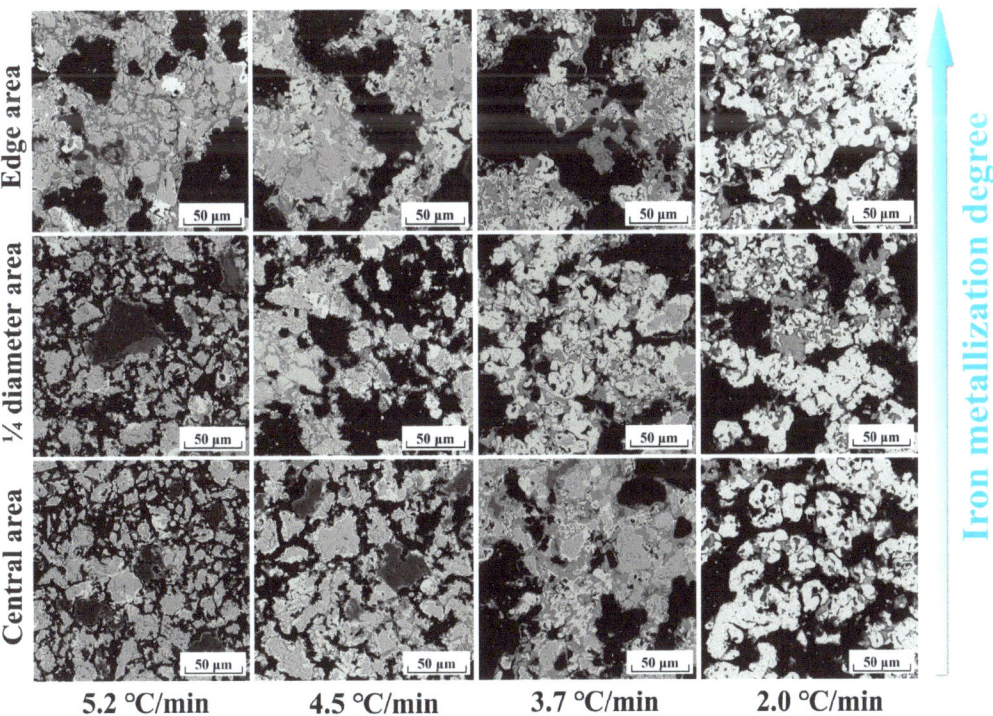

Figure 11. Microstructures of different areas of the metallized pellets obtained by reduction with different cooling rates.

In addition to the regional differences in metallization degree, the changes in porosity can be clearly observed in Figure 11. There was a similar trend to that of iron metallization degree. In other words, the porosity increased with increasing iron metallization degree. When the cooling rate decreased to 2.0 °C/min, the metallized pellets became more porous, lowering their compressive strength.

3.2. Effect of Cooling Rate on the Porosity

As aforementioned, the porosity varied with the cooling rate during the reduction process. To describe the trend more accurately, the distribution of porosity inside different metallized pellets was analyzed. Figure 12a shows the pore distributions of different pellets from the edge to the center. More pores could be clearly observed in the left side of the green line. The porosity of the metallized pellets increased considerably with decreasing cooling rate. At the same time, this change progressed inward. The interior of the metallized pellets could be divided into the high-porosity region and low-porosity region. By combining with the line-scan fitting curve of elemental carbon from epoxy resin in Figure 12b, it was found that the carbon content of the pellets after reduction increased significantly, indicating the increase in porosity of the pellets via reduction. It was related to the migration and aggregation of metallic iron particles and emission of the reduction product, H_2O, during the reduction process. Meanwhile, as the reduction process proceeded, the fluctuations in carbon content along the radial direction gradually became smaller, indicating the relatively uniform pore distribution.

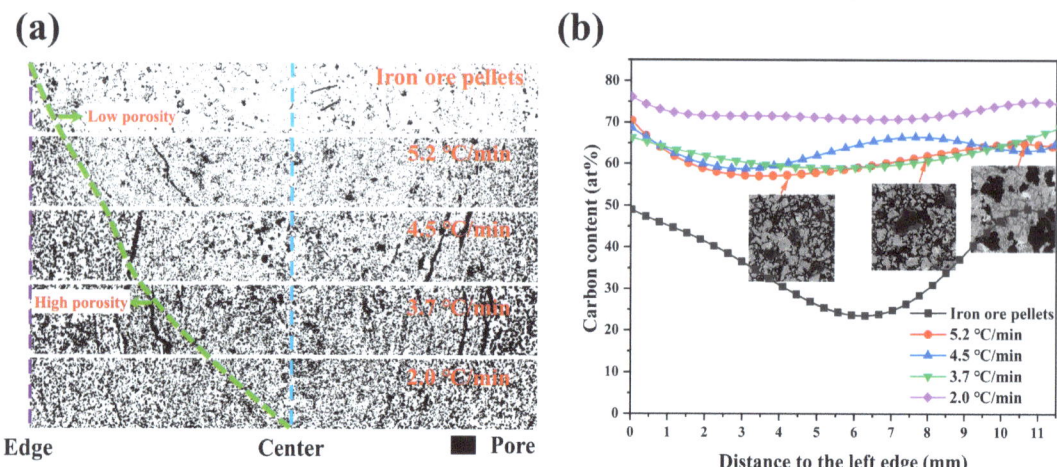

Figure 12. (a) Pore distributions and (b) fluctuations in carbon content along the radial direction in the metallized pellets obtained by reduction with different cooling rates.

Figure 13 shows the porosities of the metallized pellets obtained by reduction with different cooling rates. Compared to the iron ore pellets, the porosity of the metallized pellets increased by 29.7%, 50.2%, 72.0%, and 127.2% when the cooling rate was 5.2 °C/min, 4.5 °C/min, 3.7 °C/min, and 2.0 °C/min, respectively. Although a larger porosity indicated a higher reduction degree and iron metallization degree, it was detrimental to the compressive strength. Therefore, selecting a proper cooling rate would be critical for HCR.

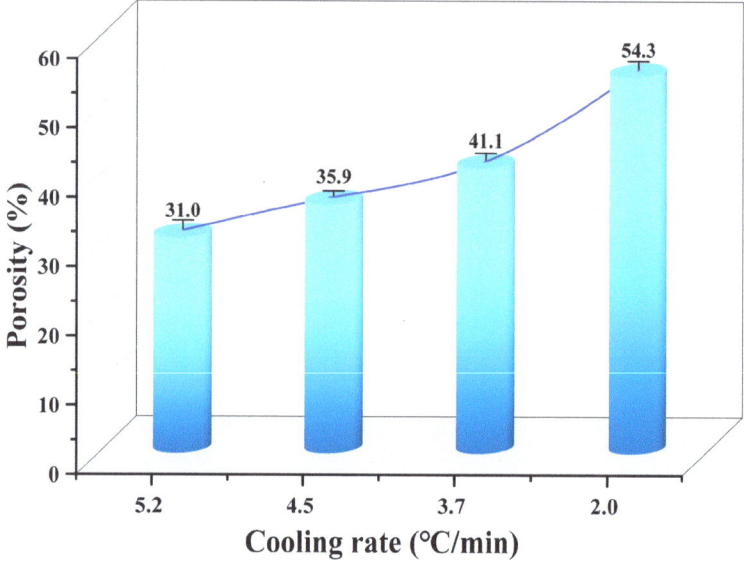

Figure 13. Porosities of the metallized pellets obtained by reduction with different cooling rates.

3.3. Effect of Cooling Rate on the Generation of Metallic Iron Particles

In the HCR process, metallic iron particles would form with variable morphological features. Figure 14 shows the micro-morphologies of the metallized pellets obtained by reduction with different cooling rates.

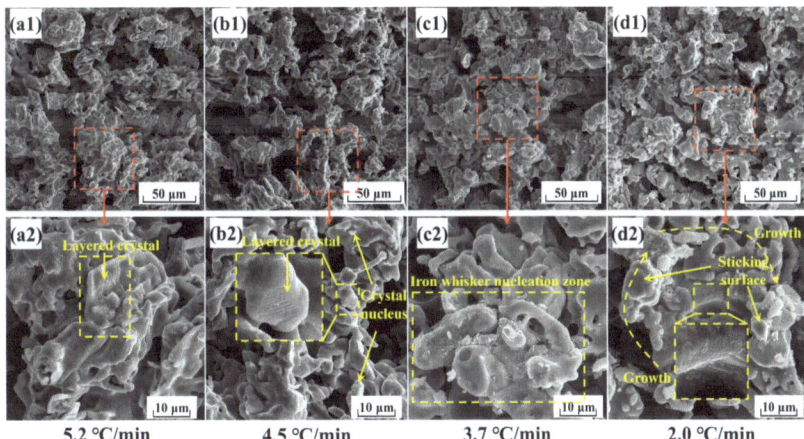

Figure 14. Micro-morphologies of the metallized pellets obtained by reduction with different cooling rates.

According to Figure 14, when the cooling rate was 5.2 °C/min, there were a number of layered iron particles with relatively smooth surfaces, consistent with the iron precipitation morphology found below 800 °C [38]. It could be attributed to the rapid temperature decrease and the short dwell time at high temperatures.

When the cooling rate decreased to 4.5 °C/min, more nascent iron particles formed on the surfaces of layered iron particles. Because the dwell time of the reaction system in the high temperature range (>900 °C) was extended with this cooling rate, the precipitation rate of metallic iron increased and the mode of precipitation shifted from the steady state to the unsteady state [39]. In this case, both layered and small spherical iron particles co-existed.

Figure 14 also shows the morphologies of the metallized pellets when the cooling rate was 3.7 °C/min. Because of the prolonged dwell time in the high temperature range, most of metallic iron nucleated at points and grew into iron whiskers. At the same time, due to the high reduction potential on the surface of the pellets, the reduction reaction proceeded rapidly and the migration rate of oxygen atoms was larger than the precipitation rate of iron atoms, leading to the precipitation of a small portion of metallic iron in the form of porous reticular structure. With this cooling rate, non-steady state precipitation became dominant.

When the cooling rate decreased to 2.0 °C/min, the iron whiskers on the surfaces of the metallized pellets had longer lengths. Meanwhile, more metallic iron precipitated in the porous reticular form. It microscopically explained the huge increase in porosity and the considerable decrease in compressive strength.

To further assess the effect of the cooling rate on the growth behavior of metallic iron particles, the iron particle sizes in different areas of the metallized pellets were characterized. The results are shown in Figures 15–18.

Figures 15–18 show that the mean size of metallic iron particles at the edge area of the metallized pellets was larger than those at 1/4 diameter and the central areas of the pellets. This was particularly evident at high cooling rate. Moreover, the d_{10} values of metallic iron particles of different metallized pellets were all around 1.5 μm. However, the d_{50} value of metallic iron particles in the metallized pellets increased significantly with decreasing cooling rate due to longer dwell time in the high temperature range.

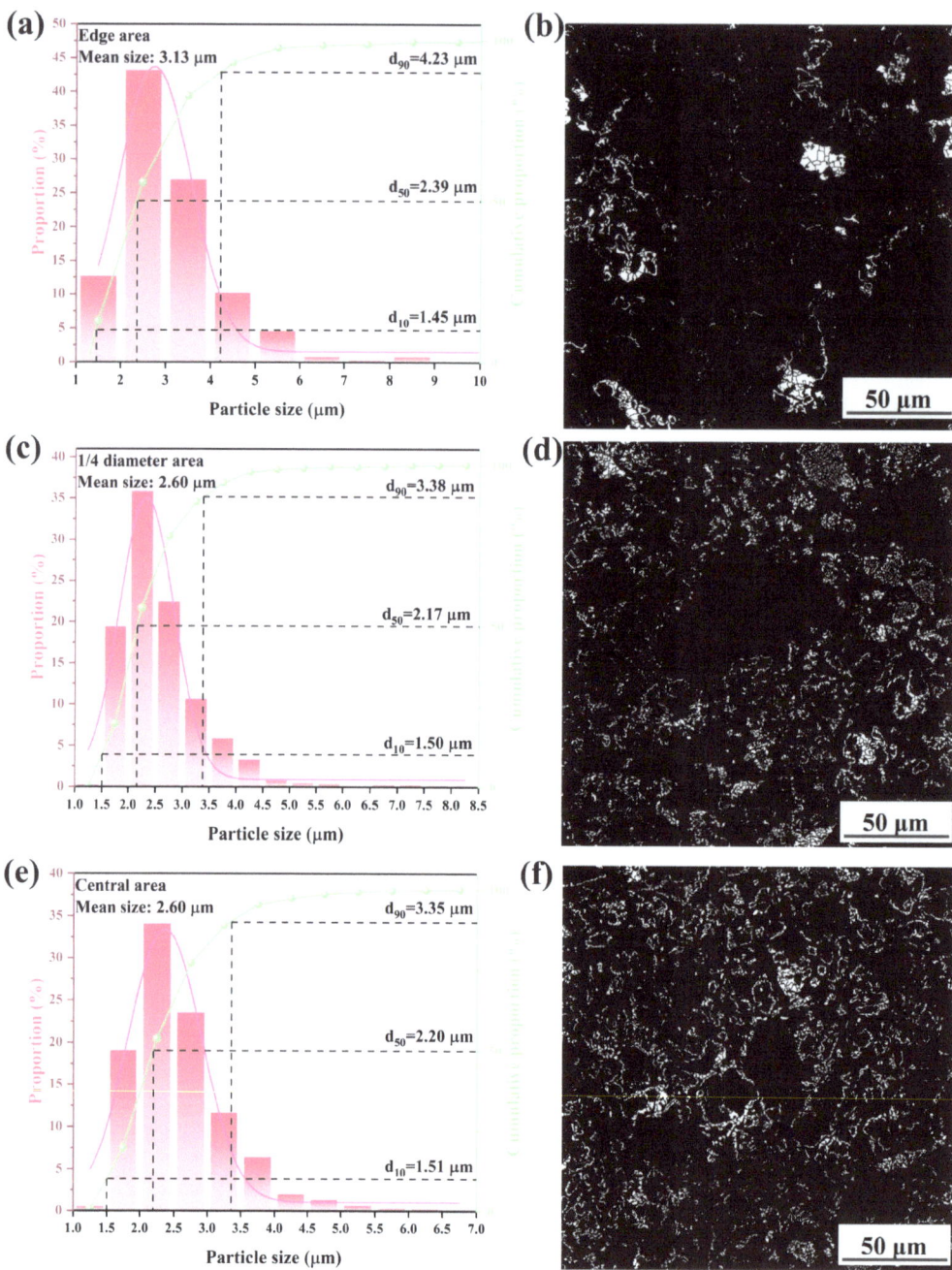

Figure 15. Histograms and microscopic features of iron particle size distributions in different areas of the metallized pellets obtained by reduction with the cooling rate of 5.2 °C/min: (**a**,**b**) edge area, (**c**,**d**) 1/4 diameter area, and (**e**,**f**) central area.

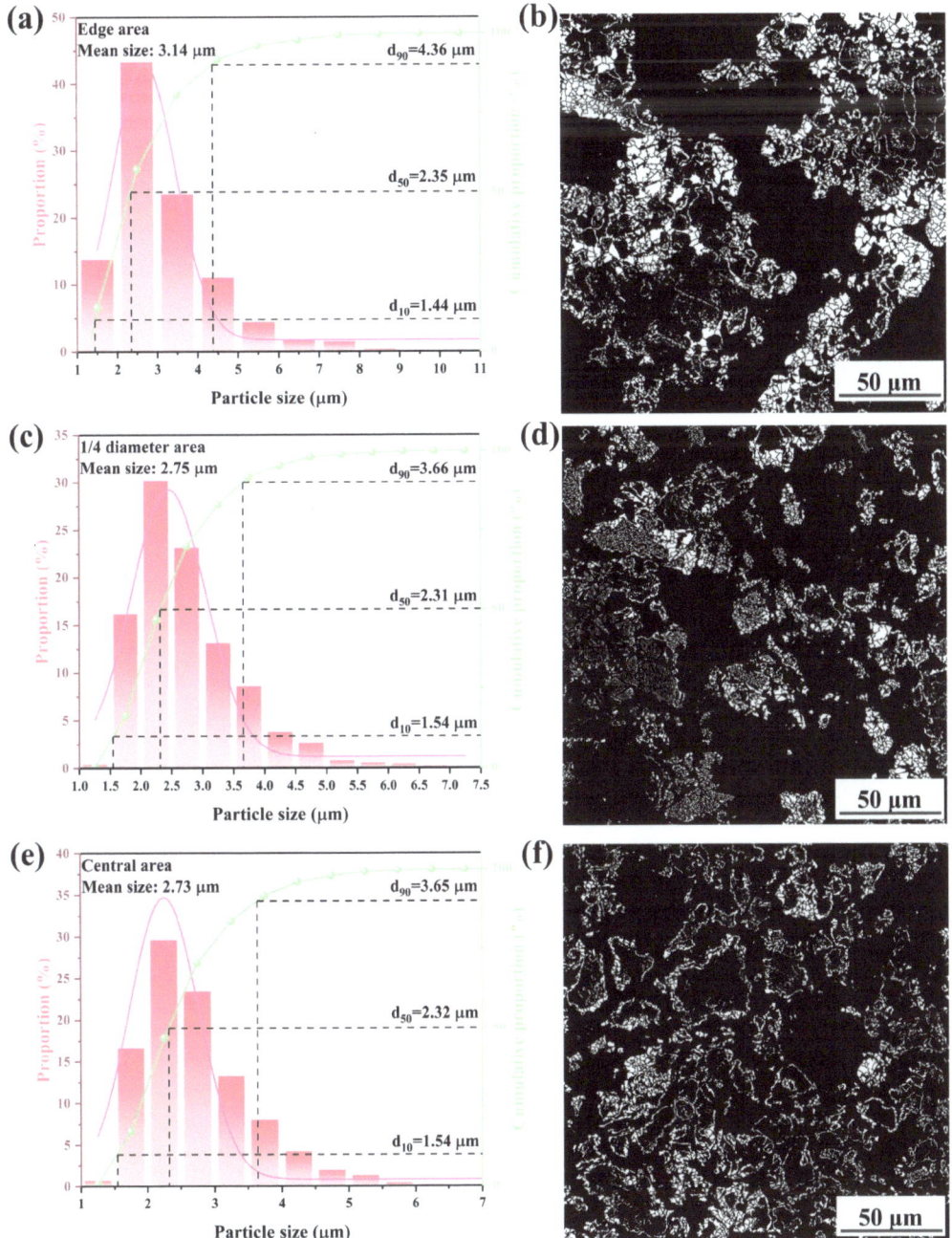

Figure 16. Histograms and microscopic features of iron particle size distributions in different areas of the metallized pellets obtained by reduction with the cooling rate of 4.5 °C/min: (**a**,**b**) edge area, (**c**,**d**) 1/4 diameter area, and (**e**,**f**) central area.

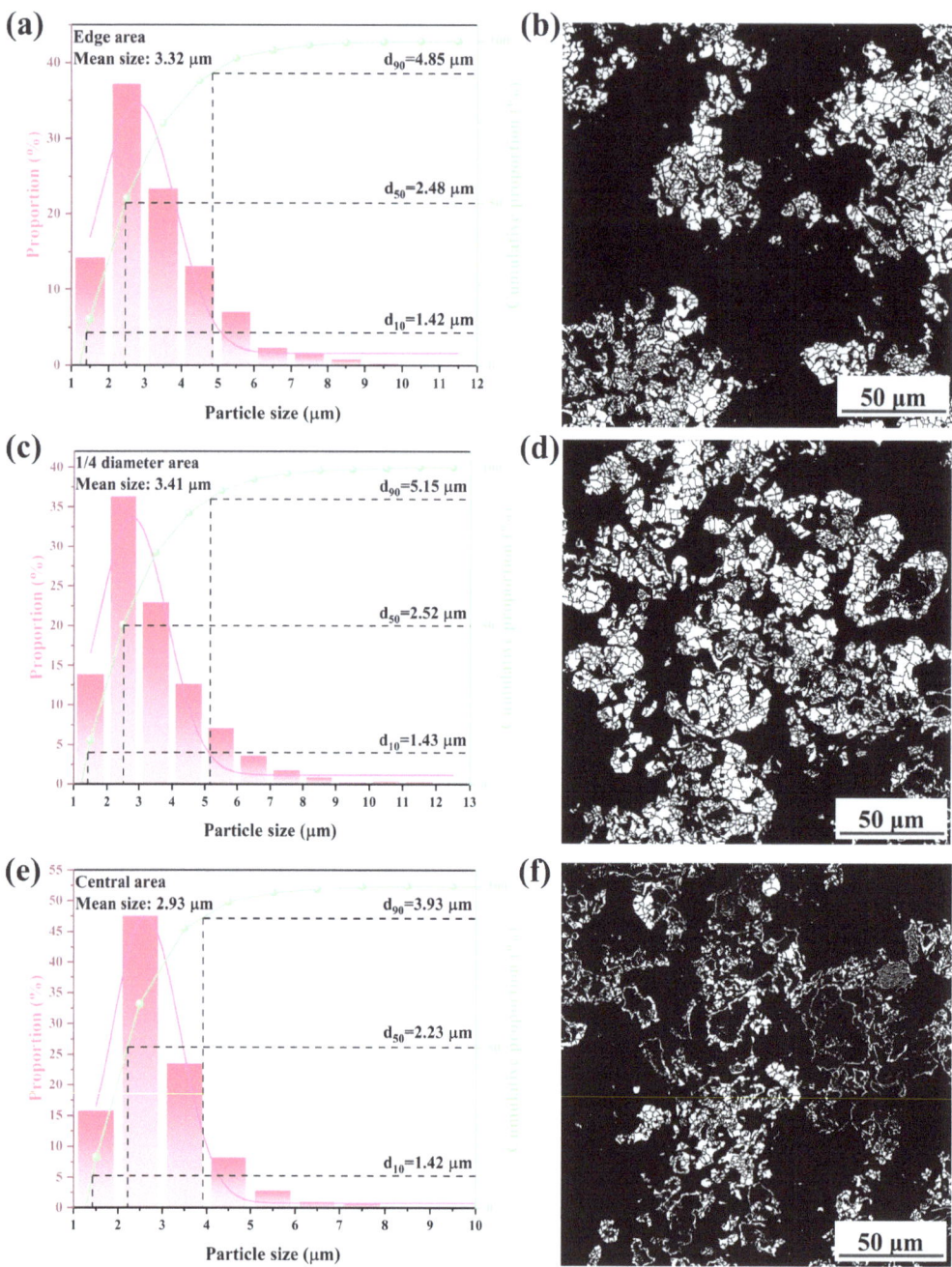

Figure 17. Histograms and microscopic features of iron particle size distributions in different areas of the metallized pellets obtained by reduction with the cooling rate of 3.7 °C/min: (**a**,**b**) edge area, (**c**,**d**) 1/4 diameter area, and (**e**,**f**) central area.

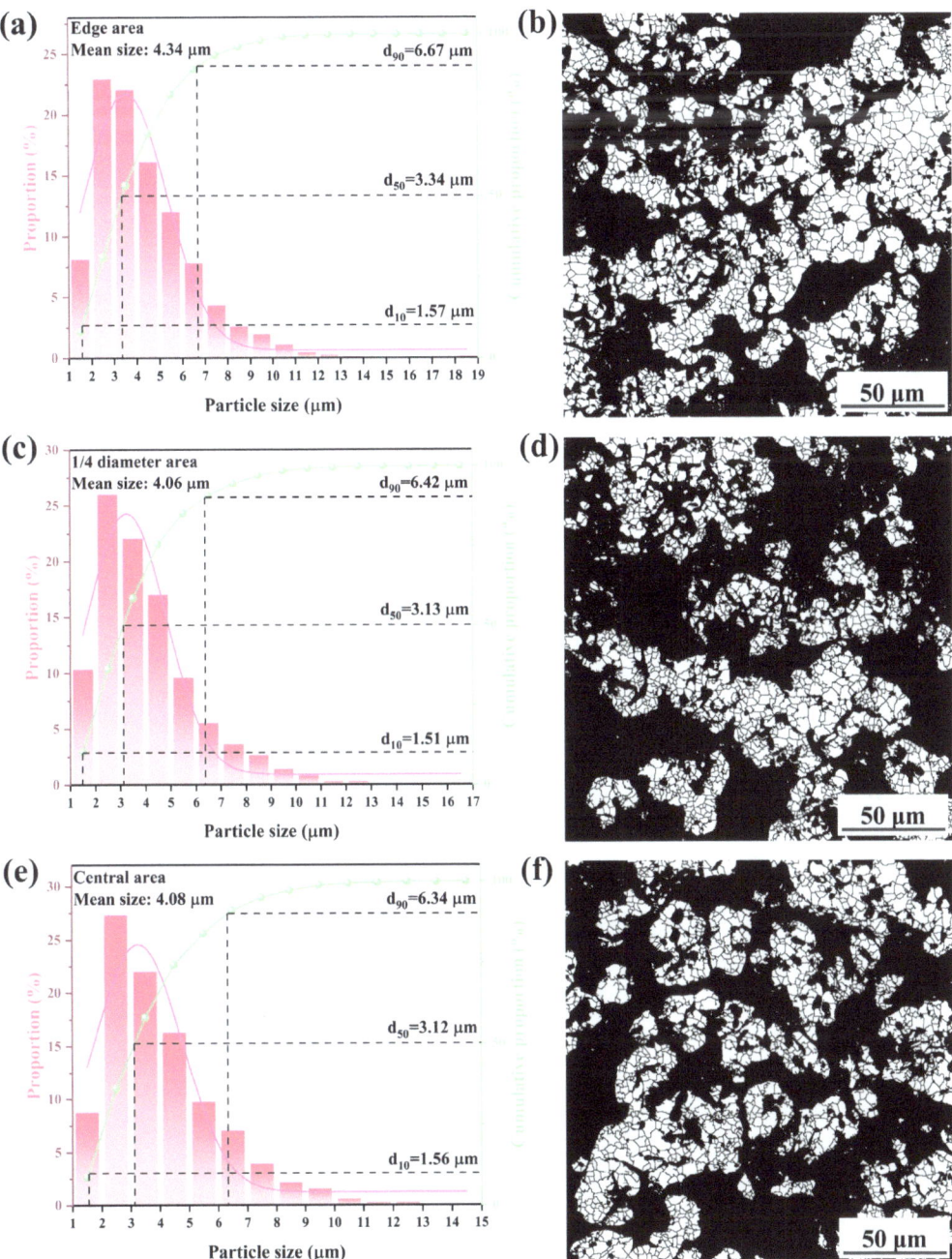

Figure 18. Histograms and microscopic features of iron particle size distribution in different areas of the metallized pellets obtained by reduction with the cooling rate of 2.0 °C/min: (**a**,**b**) edge area, (**c**,**d**) 1/4 diameter area, and (**e**,**f**) central area.

As shown in Figure 15, only a small quantity of metallic iron particles larger than 5 μm were observed at the edges. The size distribution of metallic iron particles at 1/4 diameter area of the pellets was approximately the same as that at the central area of the pellets, with more than 90% of the particles being smaller than 3.4 μm. According to Figure 16, when the cooling rate decreased to 4.5 °C/min, the metallic iron particles grew obviously along the edges of FeO particles and spread inward. The mean size increased from 3.13 μm (edge area), 2.60 μm (1/4 diameter area), and 2.60 μm (central area) to 3.14 μm (edge area), 2.75 μm (1/4 diameter area), and 2.73 μm (central area). When the cooling rate decreased to 3.7 °C/min, as shown in Figure 17, the mean sizes of metallic iron particles were 3.32 μm (edge area), 3.41 μm (1/4 diameter area), and 2.93 μm (central area). It indicated that the reduction performance was comparable at the edge and 1/4 diameter areas of the pellets, while the reduction at the central area was relatively incomplete. As expected, further reducing the cooling rate to 2.0 °C/min promoted the reduction of the pellets. As shown in Figure 18, the mean sizes of metallic iron particles reached 4.34 μm (edge area), 4.06 μm (1/4 diameter area), and 4.08 μm (central area). In addition, the distribution range and quantity of metallic iron particles in these three areas were nearly the same. It revealed that the reduction of the pellets became more uniform.

4. Conclusions

In this study, the effect of cooling rate on the HCR of iron ore pellets was investigated. When the cooling rate decreased from 5.2 °C/min to 2.0 °C/min, the total iron content, reduction degree, and iron metallization degree of the pellets increased from 74.0 wt%, 52%, and 31.1% to 84.9 wt%, 93.4%, and 89.2%, respectively. However, the compressive strength of the pellets increased initially from 2100 N/p to 2436 N/p and subsequently decreased significantly to 841 N/p. A too fast or too slow cooling rate would negatively affect the compressive strength of the pellets. During the reduction process, the main phase of the pellets, Fe_2O_3, had stepwise reactions to Fe, with diminishing FeO and Fe_2SiO_4. As the cooling rate declined, the porosity of the pellets increased from 23.9% to 54.3%, with a higher distribution uniformity of pores. The morphology of metallic iron particles transited from a layered form to a spherical form and finally to a porous reticular form. Meanwhile, the metallic iron particles in the pellets grew evidently and distributed more uniformly based on the analysis of metallic iron particle sizes in different areas of the pellets, namely edge area, 1/4 diameter area, and central area. After reduction with the cooling rate of 3.7 °C/min, the resulting metallized pellets had the reduction degree of 74.2%, iron metallization degree of 66.9%, and the highest compressive strength of 2436 N/p, mainly associated with the spherical morphology and relatively large size of metallic iron particles. Further efforts will be made to obtain the balance between iron metallization degree and compressive strength of the pellets.

Author Contributions: Conceptualization: G.L. and Z.P.; data curation, G.L.; formal analysis, G.L., Z.P., R.T. and L.Y.; funding acquisition, Z.P.; investigation, G.L. and K.G.; methodology, G.L., Z.P., W.F. and R.T.; project administration, Z.P.; resources, Z.P.; software, K.G., W.F. and M.R.; supervision, Z.P.; visualization, G.L., K.G., L.Y. and M.R.; writing—original draft, G.L.; writing—review & editing, G.L. and Z.P. All authors have read and agreed to the published version of the manuscript.

Funding: This work was partially supported by China Baowu Low Carbon Metallurgy Innovation Foundation under Grant BWLCF202103, the Hunan Provincial Natural Science Foundation of China under Grant 2023JJ10073, and Hunan Provincial Key Research and Development Program under Grant 2023SK2079.

Institutional Review Board Statement: Not applicable.

Informed Consent Statement: Not applicable.

Data Availability Statement: The datasets presented in this article are not readily available because the data are part of an ongoing study. Requests to access the datasets should be directed to zwpeng@csu.edu.cn.

Conflicts of Interest: The authors declare no conflict of interest.

References

1. Martins, T.; Barreto, A.C.; Souza, F.M.; Souza, A.M. Fossil fuels consumption and carbon dioxide emissions in G7 countries: Empirical evidence from ARDL bounds testing approach. *Environ. Pollut.* **2021**, *291*, 7. [CrossRef]
2. Zhao, J.; Zuo, H.B.; Wang, Y.J.; Wang, J.S.; Xue, Q.G. Review of green and low-carbon ironmaking technology. *Ironmak. Steelmak.* **2020**, *47*, 296–306. [CrossRef]
3. Tang, C.M.; Guo, Z.Q.; Pan, J.; Zhu, D.Q.; Li, S.W.; Yang, C.C.; Tian, H.Y. Current situation of carbon emissions and countermeasures in China's ironmaking industry. *Int. J. Miner. Metall. Mater.* **2023**, *30*, 1633–1650. [CrossRef]
4. Ren, L.; Zhou, S.; Peng, T.D.; Ou, X.M. A review of CO_2 emissions reduction technologies and low-carbon development in the iron and steel industry focusing on China. *Renew. Sustain. Energy Rev.* **2021**, *143*, 23. [CrossRef]
5. Zhang, S.H.; Yi, B.W.; Guo, F.; Zhu, P.Y. Exploring selected pathways to low and zero CO_2 emissions in China's iron and steel industry and their impacts on resources and energy. *J. Clean. Prod.* **2022**, *340*, 18. [CrossRef]
6. Na, H.M.; Yuan, Y.X.; Du, T.; Zhang, T.B.; Zhao, X.; Sun, J.C.; Qiu, Z.Y.; Zhang, L. Multi-process production occurs in the iron and steel industry, supporting 'dual carbon' target: An in-depth study of CO_2 emissions from different processes. *J. Environ. Sci.* **2024**, *140*, 46–58. [CrossRef] [PubMed]
7. Chen, Y.B.; Zuo, H.B. Review of hydrogen-rich ironmaking technology in blast furnace. *Ironmak. Steelmak.* **2021**, *48*, 749–768. [CrossRef]
8. Shao, L.; Zhang, X.N.; Zhao, C.X.; Qu, Y.X.; Saxén, H.; Zou, Z.S. Computational analysis of hydrogen reduction of iron oxide pellets in a shaft furnace process. *Renew. Energy* **2021**, *179*, 1537–1547. [CrossRef]
9. Tang, J.; Chu, M.S.; Li, F.; Feng, C.; Liu, Z.G.; Zhou, Y.S. Development and progress on hydrogen metallurgy. *Int. J. Miner. Metall. Mater.* **2020**, *27*, 713–723. [CrossRef]
10. Li, W.; Fu, G.Q.; Chu, M.S.; Zhu, M.Y. Reduction behavior and mechanism of Hongge vanadium titanomagnetite pellets by gas mixture of H_2 and CO. *J. Iron. Steel Res. Int.* **2017**, *24*, 34–42. [CrossRef]
11. Behera, P.R.; Bhoi, B.; Paramguru, R.K.; Mukherjee, P.S.; Mishra, B.K. Hydrogen Plasma Smelting Reduction of Fe_2O_3. *Metall. Mater. Trans. B* **2019**, *50*, 262–270. [CrossRef]
12. Souza Filho, I.R.; Ma, Y.; Kulse, M.; Ponge, D.; Gault, B.; Springer, H.; Raabe, D. Sustainable steel through hydrogen plasma reduction of iron ore: Process, kinetics, microstructure, chemistry. *Acta Mater.* **2021**, *213*, 116971. [CrossRef]
13. Pauna, H.; Ernst, D.; Zarl, M.; Souza Filho, I.R.D.; Kulse, M.; Büyükuslu, Ö.; Jovičević-Klug, M.; Springer, H.; Huttula, M.; Schenk, J.; et al. The Optical Spectra of Hydrogen Plasma Smelting Reduction of Iron Ore: Application and Requirements. *Steel Res. Int.* **2024**, *95*, 2400028. [CrossRef]
14. Sohn, H.Y.; Fan, D.-Q.; Abdelghany, A. Design of Novel Flash Ironmaking Reactors for Greatly Reduced Energy Consumption and CO_2 Emissions. *Metals* **2021**, *11*, 332. [CrossRef]
15. Moriya, K.; Takahashi, K.; Murao, A.; Sato, T.; Fukada, K. Prediction of Pulverized Coal Combustibility by Measuring Chemiluminescence of Radical Species around Tuyere. *ISIJ Int.* **2022**, *62*, 1371–1380. [CrossRef]
16. Zhang, X.; Jiao, K.; Zhang, J.; Guo, Z. A review on low carbon emissions projects of steel industry in the World. *J. Clean. Prod.* **2021**, *306*, 127259. [CrossRef]
17. De Castro, J.A.; De Medeiros, G.A.; Da Silva, L.M.; Ferreira, I.L.; De Campos, M.F.; De Oliveira, E.M. A Numerical Study of Scenarios for the Substitution of Pulverized Coal Injection by Blast Furnace Gas Enriched by Hydrogen and Oxygen Aiming at a Reduction in CO_2 Emissions in the Blast Furnace Process. *Metals* **2023**, *13*, 21. [CrossRef]
18. Jiang, X.; Wang, L.; Shen, F.M. Shaft Furnace Direct Reduction Technology—Midrex and Energiron. *Adv. Mater. Res.* **2013**, *805*, 654–659. [CrossRef]
19. Wang, R.R.; Zhao, Y.Q.; Babich, A.; Senk, D.; Fan, X.Y. Hydrogen direct reduction (H-DR) in steel industry—An overview of challenges and opportunities. *J. Clean. Prod.* **2021**, *329*, 129797. [CrossRef]
20. Scharm, C.; Küster, F.; Laabs, M.; Huang, Q.; Volkova, O.; Reinmöller, M.; Guhl, S.; Meyer, B. Direct reduction of iron ore pellets by H_2 and CO: In-situ investigation of the structural transformation and reduction progression caused by atmosphere and temperature. *Miner. Eng.* **2022**, *180*, 107459. [CrossRef]
21. Man, Y.; Feng, J.X.; Li, F.J.; Ge, Q.; Chen, Y.M.; Zhou, J.Z. Influence of temperature and time on reduction behavior in iron ore–coal composite pellets. *Powder Technol.* **2014**, *256*, 361–366. [CrossRef]
22. Feng, J.; Tang, J.; Chu, M.; Liu, P.; Zhao, Z.; Zheng, A.; Wang, X.; Han, T. Sticking Behavior of Pellets During Direct Reduction Based on Hydrogen Metallurgy: An Optimization Approach Using Response Surface Methodology. *J. Sustain. Metall.* **2023**, *9*, 1139–1154. [CrossRef]
23. *GB/T 6730.5-2022*; Iron Ores—Determination of Total Iron Content—Titrimetric Methods after Titanium (III) Chloride Reduction. State Administration for Market Regulation and Standardization Administration of China & Standardization Administration of China: Beijing, China, 2022.
24. *GB/T 24236-2009*; Iron Ores for Shaft Direct-Reduction Feedstocks—Determination of the Reducibility Index, Final Degree of Reduction and Degree of Metallization. State General Administration of China for Quality Supervision and Inspection and Quarantine & Standardization Administration of China: Beijing, China, 2009.

25. *GB/T14201–2018*; Iron Ore Pellets for Blast Furnace and Direct Reduction Feedstocks—Determination of the Crushing Strength. State Administration for Market Regulation and Standardization Administration of China & Standardization Administration of China: Beijing, China, 2018.
26. Zou, G.; She, J.; Peng, S.; Yin, Q.; Liu, H.; Che, Y. Two-dimensional SEM image-based analysis of coal porosity and its pore structure. *Int. J. Coal Sci. Technol.* **2020**, *7*, 350–361. [CrossRef]
27. Otsu, N. A Threshold Selection Method from Gray-Level Histograms. *IEEE Trans. Syst. Man Cybern.* **1979**, *9*, 62–66. [CrossRef]
28. *GB/T24586-2009*; Iron Ores—Determination of Apparent Density, True Density and Porosity. State Administration for Market Regulation and Standardization Administration of China & Standardization Administration of China: Beijing, China, 2009.
29. Zinoveev, D.; Grudinsky, P.; Zakunov, A.; Semenov, A.; Panova, M.; Valeev, D.; Kondratiev, A.; Dyubanov, V.; Petelin, A. Influence of Na_2CO_3 and K_2CO_3 Addition on Iron Grain Growth during Carbothermic Reduction of Red Mud. *Metals* **2019**, *9*, 1313. [CrossRef]
30. Grudinsky, P.; Zinoveev, D.; Pankratov, D.; Semenov, A.; Panova, M.; Kondratiev, A.; Zakunov, A.; Dyubanov, V.; Petelin, A. Influence of Sodium Sulfate Addition on Iron Grain Growth during Carbothermic Roasting of Red Mud Samples with Different Basicity. *Metals* **2020**, *10*, 1571. [CrossRef]
31. Ishak, M.H.; Sazali, N.; Ghazali, M.F. A short review on different metal alloys on rapid cooling process. *IOP Conf. Ser. Mater. Sci. Eng.* **2020**, *736*, 052034. [CrossRef]
32. Ye, L.; Peng, Z.; Ye, Q.; Wang, L.; Augustine, R.; Lee, J.; Liu, Y.; Liu, M.; Rao, M.; Li, G.; et al. Preparation of Metallized Pellets from Blast Furnace Dust and Electric Arc Furnace Dust Based on Microwave Impedance Matching. In Proceedings of the 11th International Symposium on High-Temperature Metallurgical Processing, Cham, Switzerland, 24 January 2020; pp. 569–579. [CrossRef]
33. Bersenev, I.S.; Pokolenko, S.I.; Sabirov, E.R.; Spirin, N.A.; Borisenko, A.V.; Kurochkin, A.R. Influence of the Iron Ore Pellets Macrostructure on Their Strength. *Steel Transl.* **2023**, *53*, 1018–1022. [CrossRef]
34. Prusti, P.; Rath, S.S.; Dash, N.; Meikap, B.C.; Biswal, S.K. Pelletization of hematite and synthesized magnetite concentrate from a banded hematite quartzite ore: A comparison study. *Adv. Powder Technol.* **2021**, *32*, 3735–3745. [CrossRef]
35. Gao, X.; Chai, Y.; Wang, Y.; Luo, G.; An, S.; Peng, J. Process and mechanism of preparing metallized blast furnace burden from metallurgical dust and sludge. *Sci. Rep.* **2024**, *14*, 9760. [CrossRef]
36. Rajshekar, Y.; Alex, T.C.; Sahoo, D.P.; Babu, G.A.; Balakrishnan, V.; Venugopalan, T.; Kumar, S. Iron ore slime as an alternate coolant in steelmaking: Performance evaluation at commercial scale. *J. Clean. Prod.* **2016**, *139*, 886–893. [CrossRef]
37. Wang, Z.; Peng, B.; Zhang, L.; Zhao, Z.; Liu, D.; Peng, N.; Wang, D.; He, Y.; Liang, X.; Liu, H. Study on Formation Mechanism of Fayalite (Fe_2SiO_4) by Solid State Reaction in Sintering Process. *JOM* **2018**, *70*, 539–546. [CrossRef]
38. Zhang, L.; Chen, H.; Deng, R.; Zuo, W.; Guo, B.; Ku, J. Growth behavior of iron grains during deep reduction of copper slag. *Powder Technol.* **2020**, *367*, 157–162. [CrossRef]
39. Guo, L.; Zhong, S.; Bao, Q.; Gao, J.; Guo, Z. Nucleation and Growth of Iron Whiskers during Gaseous Reduction of Hematite Iron Ore Fines. *Metals* **2019**, *9*, 750. [CrossRef]

Disclaimer/Publisher's Note: The statements, opinions and data contained in all publications are solely those of the individual author(s) and contributor(s) and not of MDPI and/or the editor(s). MDPI and/or the editor(s) disclaim responsibility for any injury to people or property resulting from any ideas, methods, instructions or products referred to in the content.

Article

Formability Prediction Using Machine Learning Combined with Process Design for High-Drawing-Ratio Aluminum Alloy Cups

Yeong-Maw Hwang [1], Tsung-Han Ho [1], Yung-Fa Huang [2,*] and Ching-Mu Chen [3]

1. Department of Mechanical and Electro-Mechanical Engineering, National Sun Yat-sen University, Kaohsiung 80424, Taiwan; ymhwang@mail.nsysu.edu.tw (Y.-M.H.); rf30526@gmail.com (T.-H.H.)
2. Department of Information and Communication Engineering, Chaoyang University of Technology, Taichung 413310, Taiwan
3. Department of Electrical Engineering, National Penghu University of Science and Technology, Magong 88046, Taiwan; t20136@gms.npu.edu.tw
* Correspondence: yfahuang@cyut.edu.tw

Citation: Hwang, Y.-M.; Ho, T.-H.; Huang, Y.-F.; Chen, C.-M. Formability Prediction Using Machine Learning Combined with Process Design for High-Drawing-Ratio Aluminum Alloy Cups. *Materials* **2024**, *17*, 3991. https://doi.org/10.3390/ma17163991

Academic Editors: Jan Haubrich and Daniela Kovacheva

Received: 20 May 2024
Revised: 13 July 2024
Accepted: 8 August 2024
Published: 11 August 2024

Copyright: © 2024 by the authors. Licensee MDPI, Basel, Switzerland. This article is an open access article distributed under the terms and conditions of the Creative Commons Attribution (CC BY) license (https://creativecommons.org/licenses/by/4.0/).

Abstract: Deep drawing has been practiced in various manufacturing industries for many years. With the aid of stamping equipment, materials are sheared to different shapes and dimensions for users. Meanwhile, through artificial intelligence (AI) training, machines can make decisions or perform various functions. The aim of this study is to discuss the geometric and process parameters for A7075 in deep drawing and derive the formable regions of sound products for different forming parameters. Four parameters—forming temperature, punch speed, blank diameter and thickness—are used to investigate their effects on the forming results. Through finite element simulation, a database is established and used for machine learning (ML) training and validation to derive an AI prediction model. Importing the forming parameters into this prediction model can obtain the forming results rapidly. To validate the formable regions of sound products, several experiments are conducted and the results are compared with the prediction results to verify the feasibility of applying ML to deep drawing processes of aluminum alloy A7075 and the reliability of the AI prediction model.

Keywords: deep drawing; machine learning; aluminum alloy A7075; finite element analysis

1. Introduction

As a widely used metal forming technology, deep drawing processes have been applied to manufacturing circular or square cups in various industries for many years. With the help of stamping equipment, blanks are sheared, bent, and shaped into desired geometries and dimensions. Artificial intelligence (AI) is an emerging field of science and technology development. The purpose of AI is to allow computers to learn like humans, and through training with a database to make decisions or perform various functions. The objective of this study is to explore the influence of forming temperature, punch speed, blank diameter, and blank thickness on the forming results using finite element simulations with the variation of the four parameters and obtain a possible forming range for sound-drawn products in A7075 deep drawing processes using machine learning (ML) algorithm. A dataset for formable ranges of the parameters is constructed from ML training and verification. An AI prediction model is also established. Inputting the forming parameters, the predicted forming results can be obtained quickly. Finally, deep drawing experiments are conducted and the results obtained are used to verify the predicted formable regions of sound products and the proposed AI model.

Deep drawing is a long-established technology. Using a stamping machine, a circular sheet can be formed into circular cups or more complex shape parts used in automotive or aircraft industries [1]. Due to long-term development, deep drawing and stamping have become a stable and well-known technology. However, these methods are limited to forming ductile metal materials. For some brittle metal materials with low formability and high hardness, such as aluminum alloy A7075, cracking or fractures are likely to occur at the cup corner during deep drawing processes. To raise the formability of the products, an elevated temperature is usually used, and other forming parameters have to be adjusted appropriately. It is necessary to obtain the formable ranges for various forming parameters. The finite element analysis has been widely applied to construct a comprehensive model for many different cases. Nevertheless, a lot of time is needed to complete all the simulations with various cases. Therefore, a new approach based on ML is proposed to replace the traditional finite element simulation-based method. The ML system learns from input data and improves the output results by adjusting internal weights. After training, it can be used to predict results or assist decision-making. After learning, the predicted forming results can be obtained quickly under different forming conditions by inputting relevant information later. This approach can save time and cost in developing various deep-drawn cup products.

Regarding the mechanical properties of aluminum alloy A7075, Tajally and Emadoddin [2] and Cerri et al. [3] investigated the mechanics and anisotropic behavior of A7075 plates. The former conducted annealing of A7075 plates to reduce the yield strength and hardness and to improve its formability. A series of experiments verified that annealing at 350~400 °C has obvious effects. The latter conducted high-temperature torsion testing. As the temperature increases, the ductility between 250 °C and 300 °C increases due to changes in the microstructure. In view of the influence of various forming parameters of deep drawing, Colgan and Monaghan [1] examined some forming factors influencing the deep drawing process, utilizing an experimental rig design and statistical analysis. The parameters include the punch and die radii, the punch velocity, clamping force, friction, and drawing depth. From finite element analysis and experimental results, it seems that the punch/die radii have the greatest effect on the thickness of the deformed mild steel cups compared to blank-holder force or friction. Chen et al. [4] used finite element analysis to investigate the effects of the gap between the punch and die, the corner radius of the punch, and the corner radius of the die shoulder in deep drawing with a circular and square die design. They found that the defects of cracks occurred due to excessive stretching of the blank. To avoid cracking to improve the formability, the die geometrical dimensions and forming parameters have to be set approximately. Gowtham et al. [5] investigated the effects of the die radius on deep drawing results and found that reducing the die radius can cause stretch marks on the blank and uneven height of the cup.

Concerning the studies on AI, a few papers have been published to examine network algorithms to combine metal-forming technology with ML. For example, Liu et al. [6] used a long short-term memory (LSTM) network to predict bearing failure and fracture and proposed a new LSTM model. This model combines the advantages of an LSTM network and statistical analysis to predict aviation engine bearing failure. The results show that this method has higher accuracy than recurrent neural network (RNN), support vector machine (SVM), and LSTM, thereby improving the prediction accuracy of bearing performance degradation trend and remaining service life. ML facilitates computers to read and interpret from the previously present data automatically and makes use of multiple algorithms to build models, mathematical in nature, and then makes predictions for the new data using the past data and knowledge. Lately, it has been adopted for text detection, hate speech detection, recommender system, face detection, and more. In [7], the aspects concerning ML algorithms, K-nearest neighbor (KNN), genetic algorithm (GA), SVM, decision tree (DT), and LSTM network have been investigated.

In [8], the authors calculated the accuracy of ML algorithms for predicting heart disease, with KNN, DT, linear regression, and SVM by using the University of California, Irvine (UCI) repository dataset for training and testing. In [9], the previous works proposed a hybrid traffic classification method based on ML using the packet-multilayer perceptron (P-MLP) model and majority voting method to effectively classify the encrypted traffic in the network. In previous research [10], a FaceNet training method is used to train the mask face recognition (MFR) model using migration learning combined with a CNN model and a fully connected SoftMax output classifier to dynamically update the optimizer's learning rate (LR). The proposed ML models combined with fingerprinting for indoor localization using channel side information (CSI) [11]. Experiments were conducted on the positioning accuracy performance of RF and backpropagation neural network (BPNN) ML models using received signal strength indicator (RSSI) and CSI information, respectively.

Shinomiya et al. [12] used ML to control the movement of a slider and applied a convolutional neural network (CNN) with consideration of the elastic strain of the punch to predict the quality of the extruded products. Using this intelligent slider motion control, the defects can be prevented, and sound products can be obtained. Accordingly, the defect rate can be reduced significantly. Crystal plasticity analysis in sheet metal forming consumes a lot of calculation time. Cancemi et al. [13] proposed deep learning (DL) in the investigation of the safety behavior of nuclear plant items. The proposed innovative methodology is based on an unsupervised neural network (NN) to predict potential anomalies in the cooling system of a pressurized water reactor. Yamanaka [14] used ML models to replace complicated formula calculations and used simpler dimensionality reduction models to calculate the data from sheet-forming experiments. Thus, the efficiency of sheet metal forming simulations is improved by combining with the ML learning models.

Due to the shortage of manpower, many manufacturing companies are developing their production toward automation and intelligence. Therefore, in recent years, many companies introduced metal-forming technology combined with AI. In this paper, a new approach based on ML is proposed to predict the formable regions of a sound product for various forming parameters in the deep drawing process of an aluminum ally A7075 circular cup. By this method, the development cycle of a newly drawn product can be shortened and the production efficiency can be improved.

2. Finite Element Simulations

2.1. Geometric Configurations in the Deep Drawing Process

DEFORM (v11.0.2, Scientific Forming Technologies Corporation (SFTC), Columbus, OH, USA) [15], which is a finite element simulation software, was used in this paper for a series of simulations. It was used to carry out deep drawing simulations to analyze the plastic deformation and heat transfer of the metal material and exhibit the simulation results of drawing force and product geometries.

The schematic diagrams of before and after deep drawing forming processes are shown in Figure 1, and the relevant geometrical dimensions and forming parameters used in finite element simulations are shown in Table 1. The simulation objects are mainly composed of four parts: a punch, a blank holder, a blank, and a die. During the simulation process, the punch moves downward at a constant speed. After contacting the blank material, punch force is applied to bend the blank downward and press it into the die. During this period, a blank holder is used to compress the blank. In the actual experiments, a spring is used as the source of the compression force; thus, the compression force is set as the function of punch stroke to ensure the simulation conditions are the same as those in the experiments. The circular blank material is aluminum alloy A7075. Initially, the blank is placed on the top of the die and compressed by the blank holder. It is bent by the downward punch and drawn into the die cavity to form a cup-shaped product. The gap between the punch and the die is an important geometric parameter, which is usually set as 1.1–1.3 times the blank thickness. If the gap is smaller, a longer cup can be obtained, but it may result in breakage at the cup wall. If the gap is too large, wrinkles or uneven material stretching may occur. In this study, the die gap is set at 2.2 mm, 1.1 times the blank thickness. The blank material of aluminum alloy A7075 is set as an elastoplastic body, whereas the punch, die and blank holder are regarded as rigid bodies for saving the simulation time. The material properties of aluminum alloy A7075 used in the finite element simulations are shown in Table 2.

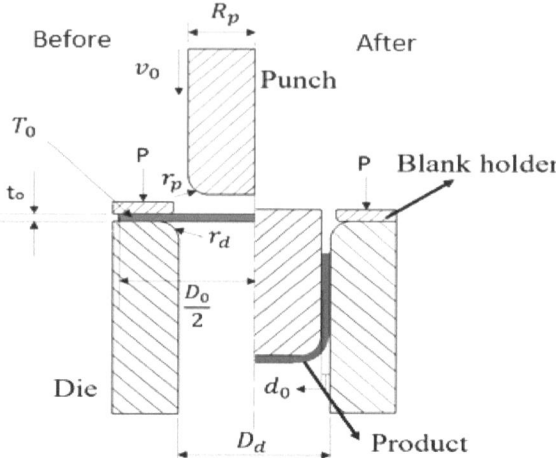

Figure 1. Schematic diagrams of before and after deep drawing forming processes.

Table 1. Geometric and forming parameters used in deep drawing simulations.

Variable	Description	Dimension
R_p	Punch diameter (mm)	16
v_0	Punch speed (mm/s)	12.08
r_p	Punch fillet radius (mm)	5
P	Compression force (N)	Stroke function
T_0	Forming temperature (°C)	250
t_l	Blank thickness (mm)	2
D_k	Blank diameter (mm)	65
r_d	Die fillet radius (mm)	5
d_0	Gap between punch and die (mm)	2.2
D_d	Die inner diameter (mm)	36.4

Table 2. Material properties of A7075 used in finite element simulations.

Material Parameters	Values and Unit
Elastic modulus	71,700 (MPa)
Poisson's ratio	0.33
Thermal expansion coefficients	$2.2 \times 10^{-5} \left(\frac{1}{°C}\right)$
Thermal conductivity	$41.7 \left(\frac{N}{sec \times K}\right)$
Heat capacity	$0.96 \left(\frac{J}{g \times °C}\right)$
Mass density	$2.81 \times 10^{-5} \left(\frac{kg}{mm^3}\right)$
Yield criterion	von Mises
Failure criteria	Normalized C&L [16]
Friction coefficient	0.1
Heat transfer coefficient	145 W/(m × K)

2.2. Convergence Analysis for Finite Element Simulations of Deep Drawing Process

To discuss the maximal deviation of the simulation results, convergence analysis of maximal load variation in deep drawing processes was implemented. The effects of the total element number on the maximal load variation in deep drawing processes are shown in Figure 2. Clearly, as the element number increases from 1000 elements (8-layer elements in the thickness direction) to 10,000 elements (25-layer elements in the thickness direction), the maximal load converges gradually from 1.72 tons to 1.74 tons and the maximal variations decrease by 0.5% as the element number exceeds 10,000. Accordingly, an element number of 10,000 is adopted in the subsequent finite element simulations.

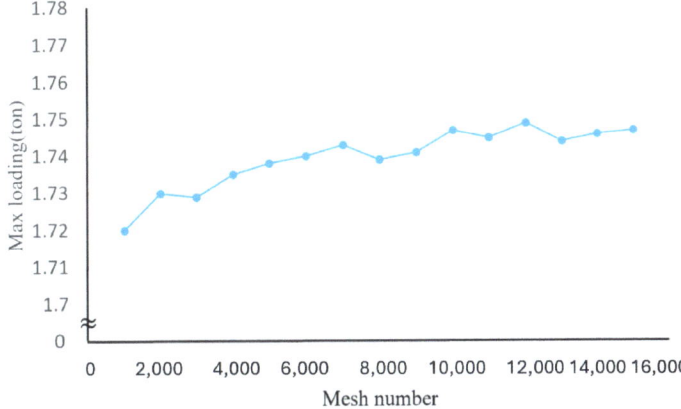

Figure 2. Effects of the element number on maximal load variation in the deep drawing process.

2.3. Compression Tests of A7075

In deep drawing processes, the blank material of A7075 undergoes plastic deformation. Thus, compression tests of aluminum alloy A7075 using a 160-ton servo press were conducted to obtain its flow stresses. In compression test experiments, the load and displacement of the upper die were recorded. The load is divided by the cross-section area of the specimen to obtain engineering stress and the top die displacement is divided by the initial length of the specimen to obtain engineering strain. Engineering stress and engineering strain can be converted into true stress σ_t and true strain ε_t through the equations by

$$\sigma_t = \sigma_e(1 + \varepsilon_e), \tag{1}$$

and

$$\varepsilon_t = ln(1 + \varepsilon_e) \tag{2}$$

respectively, where σ_e is the engineering stress and ε_e is the engineering strain [17]. The obtained A7075 flow stress curves are used in the finite element simulations. To ensure the accuracy of the flow stress obtained, finite element simulations of compression tests using the calculated flow stress were conducted. The comparisons between simulated and experimental loads are shown in Figure 3. The maximal difference between the two loads is about 10%, which validates the flow stress curve used in the finite element simulations. Compression tests at different temperatures were conducted. The flow stresses at different temperatures were input into DEFORM for the simulations at different forming temperatures. Heat conduction among the punch, blank and die was set. The related coefficients are given in Table 2.

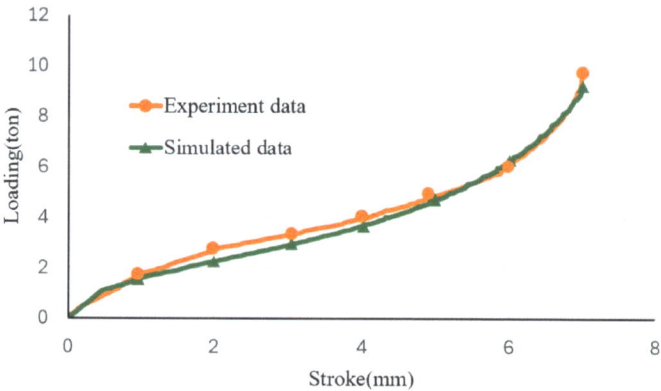

Figure 3. Comparisons of load-stroke curves between simulations and compression tests.

The critical fracture value of the material varies with the forming temperature. Normalized Cockcroft–Latham ductile fracture criteria [16] are used to calculate the fracture value of each point in the forming blank. During deformation, as the damage value at any point inside the blank is larger than the critical damage value, the point or element will disappear, which means a crack or fracture may occur. The critical damage values of A7075 at different temperatures shown in Table 3 were obtained by comparing the simulated compression tests with the actual compression tests. A7075 has tiny holes in the internal crystal lattice at higher temperatures, thus, the critical fracture values decrease slightly at higher temperatures [18].

Table 3. Critical damage values of A7075 at different forming temperatures.

Forming Temperature (°C)	Critical Fracture Value
250	0.35
275	0.345
300	0.34
325	0.335
350	0.33
375	0.315
400	0.3
425	0.285
450	0.27

2.4. Simulation Results and Discussion

A series of static-implicit finite element simulations with variations of the four forming parameters: forming temperature, punch speed, blank thickness, and blank diameter were conducted. Several levels for each parameter are chosen for the finite element simulations. The values or levels for the four forming parameters are listed in Table 4.

Table 4. The level setting for four forming parameters.

Parameters	Level Setting	Number of Levels
Forming temperature (°C)	50, 100, 150, 200, 225, 250, 275, 300, 325, 350, 375, 400, 425, 450	13
Average punch speed (mm/s)	6.0, 12.1, 18.1, 24.2, 30.2, 36.3	6
Blank diameter (mm)	51, 54, 57.5, 65	4
Blank thickness (mm)	0.5, 1.0, 1.5, 2.0, 2.5	5

In case 1, the punch speed, blank diameter, and thickness are fixed at 12.1 mm/s, 57.5 mm, and 2.0 mm, respectively, and finite element simulations with variations of forming temperature were conducted. The simulation results with temperatures of 250 °C and 375 °C are shown in Figure 4a and Figure 4b, respectively, where the product is shown by light yellow color for easily distinguished. Clearly, a sound product was obtained at a temperature of 250 °C, whereas fracture occurred around the corner of the cup at the temperature of 375 °C. From a series of simulation results, it is known that at either too low or too high temperatures necking or fracture occurred in the product. The blank at lower temperatures is subjected to greater stress during deformation and is prone to cracking at the fillets or corners of the deep-drawn cup. Because of smaller critical damage values at higher temperatures, failure or cracks are easier to occur during the drawing process. Therefore, deep drawing processes should be conducted within a certain temperature range for better-quality products.

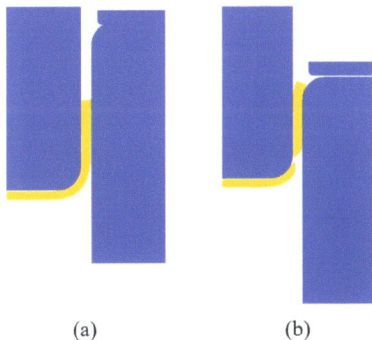

(a) (b)

Figure 4. Simulation results with different temperatures: (**a**) 250 °C and (**b**) 375 °C.

In case 2, the forming temperature, blank diameter, and thickness are fixed at 375 °C, 57.5 mm, and 2.0 mm, respectively, and finite element simulations with variations of punch speed were conducted. The simulation results with punch speeds of 36.3 mm/s and 12.1 mm/s are shown in Figure 5a and Figure 5b, respectively. Clearly, a sound product was obtained at a punch speed of 36.3 mm/s, whereas a fracture occurred around the corner of the cup at a punch speed of 12.1 mm/s. From a series of simulation results, it is known that at a faster punch speed, a sound-drawn cup can be obtained, whereas, at a slower punch speed the blank is easier to break during the drawing process. That is probably because a slower punch speed may limit the plastic deformation of the blank and a too low strain rate may result in uneven product surfaces.

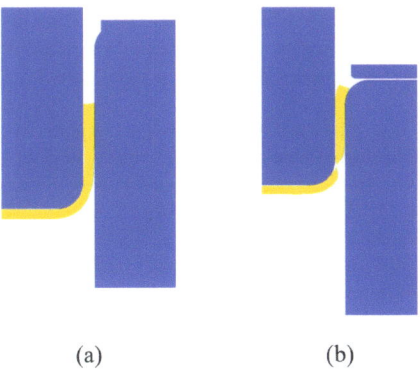

(a) (b)

Figure 5. Simulation results with different punch speeds: (**a**) 36.3 mm/s and (**b**)12.1 mm/s.

The blank diameter affects greatly the drawing ratio of the deep-drawn cup. The limited drawing ratio is expressed by the following:

$$LDR = \frac{D_m}{D_p} \quad (3)$$

where D_p is the punch diameter and D_m is the blank diameter [19]. In case 3, the forming temperature, punch speed, and blank diameter are fixed at 375 °C, 29.7 mm/s, and 65 mm, respectively, and finite element simulations with variations of blank thickness were conducted. The simulation results with blank thicknesses of 2.0 mm and 2.5 mm are shown in Figure 6a and Figure 6b, respectively, where the product is shown by yellow color for easily distinguished. Clearly, a sound product was obtained at a blank thickness of 2.0 mm, whereas fracture occurred at a blank thickness of 2.5 mm. From a series of simulation results, it is known that it is difficult to form a too-thick or too-thin blank. For thicker blanks, it is difficult to bend the blank at the round corner, which results in a bad cup shape. For thinner blank, although it is easier to bend at the round corner, however, its strength decreases, which may lead to cracking at the cup wall. Therefore, the blank thickness should be within a range to achieve better-forming results.

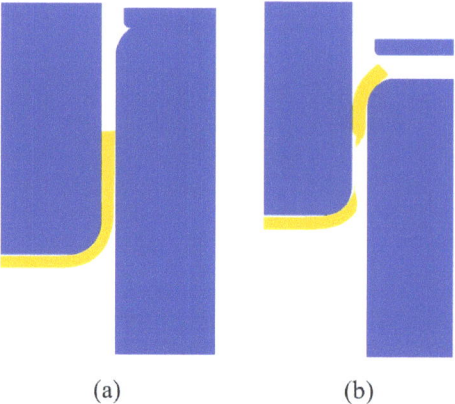

(a)   (b)

Figure 6. Simulation results with different blank thicknesses of (**a**) 2.0 mm and (**b**) 2.5 mm.

3. Machine Learning Classifier

A classifier is an ML model that assigns data points to different categories or labels, using supervised learning methods to learn labeled data to predict the category of subsequent data [20]. With the development of modern AI, its advantages are gradually valued by the manufacturing industry. In terms of improving production efficiency, AI's high degree of learning and efficient computing power can make the most appropriate decisions quickly, or by establishing a dataset for AI to learn, subsequent data can be processed in various tasks [12].

This study uses the finite element software simulation results explained in the previous section as input data to write a prediction model and create an input dataset, allowing the model to learn the input dataset, predict subsequent inputs, and obtain the forming range of aluminum alloy 7075. In this study, the classifier function in ML is used to input data such as material temperature, material diameter, forming speed, and other parameters and forming results to classify, thereby predicting whether the input forming parameters can be successfully formed in the future. The classifiers used in this study can be divided into two categories: weak classifiers and integrated classifiers. Weak classifiers are mainly composed of a single classifier model. Weak classifiers are faster in learning speed but have lower learning rates or relatively poor accuracy.

The weak classifiers used in this study include Logistic regression, KNN, SVM, DT, etc. Integrated learning mainly consists of multiple or multiple weak classifiers, which are systematically integrated, and then the final classification prediction is obtained through weighted calculation or voting calculation. Compared with weak classifiers, due to the use of multiple weak classifiers for combined learning, the accuracy and learning effect of integrated learning classifiers are usually better than those of general weak classifiers.

AdaBoost, the full name of adaptive boosting, also known as adaptive enhancement, is an integrated learning classifier that uses the Boosting concept [20]. The basic concept of Boosting is to combine multiple weak classifiers to form a strong classifier with a strong classification effect. The main idea is to improve prediction results through iteration. Boosting uses a weighting method to process the original data. The data of each training input set is given an initial weight by the model. In each round, a new weak classifier is used to train the training set, and each classifier is also given initial weights. After each round of learning, the data are divided into two categories: correctly classified and incorrectly classified. The weights of the two data types are updated, increasing the weight of the incorrectly classified data and reducing the weight of the correctly classified data so that the model can respond to the incorrectly classified data. The data are learned again, and the updated weights are given to the next round of classifiers for re-training.

Since the efficiency and accuracy of each classifier are different, all classifiers need to be integrated at the end of all training. At this time, the weight of the classifier with a higher error rate and poor accuracy is reduced, while the weight of the classifier with better performance and poor accuracy is reduced. The weight of higher classifiers increases, and finally, the updated weights of all classifiers and their corresponding classification results are statistically calculated to obtain the prediction results of the Boosting model. The structure diagram of AdaBoost is shown in Figure 7 [21]. After iterative training of the classifier, the overall model is gradually improved, and finally, excellent prediction results are obtained.

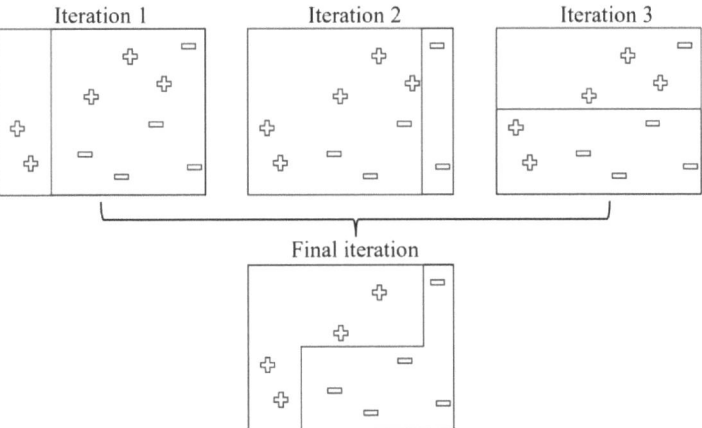

Figure 7. AdaBoost algorithm iteration structure.

In terms of programming, Python is used to write research-related programs and establish a virtual environment for running classification prediction models. The input data learned by the program are the four forming parameters and forming results set in the Deform simulation. The four parameters are forming temperature, punch speed, material diameter, and material thickness. A total of 624 pieces of the input dataset are recorded for model training, as shown in Table 5, where the forming results are represented by 0 and 1, where 1 represents the material being formed smoothly, and 0 represents material rupture.

Table 5. Parts of input dataset for CSV file.

Thickness	Diameter	Temperature	Speed	Forming
1.5	51	250	6.0	1
1.5	51	250	12.1	1
1.5	51	250	18.1	1
1.5	51	250	24.2	1
1.5	51	250	30.2	1
1.5	51	250	36.3	1
1.5	51	250	6.0	1
1.5	51	275	12.1	1
1.5	51	275	18.1	1

Input the dataset into the program and use it. The data are divided into two categories: training and test sets, with the number of records being 8:2. Then, write the classifier used in the study. The operation flow chart of the program is shown in Figure 8. The input data are imported into the specified folder at the model's front end. It is necessary to check whether the input dataset has missing values or other incorrect data formats before it can be input into the classifier for internal learning. Otherwise, it needs to be re-entered. Confirm the content of the dataset. The program will pop up an error message if an error occurs during the classifier learning process. You must reconfirm the input dataset or modify the classification model to avoid program code or logic errors.

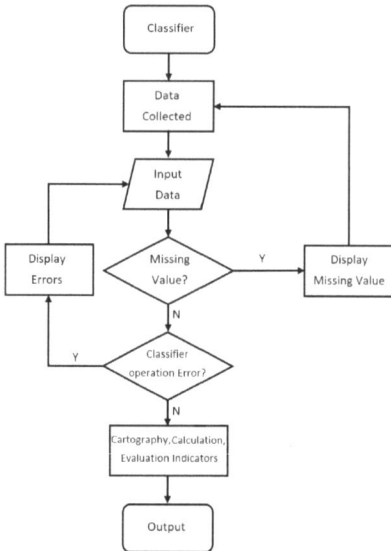

Figure 8. Flow chart for the classification prediction model.

After the input dataset is confirmed to be correct, the model will be divided into training and test datasets. In ML, to evaluate the accuracy of the prediction model, the model will use the data of the training set to learn and then predict the output of the test set. The accuracy is obtained by comparison. The ratio of training and test sets is 8:2. After obtaining the test set, a program must be set up to distinguish the data used for prediction and the output result data. In this study, the four data types were used for prediction. The parameter factor, the output result data, is the material forming result. After the data are learned by the classifier, the results of model learning need to be tested.

The trained model predicts the output of the test set, and the results will be compared with the original correct answers. Evaluation indicators such as accuracy and error will be recorded. After training and testing, the model will summarize and export the model evaluation indicators of all classifiers and present the prediction results in charts, drawings, etc., to facilitate the researchers' interpretation. Model evaluation indicators can be used to determine which classifier has the best performance and the most accurate prediction results, and the classifier with the best results can be selected as a classification prediction system. In the future, one only needs to enter relevant forming parameters, and the model will provide prediction results for reference.

The confusion matrix is a 2 × 2 matrix. The horizontal and vertical axes represent the actual and predicted results, respectively. The four blocks of the matrix represent different meanings, as shown in Figure 9 [22].

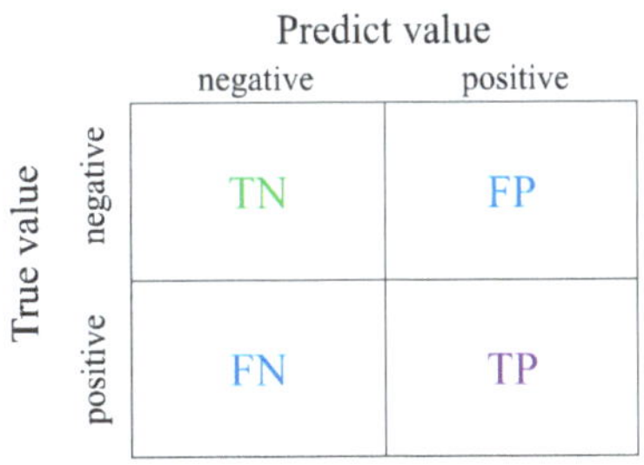

	Predict value	
	negative	positive
True value — negative	TN	FP
True value — positive	FN	TP

Figure 9. Confusion matrix for the classification prediction model.

Some results of the confusion matrix of classifiers are shown in Figure 10, where the dark color is used for highlighting the values. From Figure 10a, it is observed that the weak classifier of DT suffers a higher FN with 12. From Figure 10b, it is observed that the weak classifier of KNN suffers higher FP with 19. However, From Figure 10c, it is observed that there are lower FP with 2 and FN with 0 for the strong classifier of AdaBoost.

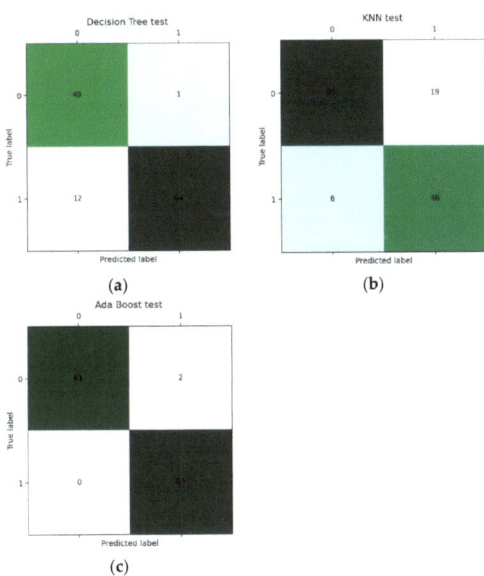

Figure 10. Some confusion matrices of classifiers: (**a**) DT, (**b**) KNN, and (**c**) AdaBoost.

Other evaluation indicators can be derived from the confusion matrix, and various model functions can be analyzed. The accuracy rate indicates the accuracy of the model's incorrect prediction. The calculation formula is shown as follows:

$$Accuracy = \frac{TP+TN}{TP+FP+FN+TN}. \tag{4}$$

The highest accuracy may not always mean that the model's prediction effect is the best, so it needs to be combined with other evaluations. Indicators are discussed together. The precision rate represents the model's accuracy in predicting positive examples. The calculation formula is shown as follows:

$$Precision = \frac{TP}{TP+FP} \tag{5}$$

where the higher the ratio of TP, the higher the model's accuracy. A high accuracy rate means that the model is less likely to make mistakes when its prediction is positive. The recall rate represents the rate of correct samples being classified as positive. The calculation formula is shown as follows:

$$recall = \frac{TP}{TP+FN}. \tag{6}$$

A high recall rate means that the higher the proportion of correct samples that are judged as positive in the model, the less likely it is that the correct sample will be judged as wrong. The calculation formula of the F1 score is shown as follows:

$$F1score = 2\frac{precision \times recall}{precision + recall} \tag{7}$$

which requires using two widely referenced values, precision and recall because they can reflect the algorithm's accuracy. The receiver operating characteristic curve (ROC curve) is a visual accuracy indicator, as shown in Figure 11. The ROC curve's horizontal and vertical axes represent the model's false positive rate (FPR) and true positive rate (TPR). The farther the curve result deviates from the diagonal, the higher the accuracy. The larger the area under the curve (AUC, Area under the curve), the higher the accuracy and the better the model performance. The AUC value can compare the performance of different models, and models with larger AUC have better classification performance [23].

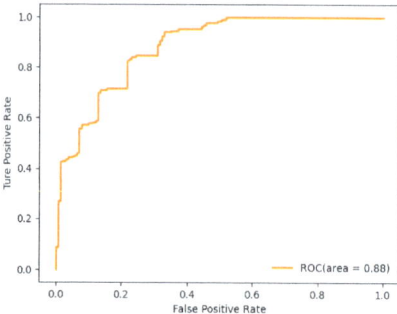

Figure 11. ROC curve diagram.

The numerical values of the evaluation indicators can be used to determine which classifier model is better. Compare the accuracy, precision, recall, F1 score, and area under the ROC curve AUC, as shown in Table 6. The closer the values of these five indicators are to 1, the better the performance and effect of the model. From the values in the table, we know that the first 6 rows are the results of weak classifiers. In terms of accuracy, some weak classifiers have higher accuracy. The precision and recall rate values are also the same as the accuracy rates. The performance of weak classifiers is mostly between 0.84 and 0.94. The prediction effect needs to be strengthened. Since the values of the precision rate and recall rate also affect the performance, the F1 score's value of the F1 score also fluctuates wildly. The value of AUC reflects the predictive value of the model.

Table 6. Integration of various classifier evaluation indicators.

Model Name	Accuracy	Precision	Recall	F1 Score	AUC
Logistic regression	0.79	0.88	0.69	0.77	0.94
KNN	0.8	0.88	0.71	0.78	0.87
SVM linear	0.75	0.83	0.66	0.73	0.87
SVM Poly	0.73	0.66	1	0.79	0.85
SVM RBF	0.73	0.7	0.85	0.76	0.84
SVM sigmoid	0.33	0.35	0.34	0.34	0.3
Decision tree	0.9	0.84	0.98	0.9	0.96
Random forest	0.98	0.98	0.98	0.98	1
XGBoost	0.98	0.98	0.98	0.98	1
AdaBoost	0.98	1	0.97	0.98	1
Stacking	0.98	1	0.97	0.98	1
LGBM	1	1	1	1	1

In Table 6, the evaluation index results of the integrated learning classifiers are shown in the 7th to 11th rows. From the numerical comparison, the integrated learning's accuracy, precision, and recall have greater progress than the weaker classifiers. The accuracy, precision, and recall rate even have a performance of 1, which improves prediction accuracy. The closer the AUC is to 1, the higher the prediction value of the model. Therefore, the ensemble learning classifier has a better prediction effect and higher classification accuracy than the weak classifier. The model has better predictive value, so this study will use an integrated learning classifier as the main architecture of the classification prediction model, in which the Ada Boost classifier is used to predict the material formability range of aluminum alloy A7075.

After the model completes learning, the subsequent input of forming parameters can predict the results. From this, different forming parameters can be input to obtain the formable range of the material. The material thickness is fixed at 2 mm, and the forming ranges under different forming temperatures, punch speeds, and material diameters are compared, as shown in Figure 12. The three lines in the figure represent the three diameters of aluminum alloy 7075 discs, respectively, with diameters of 55, 58, and 61 mm. The area under the curve is the range in which the material can be successfully formed. The forming range of the material with a diameter of 61 mm is compared to the diameters of 55 and 58 mm. The range is small and difficult to form at low or too-high temperatures. The forming range of a diameter of 61 mm is approximately above 180 °C and below 350 °C. Forming in this range will have better results. The forming range of a material with a diameter of 58 mm is larger than that of 61 mm. However, it is also difficult to form at room (or lower) temperatures due to excessive material stress. Too high temperatures will also cause poor forming effects. Therefore, the formable range is from approximately 170 °C to 380 °C. The forming range of the material with a diameter of 55 mm is relatively loose, and the forming range can be smoothly formed from 250 °C to as high as 450 °C. Since the deep extension cup formed by the material disc with a diameter of 55 mm has a relatively low extension, the specimen material has a relatively low extension. The deformation is not large, and the forming conditions are relatively loose, so the impact of temperature and stamping speed is small.

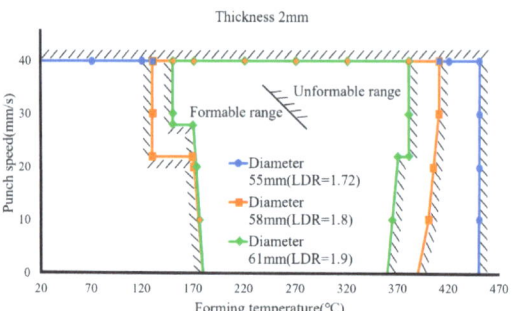

Figure 12. Formable ranges for different blank diameters.

The material diameter is controlled to 61 mm and the stretch ratio to 1.9. We change the material thickness, forming temperature, and punch speed for comparison, as shown in Figure 13. The three lines represent the material forming range with material thicknesses of 1, 1.5, and 2 mm, respectively. The forming curve range of a material with a thickness of 1 mm is small. The forming temperature is about 170 °C to 270 °C. It can be formed smoothly. After 270 °C, the punch speed needs to be controlled. It cannot be formed after 340 °C. Therefore, the forming range of a 1 mm material is more suitable for forming between 200 °C and 250 °C. The forming range of a material with a thickness of 1.5 mm is larger than that of a material with a thickness of 1 mm. The formable temperature range falls between 140 °C and 410 °C. The blank is not easy to form at a too-low temperature, so deep processing should be carried out between 170 °C and 370 °C. The extension-forming effect is better. The forming range of a material with a thickness of 2 mm is larger than that of 1 mm but smaller than that of 1.5 mm. The forming temperature range is about 170 °C to 370 °C. Forming at low or high temperatures is difficult, so it is suitable. The forming temperature is about 200 °C to 350 °C.

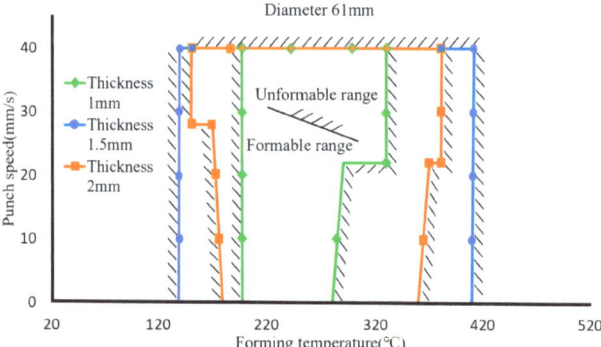

Figure 13. Formable ranges for different blank thicknesses.

To sum up, under the conditions of the same material diameter and the same draw ratio, the suitable thickness of the extension cup falls within a range rather than excessively increasing or reducing the thickness. If you want to produce thinner or thicker plates for a deep extension cup, the extension ratio of the extension cup needs to be reduced; that is, the material diameter can be reduced to form effectively. From the previous results, it can be concluded that the three parameters—forming temperature, material diameter, and material thickness—will have a greater impact on the forming of deep extension cups, while the impact of punch speed is relatively small. The punch speed is fixed at 24 mm/s, and the effects of the other three parameters are compared, as shown in Figure 14.

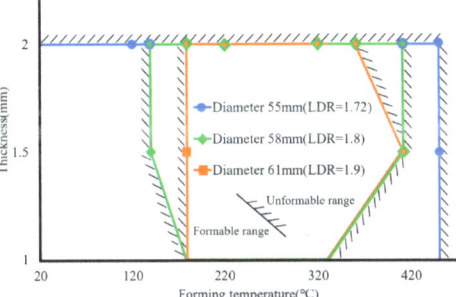

Figure 14. Formable ranges of temperature for different blank thicknesses and diameters.

As the blank diameter increases or the draw ratio increases, the formable ranges become narrow. Therefore, the forming temperature must be controlled within a certain range to avoid bad forming results. Forming at too low or too high temperatures may lead to bad forming results. The correlation function of forming temperature with respect to blank diameter and thickness can be calculated from the above figures. When the drawing ratio is between 1.72 and 1.8, the blank thickness is 1.5 to 2 mm. The minimum and maximum forming temperatures can be expressed by the following:

$$T_{min} = 20 + (LDR - 1.72) \cdot \frac{120}{0.08} \tag{8}$$

and

$$T_{max} = 450 - (LDR - 1.72) \cdot \frac{40}{0.08}, \tag{9}$$

respectively, where LDR (limiting drawing ratio) is the draw ratio of the drawn cup [19]. When the draw ratio is between 1.72 and 1.8, and the thickness is between 1 mm and 1.5 mm, the minimum and maximum forming temperatures are expressed by the following:

$$T_{min} = 20 + (LDR - 1.72) \cdot \frac{120}{0.08} + (1.5 - t) \cdot \frac{40}{0.5} \tag{10}$$

and

$$T_{max} = 450 - (LDR - 1.72) \cdot \frac{40}{0.08} - (1.5 - t) \cdot \frac{80}{0.5}, \tag{11}$$

respectively, where t is the blank thickness. If the draw ratio is 1.8 to 1.9 and the thickness is 1.5 to 2 mm, the minimum and maximum forming temperatures are expressed by the following:

$$T_{min} = 140 + (LDR - 1.8) \cdot \frac{40}{0.1} \tag{12}$$

and

$$T_{max} = 410 - (t - 1.5) \cdot \frac{50}{0.5}, \tag{13}$$

respectively. If the draw ratio is 1.8 to 1.9 and the thickness is 1 to 1.5 mm, the minimum and maximum forming temperatures can be expressed by the following:

$$T_{min} = 140 + (1.5 - t) \cdot \frac{40}{0.5} \tag{14}$$

and

$$T_{max} = 410 - (1.5 - t) \cdot \frac{80}{0.5}, \tag{15}$$

respectively. It can be known from the above formable ranges and related functions under the same blank thickness, materials with smaller blank diameters have advantages in forming compared to materials with larger blank diameters. Different blank thicknesses affect the formable range differently. The suitable blank thickness is inside the formable range. A thinner blank may not withstand the material's stretching during deep drawing and make cracks occur. A thicker blank may be difficult to bend, thus, breakage may occur. Therefore, it is important to control the forming parameter appropriately in the deep drawing process of aluminum alloy A7075.

4. Deep Drawing Experiments

4.1. Experimental Procedures

To validate the finite element modeling and prediction functions proposed, a serious of deep drawing experiments were conducted. A SEYI SD1-160 servo press was used to conduct the deep drawing experiments of aluminum alloy A7075. The SEYI SDI-160 servo press was manufactured by Shieh Yih machinery industry Co., LTD. in Taoyuan, Taiwan. The servo press can control the movement of the upper die precisely. It is equipped with a monitor to show the load variations and punch displacements in real-time. This function makes the user easily set the forming parameters and operate this machine. The appearance of the servo press is shown in Figure 15a and the assembly drawing of the die set for deep drawing processes is shown in Figure 15b. The components in the die set are given in Table 7. The die (no. 4) was designed to be positioned above the punch (no. 7); thus, the relative movement of the punch was opposite to that shown in Figure 1. The blank holder (no. 6) is pressed by four springs (no. 8) with an initial force of 448 N. As the punch moves forward to form the blank (no. 5), the pressing force provided by the springs increases and reaches 2.0 KN at the end of the stroke. Before forming, the workpieces of blank A7075 (no. 5) were heated to a set temperature with a heater, and the punch (no. 7) and die (no. 4) were also heated to the same temperature with a gas torch. Two heat shields were used to prevent heat transfer from the die or punch to the press machine, which may make the machine not function well. After deep drawing experiments, the cup product was taken out of the die set. Its dimensions were measured and some defects such as necking or fracture were checked.

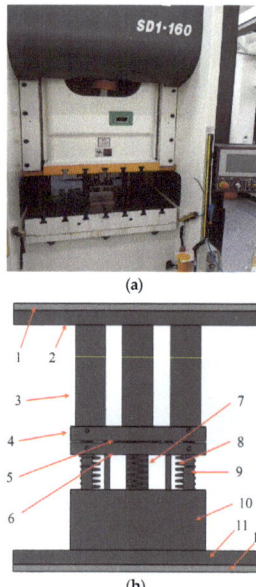

Figure 15. The servo press machine and the deep drawing die set for the experimental procedures. (**a**) Front appearance of servo press machine. (**b**) Assembly drawing of the deep drawing die set.

Table 7. Components in the deep drawing experimental die set.

no.	Components	no.	Components
1	Upper heat shield	7	Punch
2	Top plate	8	Spring
3	Die bracket	9	Guide post
4	Die	10	Punch fixture block
5	Blank	11	Lower heat shield
6	Blank holder	12	Bottom plate

The blank material is aluminum alloy A7075 with a thickness of 2.0 mm. Four kinds of circular blank diameters of 55, 58, 61, and 65 mm were prepared for the deep drawing experiments. The forming temperatures and punch speeds for the experiments are shown in Table 8.

Table 8. Forming parameters for deep drawing experiments.

Forming Parameters	Values
Forming temperature T_i (°C)	250, 350, 450
Punch speed v_j ($\frac{mm}{s}$)	12.1, 24.2, 36.3
Blank diameter D_k (mm)	55, 58, 61, 65
Blank thickness t_l (mm)	2.0

4.2. Experimental Results

The experimental results with variations of forming temperatures and punch speeds are shown in Table 9. The initial blank diameter is 55 mm. There is a slightly uneven thickness distribution occurring at the cup rims at lower punch speeds. However, generally, the blanks were formed into circular cups successfully. The forming parameters used in the experiments were inputted into the prediction model to predict the forming results. Table 10 shows the predicted forming results for the blank diameter of 55 mm. Clearly, the predicted results are consistent with the actual experimental results.

Table 9. Experimental results of the drawn products with a blank diameter of 55 mm.

Temperature \ Speed	12.08 mm/s	24.16 mm/s	36.25 mm/s
250 °C			
350 °C			
450 °C			

Table 10. Prediction results for a blank diameter of 55 mm.

Blank Thickness (mm)	Blank Diameter (mm)	Forming Temperature (°C)	Punch Speed (mm/s)	Prediction
2	55	250	12.1	O
2	55	250	24.2	O
2	55	250	36.3	O
2	55	350	12.1	O
2	55	350	24.2	O
2	55	350	36.3	O
2	55	450	12.1	O
2	55	450	24.2	O
2	55	450	36.3	O

O: Sound products

The experimental results of the drawn products for a blank diameter of 65 mm are shown in Table 11. Clearly, the drawn products are failures. All the cups are broken under all conditions, which means aluminum alloy A7075 is difficult for deep drawing with a larger drawing ratio. The forming parameters used in the experiments with a blank diameter of 65 mm were inputted into the prediction model to predict the forming results. The prediction results are shown in Table 12. The predicted results show the forming results are failures, which is consistent with the actual experimental results. From the comparisons, it can be said that the prediction models proposed by this study can predict deep drawing results of aluminum alloy A7075 rapidly and effectively.

Table 11. Experimental results of drawn products with a blank diameter of 65 mm.

Temperature \ Speed	12.08 mm/s	24.16 mm/s	36.25 mm/s
250 °C			
350 °C			
450 °C			

Table 12. Prediction results for a blank diameter of 65 mm.

Thickness	Diameter	Temperature	Speed	Predict
2	65	350	12.1	X
2	65	350	24.2	X
2	65	350	36.3	X
2	65	350	12.1	X
2	65	350	24.2	X
2	65	350	36.3	X
2	65	350	12.1	X
2	65	350	24.2	X
2	65	350	36.3	X
		X: not ok for products		

4.3. Comparisons between Experimental and Predicted Results

The above experimental results have verified the correctness of the prediction model for blank diameters of 58 and 65 mm. Some more experiments were conducted to verify the correctness of the formable regions obtained by the prediction model. The forming conditions and experimental results for verification of predicted formable regions are shown in Table 13. The product appearance for each case is also shown in Table 13. The experimental results and predicted formable regions for different blank diameters and forming temperatures are shown in Figure 16. The inner space is the predicted safe region between forming temperature and punch speed. The outer space with hatched lines is the predicted failure region. Clearly, the predicted formable regions are affected significantly by the forming temperature. The formable region for a blank diameter of 58 mm (drawing ratio 1.8) is larger slightly than that for a blank diameter of 61 mm (drawing ratio 1.9). Symbols ∆ and o denote experimental results of sound products for blank diameters of 58 and 61 mm, respectively. Symbols ▲ and ● denote experimental failure results for blank diameters of 58 and 61 mm, respectively. From Figure 16, it is clear that the successful experimental results were located within formable regions, while the failed experimental results were located outside the unformable regions.

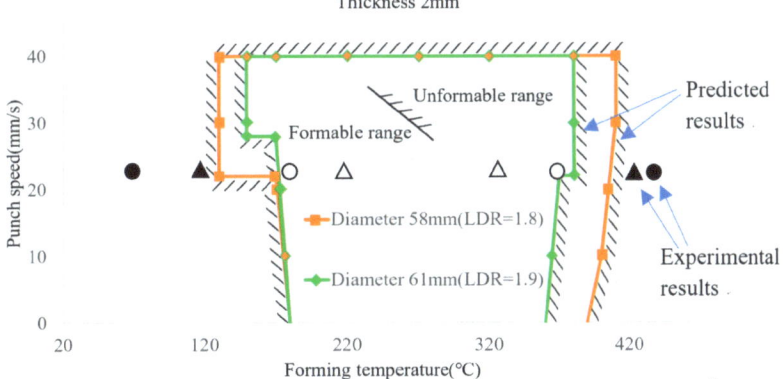

Figure 16. Experimental results and predicted formable regions for different blank diameters and forming temperatures.

Table 13. Forming conditions and experimental results for verification of formable regions.

Thickness t (mm)	Diameter D (mm)	Temperature T (°C)	Punch Speed (mm/s)	Experimental Results	Product Appearance
2	58	70	24.1	●	
2	58	170	24.1	○	
2	58	370	24.1	○	
2	58	450	24.1	●	
2	61	120	24.1	▲	
2	61	200	24.1	△	
2	61	320	24.1	△	
2	61	420	24.1	▲	

●▲: Failure; △○: Successful.

From the product appearance shown in the last column of Table 13, it is known that there are some small cracks at the cup rims or uneven thickness distribution on the cup wall for the cases of D = 58 mm, T = 70 °C and D = 61 mm, T = 120 °C. That is, some defects in the drawn cups probably occur at too low temperatures. Bad forming results were found for the cases of D = 58 mm, T = 450 °C and D = 61 mm, T = 420 °C. That is, distortion or fracture at the cup bottom probably occurs at too high temperatures. The experimental results are consistent with the predicted formable regions, which validates the effectiveness of the prediction model by the ML approach.

5. Conclusions

This paper investigated the feasibility of applying the ML approach to propose a model to predict the formability of aluminum alloy A7075 in the deep drawing process. At first, a database from the finite element simulation results with various forming parameters was collected. Then, ML classifiers were used for learning and a prediction model was established. A dataset for formable regions of the parameters was constructed from ML training and verification. Some experiments of deep drawing of aluminum alloy A7075 were conducted and the experimental results were compared with the predicted formable regions. The experimental results were consistent with the predicted formable regions and the effectiveness of the prediction model by ML approach was verified. Using this ML approach can save simulation time and shorten the development cycles of new products or processes.

Author Contributions: Conceptualization, Y.-M.H. and Y.-F.H.; methodology, Y.-M.H. and Y.-F.H.; software, T.-H.H.; validation, Y.-F.H. and C.-M.C.; formal analysis, Y.-M.H. and Y.-F.H.; investigation, Y.-M.H. and T.-H.H.; resources, Y.-M.H. and Y.-F.H.; data curation, Y.-M.H. and Y.-F.H.; writing—original draft preparation, Y.-M.H. and T.-H.H.; writing—review and editing, Y.-F.H. and C.-M.C.; visualization, Y.-F.H. and C.-M.C.; supervision, Y.-M.H.; project administration, Y.-M.H. and Y.-F.H. All authors have read and agreed to the published version of the manuscript.

Funding: This work was supported by the National Science and Technology Council of Taiwan under grant NSTC 113-2221-E-110-024.

Institutional Review Board Statement: Not applicable.

Informed Consent Statement: Not applicable.

Data Availability Statement: Data are contained within the article.

Conflicts of Interest: The authors declare no conflicts of interest.

References

1. Colgan, M.; Monaghan, J. Deep drawing process: Analysis and experiment. *J. Mater. Process. Technol.* **2003**, *132*, 35–41. [CrossRef]
2. Tajally, M.; Emadoddin, E. Mechanical and Anisotropic Behaviors of 7075 Aluminum Alloy Sheets. *Mater. Des.* **2011**, *32*, 1594–1599. [CrossRef]
3. Cerri, E.; Evangelista, E.; Forcellese, A.; McQueen, H.J. Comparative hot workability of 7012 and 7075 alloys after different pretreatments. *Mater. Sci. Eng.* **1995**, *A197*, 181–198. [CrossRef]
4. Chen, D.C.; Guo, J.Y.; Li, C.Y.; Lai, Y.Y.; Hwang, Y.M. Study of circular and square aluminum alloy deep drawing. In Proceedings of the 9th International Conference on Tube Hydroforming (TUBEHYDRO 2019), Kaohsiung, Taiwan, 18–21 November 2019.
5. Gowtham, K.; Srikanth, K.V.N.S.; Murty, K.L.N. Simulation of the Effect of Die Radius on Deep Drawing Process. *Int. J. Appl. Res. Mech. Eng.* **2012**, *2*, 12–17. [CrossRef]
6. Liu, J.; Pan, C.; Lei, F.; Hu, D.; Zuo, H. Fault prediction of bearings based on LSTM and statistical process analysis. *Reliab. Eng. Syst. Saf.* **2021**, *214*, 107646. [CrossRef]
7. Bansal, M.; Goyal, A.; Choudhary, A. A comparative analysis of K-nearest neighbor, genetic, support vector machine, decision tree, and long short-term memory algorithms in machine learning. *Decis. Anal. J.* **2022**, *3*, 100071. [CrossRef]
8. Singh, A.; Kumar, R. Heart disease prediction using machine learning algorithms. In Proceedings of the 2020 IEEE International Conference on Electrical and Electronics Engineering (ICE3), Gorakhpur, India, 14–15 February 2020.
9. Huang, Y.-F.; Lin, C.-B.; Chung, C.-M.; Chen, C.-M. Research on QoS Classification of Network Encrypted Traffic Behavior Based on Machine Learning. *Electronics* **2021**, *10*, 1376. [CrossRef]
10. Cheng, C.; Hsiao, C.; Huang, Y.-F.; Li, H. Combining Classifiers for Deep Learning Mask Face Recognition. *Information* **2023**, *14*, 421. [CrossRef]
11. Lee, S.-H.; Cheng, C.-H.; Huang, T.-H.; Huang, Y.-F. Machine learning-based indoor positioning systems using multi-channel information. *J. Eng. Technol. Sci. (JETS)* **2023**, *55*, 373–383. [CrossRef]
12. Shinomiya, N.; Tsuboi, M.; Kita, S.; Yasuki, S. Reduction of Defect Rate in Impact Extrusion by Slide Motion Control Using Machine Learning. *J. JSTP* **2023**, *64*, 87–92. [CrossRef]
13. Cancemi, S.A.; Frano, R.L.; Santus, C.; Inoue, T. Unsupervised anomaly detection in pressurized water reactor digital twins using autoencoder neural networks. *Nucl. Eng. Des.* **2023**, *413*, 112502. [CrossRef]
14. Yamanaka, A. Data Scientific Application to Numerical Material Test for Sheet Metals. *Bull. JSTP* **2022**, *5*, 203–207.
15. DEFORM. DEFORM manual, v11.0.2_System_Documentation, Scientific Forming Technologies Corporation (SFTC), Columbus, Ohio 43235, USA. Available online: https://www.deform.com/ (accessed on 18 May 2024).
16. Kvačkaj, T.; Tiža, J.; Bacsó, J.; Kováčová, A.; Kočiško, R.; Pernis, R.; Fedorčáková, M.; Purcz, P. Cockroft-Latham ductile fracture criteria for nonferrous materials. *Mater. Sci. Forum* **2014**, *782*, 373–378. [CrossRef]
17. Faridmehr, I.; Osman, M.H.; Adnan, A.B.; Nejad, A.F.; Hodjati, R.; Azimi, M. Correlation between Engineering Stress-Strain and True Stress-Strain Curve. *Am. J. Civ. Eng. Archit.* **2014**, *2*, 53–59. [CrossRef]
18. Ji, H.; Ma, Z.; Huang, X.; Xiao, W.; Wang, B. Damage evolution of 7075 aluminum alloy basing the Gurson Tvergaard Needleman model under high temperature conditions. *J. Mater. Res. Technol.* **2022**, *16*, 398–415. [CrossRef]
19. Pernis, R.; Barényi, I.; Kasala, J.;Ličková, M. Evaluation of Limiting Drawing Ratio (LDR) in Deep Drawing Process. *Acta Metall. Slovaca* **2015**, *21*, 258–268. [CrossRef]
20. Pereira, F.; Mitchell, T.; Botvinick, M. Machine learning classifiers and fMRI: A tutorial overview. *NeuroImage* **2009**, *45*, 199–209. [CrossRef] [PubMed]
21. Russell, S.J.; Norvig, P. *Artificial Intelligence: A Modern Approach*; Pearson: London, UK, 2009.

22. Navin, J.R.M.; Pankaja, R. Performance Analysis of Text Classification Algorithms using Confusion Matrix. *Int. J. Eng. Tech. Res.* **2016**, *6*, 75–78.
23. Hanley, J.A.; McNeil, B.J. The meaning and use of the area under a receiver operating characteristic (ROC) curve. *Radiology* **1982**, *143*, 29–36. [CrossRef] [PubMed]

Disclaimer/Publisher's Note: The statements, opinions and data contained in all publications are solely those of the individual author(s) and contributor(s) and not of MDPI and/or the editor(s). MDPI and/or the editor(s) disclaim responsibility for any injury to people or property resulting from any ideas, methods, instructions or products referred to in the content.

Article

Toward Metallized Pellets for Steelmaking by Hydrogen Cooling Reduction: Effect of Gas Flow Rate

Wanlong Fan, Zhiwei Peng *, Ran Tian, Guanwen Luo, Lingyun Yi and Mingjun Rao

School of Minerals Processing and Bioengineering, Central South University, Changsha 410083, China; 225611071@csu.edu.cn (W.F.); 215601041@csu.edu.cn (R.T.); 225601020@csu.edu.cn (G.L.); yilingyun@csu.edu.cn (L.Y.); mj.rao@csu.edu.cn (M.R.)
* Correspondence: zwpeng@csu.edu.cn

Abstract: This study proposed a strategy to prepare metalized pellets for direct steelmaking by hydrogen cooling reduction (HCR) of iron ore pellets with a focus on the effect of H_2 flow rate on the process. It was demonstrated that increasing H_2 flow rate could effectively enhance the reduction performance of iron ore pellets. However, due to the influence of the countercurrent diffusion resistance of gas molecules, too high H_2 flow rate no longer promoted the reduction of the pellets when the maximum reduction rate was reached. The reduction swelling index (RSI) of the pellets initially increased and then decreased with increasing H_2 flow rate. This change was associated with the decreased content of Fe_2SiO_4 in the metalized pellets and the changes in porosity and iron particle size. The compressive strength (CS) decreased continuously, showing a sharp decline when the H_2 flow rate reached 0.6 L/min. It was attributed to the significant increases in porosity and average pore size of the metalized pellets, with the presence of surface cracks. When the H_2 flow rate was 0.8 L/min, the metalized pellets had the optimal performance, namely, reduction degree of 91.45%, metallization degree of 84.07%, total iron content of 80.67 wt%, RSI of 4.66%, and CS of 1265 N/p. The findings demonstrated the importance of controlling the H_2 flow rate in the preparation of metalized pellets by HCR.

Keywords: hydrogen cooling reduction; hydrogen flow rate; metalized pellets; porosity

Citation: Fan, W.; Peng, Z.; Tian, R.; Luo, G.; Yi, L.; Rao, M. Toward Metallized Pellets for Steelmaking by Hydrogen Cooling Reduction: Effect of Gas Flow Rate. *Materials* **2024**, *17*, 3896. https://doi.org/10.3390/ma17163896

Academic Editors: Seong-Ho Ha, Young-Ok Yoon, Young-Chul Shin and Dong-Earn Kim

Received: 5 July 2024
Revised: 25 July 2024
Accepted: 26 July 2024
Published: 6 August 2024

Copyright: © 2024 by the authors. Licensee MDPI, Basel, Switzerland. This article is an open access article distributed under the terms and conditions of the Creative Commons Attribution (CC BY) license (https:// creativecommons.org/licenses/by/ 4.0/).

1. Introduction

The iron and steel industry is an energy-intensive sector with substantial resource consumption [1]. It uses coal or coke as an energy resource which leads to significant CO_2 emission [2,3]. This is particularly serious in the long process of steel production, namely the blast furnace-basic oxygen furnace (BF-BOF) process [4,5]. For cleaner and more efficient production, the short process, namely the direct reduction-electric arc furnace (DR-EAF) process, has been receiving increasing attention. It offers the advantages of higher efficiency and better sustainability, aligning with the future development of the iron and steel industry [6].

The DR-EAF process necessitates the reduction of iron ore pellets, which are produced from iron concentrates after pelletizing, drying, preheating (500–1100 °C), and roasting (1150–1400 °C) successively to obtain direct reduced iron (DRI) or metalized pellets for EAF smelting to produce crude steel [7,8]. It relies on reducing agents such as coal, natural gas, and H_2. Among these, H_2 stands out as an exceptionally clean reducing agent with superior reduction capability [9]. In recent years, with the maturation of "green hydrogen" technologies from clean energy, such as wind and water, the industrial application of using H_2 for iron ore reduction to produce DRI has gradually become feasible [10,11]. Consequently, fostering the development and application of hydrogen-based direct reduction processes is vital for achieving "dual carbon" goals and producing high-quality steel.

For hydrogen-based direct reduction technologies, many studies have been reported [10,12, 13]. These studies primarily focused on the effects of factors such as reduction temperature, reduction time, gas concentration, and pellet porosity on the product properties [14–17]. It was reported that elevating temperature could significantly enhance the reduction rate. However, excessively high temperature could result in an increase in pellet swelling, consequently reducing pellet strength [18]. For reduction time, its selection was of paramount importance for obtaining qualified metalized pellets with high production efficiency. For reducing gas like H_2, its concentration was also found to be important because low H_2 concentration would limit the reduction rate of the pellets, while excessively high concentration might induce pellet cracking [16]. For pellet porosity, increasing it properly would enhance contact areas between reducing gas molecules and iron oxides, thereby improving reduction efficiency [17]. These reports demonstrated that many factors controlled the hydrogen reduction process. However, there are rare reports on the influence of gas flow rate on the reduction performance of iron ore pellets.

Recently, the authors' group proposed a method to produce metalized pellets by reduction of hot iron ore pellets after roasting (called roasted or oxidized pellets with temperatures higher than 1000 °C) during their cooling process in H_2 [19]. This method was named hydrogen cooling reduction (HCR) and was expected to utilize the heat of the roasted pellets for reduction to enhance energy utilization efficiency. By following this concept, this study aimed to assess the influence of H_2 flow rate on the reduction performance of iron ore pellets by examining the variations of the reduction degree, metallization degree, reduction swelling index (RSI), compressive strength (CS), phase composition, and microstructure of the pellets.

2. Materials and Methods

2.1. Raw Materials

The main raw material used in this study was iron ore pellets. The pellets were prepared by drying of green pellets produced via traditional pelletizing of magnetite concentrate with the addition of 1.5% bentonite as the binder at 105 °C for 12 h, followed by preheating at 900 °C for 10 min, and then roasting at 1150 °C for 10 min. They had the bulk density of 3.31 g/cm³, porosity of 26.87%, compressive strength of 2100 N/p, and diameter of 14–16 mm. Table 1 shows their main chemical composition. They had total iron content of 64.00 wt%, SiO_2 content of 8.51 wt%, and FeO content of 1.48 wt%. As shown by their XRD pattern in Figure 1, they were constituted by mainly hematite (Fe_2O_3), with small amounts of quartz (SiO_2) and magnetite (Fe_3O_4).

Table 1. Main chemical composition of iron ore pellets (wt%).

TFe	SiO$_2$	FeO	Al$_2$O$_3$	CaO	Na$_2$O	MgO	K$_2$O	P	S
64.00	8.51	1.48	0.42	0.18	0.18	0.15	0.10	0.045	0.0046

TFe—total iron content.

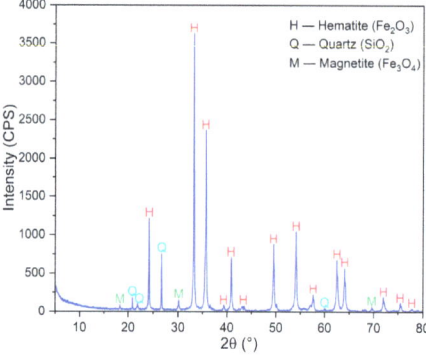

Figure 1. XRD pattern of iron ore pellets.

2.2. Experimental Methods

2.2.1. Direct Reduction

For each test, approximately 40 g of iron ore pellets were loaded in a vertical tube furnace (Figure 2) for reduction. Initially, the furnace temperature was elevated from room temperature to 1150 °C with the ramp rate of 10 °C/min. In this stage, N_2, with the flow rate of 1.0 L/min was introduced as the protective gas. After the pellets were kept at this temperature for 30 min, the gas was switched to H_2 for HCR with the cooling rate of 3.7 °C/min, which was maintained using supplementary heat provided by the furnace until the temperature of the pellets decreased to 450 °C. These parameters were selected based on the preliminary experiments. During this process, different flow rates of H_2, namely 0.2 L/min (case 1), 0.4 L/min (case 2), 0.6 L/min (case 3), 0.8 L/min (case 4), and 1.0 L/min (case 5), were selected to evaluate the effect of H_2 flow rate on the reduction performance of the pellets. After the pellet temperature reached 450 °C, the gas was switched back to N_2 with the flow rate of 1.0 L/min. The pellets after reduction, i.e., metalized pellets, were cooled naturally to room temperature for various characterizations. Note that throughout the experimental process, all gas flow rates were controlled accurately using a mass flow controller.

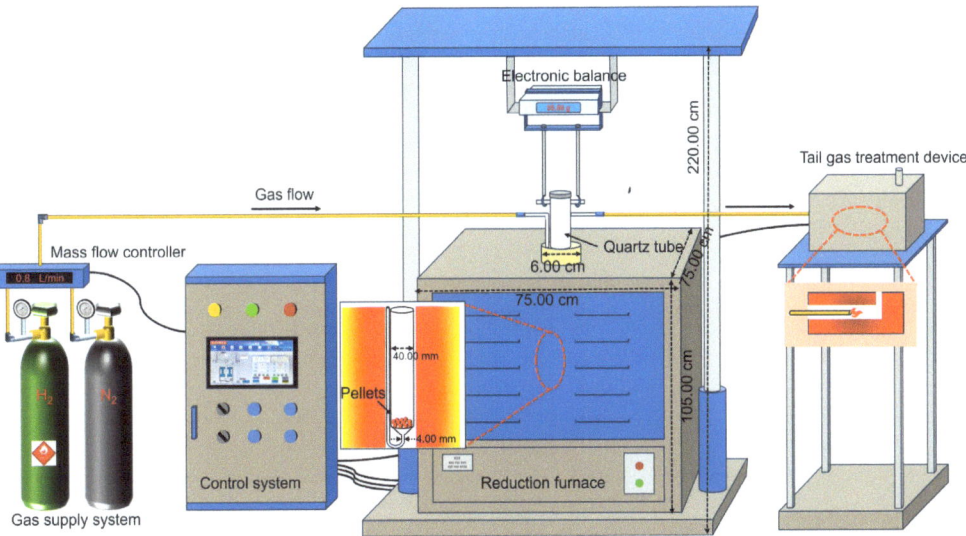

Figure 2. Schematic illustration of the experimental setup for reduction of iron ore pellets.

2.2.2. Characterizations

The reduction degree and metallization degree of the pellets were determined according to the Chinese National Standard Test Method GB/T 24236–2009 [20]. The total iron content and CS were measured according to the Chinese National Standard Test Methods GB/T 6730.5–2022 [21] and GB/T 14201–2018 [22], respectively. The RSI values of the pellets were measured using the drainage method with anhydrous ethanol as the medium [23]. The phase compositions of the pellets were determined using an X-ray diffractometer (XRD; D8 Advance, Bruker, Hanau, Germany) and analyzed using the software Jade 9.0 (Materials Data Inc., Livermore, CA, USA). The contents of different phases of the metalized pellets were determined by Rietveld full-spectrum fitting based on the XRD patterns [24]. The microstructure changes of the pellets, which preliminarily underwent cold mounting in resin and polishing, were determined using a scanning electron microscope (SEM; Sigma HD, Zeiss, Göttingen, Germany) equipped with an energy dispersive X-ray detector (EDS; EDAX Inc., Mahwah, NJ, USA). The size and distribution of metallic iron particles in the pellets were measured using the software Image J 2.0 (National Institutes of Health, Bethesda, MD, USA). The pore size distribution, average pore size, and porosity of the pellets were measured using a mercury intrusion porosimeter (Autopore IV 9500, Micromeritics, Norcross, GA, USA). The porosities of various areas of the pellets were determined by Image J 2.0, in which 10 independent SEM images were selected for each region, and the mean values were used based on the measurements.

3. Results and Discussion

3.1. Reduction Behavior

The reduction degree, metallization degree, and total iron content of the metalized pellets obtained by reduction with different H_2 flow rates are shown in Figure 3. As the H_2 flow rate increased from 0.2 L/min (case 1) to 0.8 L/min (case 4), the reduction degree, metallization degree, and total iron content of the metalized pellets increased from 57.05%, 29.81%, and 67.80 wt% to 91.45%, 84.07%, and 80.67 wt%, respectively, indicating increments of 34.40%, 54.26%, and 12.87 wt%. By increasing the H_2 flow rate to 1.0 L/min, i.e., case 5, the reduction degree, metallization rate, and total iron content of the metalized pellets increased by only 0.21%, 0.39%, and 0.27 wt%, respectively. Due to the influence of the H_2 flow rate, the reduction process was governed by both gas diffusion and chemical reaction [25,26]. Increasing the H_2 flow rate enhanced both the diffusion rate of water vapor from the pellets and the contact areas between H_2 molecules and the pellets, thereby promoting reduction of the pellets. However, the countercurrent diffusion resistance of H_2 and water vapor molecules in the pellet pores, as well as the number of reaction sites within the pellets, remain unaffected [25]. By further increasing the H_2 flow rate, the water vapor concentration within the pellets gradually approached a constant value [25]. Therefore, when the H_2 flow rate exceeded 0.8 L/min, i.e., case 5, there was no significant enhancement of reduction performance of the pellets.

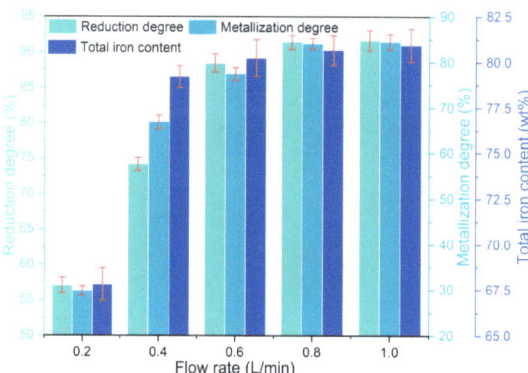

Figure 3. Effect of H_2 flow rate on the reduction degree, metallization degree, and total iron content of the metalized pellets.

Figure 4 shows the effect of H_2 flow rate on the RSI and CS of the pellets. According to the results of cases 1–5, with increasing H_2 flow rate, the RSI initially increased and then decreased. When the H_2 flow rate increased from 0.2 L/min in case 1 to 0.6 L/min in case 3, the value of RSI increased from 0.62% to 13.34%. By further increasing the H_2 flow rate to 1.0 L/min, i.e., case 5, the RSI decreased to 1.67%. During the initial reduction stage, the lattice transformation caused by the reduction of Fe_2O_3 to Fe_3O_4 resulted in an increase in pellet volume, accompanied by an increase in RSI [27,28]. As the H_2 flow rate increased, the total amount of H_2 involved in the reaction with iron ore pellets per unit time increased, leading to a higher reduction rate [16]. Therefore, the RSI increased initially and then decreased. As the H_2 flow rate increased from 0.2 L/min (case 1) to 1.0 L/min (case 5), the value of CS of the metalized pellets decreased from 2460 N/p to 1158 N/p. In particular, when the H_2 flow rate increased from 0.4 L/min (case 2) to 0.6 L/min (case 3), the CS decreased significantly by 1141 N/p, which correlated with a sharp increase in RSI [29]. The decreased CS of the metalized pellets was believed to be associated with relevant phase transformation and structural changes.

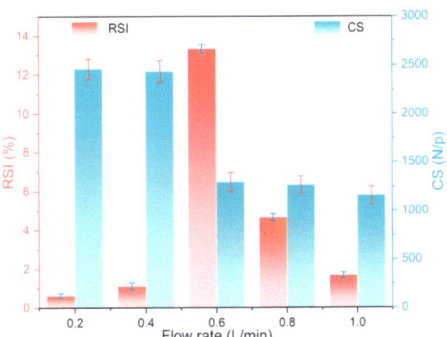

Figure 4. Effect of H_2 flow rate on the RSI and CS of the metalized pellets.

Figure 5 shows the macroscopic morphologies of iron ore pellets after reduction with different H_2 flow rates. When the H_2 flow rate was no more than 0.4 L/min, i.e., in cases 1 and 2, the resulting metalized pellets maintained intact structures and smooth surfaces. It indicated that with low H_2 flow rates, there were no significant changes in the morphologies of the pellets. In cases 3–5, as the H_2 flow rate further increased, more cracks appeared on the surfaces of the metalized pellets. The small H_2 molecules enabled rapid penetration into the interior of the pellets, facilitating the reactions with iron oxides to form metallic iron layers and thus reducing the porosity of the pellets [16]. However, when the rate of water vapor generation exceeded its rate of outward diffusion, the internal gas pressure in the pellets increased, leading to pellet cracking. Therefore, for reduction with high H_2 flow rates, the pellets were more prone to cracking, resulting in structural damage and, thus, lower strength.

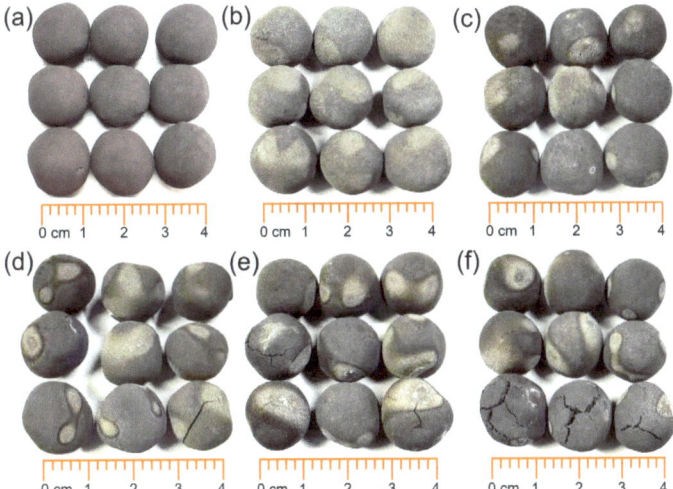

Figure 5. Macroscopic morphologies of (**a**) iron ore pellets and the metalized pellets obtained by reduction with different H_2 flow rates: (**b**)—0.2 L/min, (**c**)—0.4 L/min, (**d**)—0.6 L/min, (**e**)—0.8 L/min, and (**f**)—1.0 L/min.

3.2. Phase Transformation

The phase compositions of the metalized pellets obtained by reduction with different H_2 flow rates were determined by their XRD patterns in Figure 6. The main phases of the metallized pellets were metallic iron (Fe), wüstite (FeO), and fayalite (Fe_2SiO_4). In cases 1–5, with increasing H_2 flow rate, the diffraction peak intensity of Fe increased continuously, while the diffraction peak intensity of FeO and Fe_2SiO_4 decreased. When the H_2 flow rate reached 0.6 L/min, i.e., case 3, the diffraction peaks of FeO disappeared. It was demonstrated that increasing the H_2 flow rate promoted the reduction process.

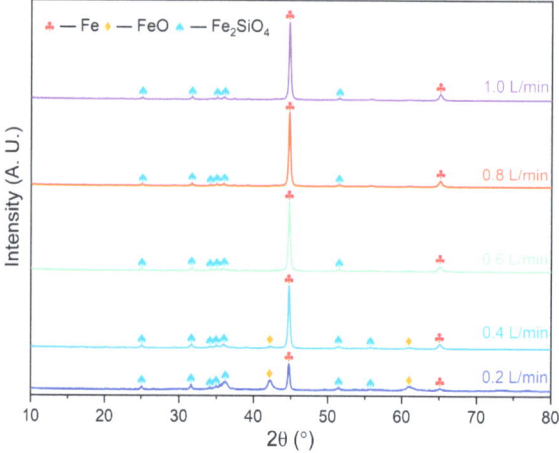

Figure 6. XRD patterns of the metalized pellets obtained by reduction with different H_2 flow rates.

Figure 7 shows the variation of contents of different phases of the metalized pellets with H_2 flow rate. As the H_2 flow rate increased from 0.2 L/min (case 1) to 0.6 L/min (case 3), the content of Fe increased from 20.21 wt% to 62.13 wt%. Concurrently, the contents of FeO and Fe_2SiO_4 decreased from 38.10 wt% and 41.69 wt% to 0 wt% and 37.87 wt%, respectively. When the H_2 flow rate further increased to 1.0 L/min, i.e., case 5, the content of Fe increased to 68.36 wt%, while that of Fe_2SiO_4 decreased to 31.64 wt%. This was because, with low H_2 flow rates, such as cases 1 and 2, there were limited reactions of iron oxides, and a part of FeO remained unreduced. With high H_2 flow rates, such as cases 3–5, the reduction of iron oxides was accelerated, producing more Fe through the reaction between FeO and H_2 and less Fe_2SiO_4 from the reaction between residual FeO and SiO_2 [30]. Consequently, with increasing H_2 flow rate, the contents of FeO and Fe_2SiO_4 decreased.

3.3. Microstructural Evolution

Along with the phase transformation, microstructural changes were expected to occur. Figures 8–12 show the SEM-EDS analysis of metalized pellets obtained by reduction with different H_2 flow rates. As shown in Figure 8, the compositions of spots 1 and 4 indicated that the predominant phase was Fe (metallic iron). At spot 2, the main elements were Fe and O, with a molar ratio of Fe to O close to 9, indicating the coexistence of both Fe and FeO in this area. At spot 3, the predominant elements were Fe, O, and Si, confirming the presence of Fe and Fe_2SiO_4. By observing the microscopic morphologies of the central, 1/4-diameter, and edge areas of the metalized pellets (denoted by A, B, and C, respectively), it was noted that the central area displayed a loose granular morphology while the edge area exhibited a relatively dense laminar structure with large pores. With a low H_2 flow rate, such as 0.2 L/min, the inward diffusion of H_2 into the pellet required a longer duration because it was impeded by the dense metallic iron layer formed in the outer layer after the reaction of H_2 with iron oxides [31–33]. Consequently, the reduction degree in the pellet interior was relatively lower, leading to a sluggish growth of metallic iron particles. Additionally, it was observed that the metallic iron in the edge area existed in a relatively smooth form. This was attributed to the reduction by H_2 in which metallic iron tended to precipitate in the form of a laminar structure rather than a whisker-like structure [34]. The laminar structure of metallic iron also contributed to a relatively low RSI of the metalized pellets, as reported before [35]. Furthermore, the formed dense metallic iron layer would help to increase CS of the pellets compared to the original pellets (2460 N/p vs. 2100 N/p).

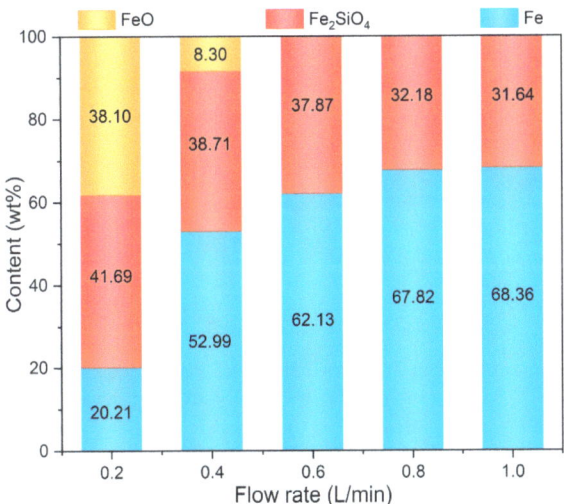

Figure 7. Contents of the main phases of the metalized pellets obtained by reduction with different H_2 flow rates.

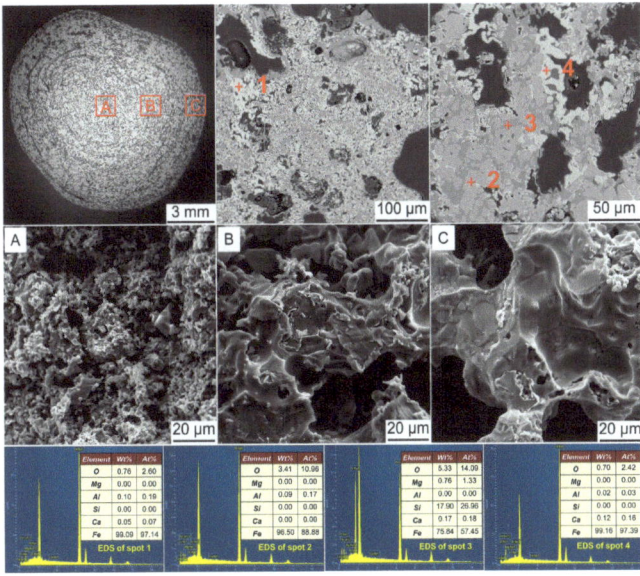

Figure 8. SEM-EDS analysis of the metalized pellets obtained by reduction with the H_2 flow rate of 0.2 L/min: (**A**)–central area, (**B**)–1/4-diameter area, and (**C**)–edge area.

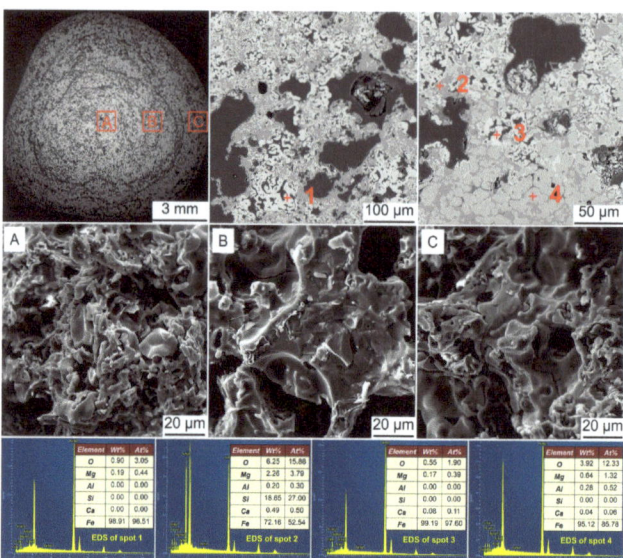

Figure 9. SEM-EDS analysis of the metalized pellets obtained by reduction with the H_2 flow rate of 0.4 L/min: (**A**)–central area, (**B**)–1/4-diameter area, and (**C**)–edge area.

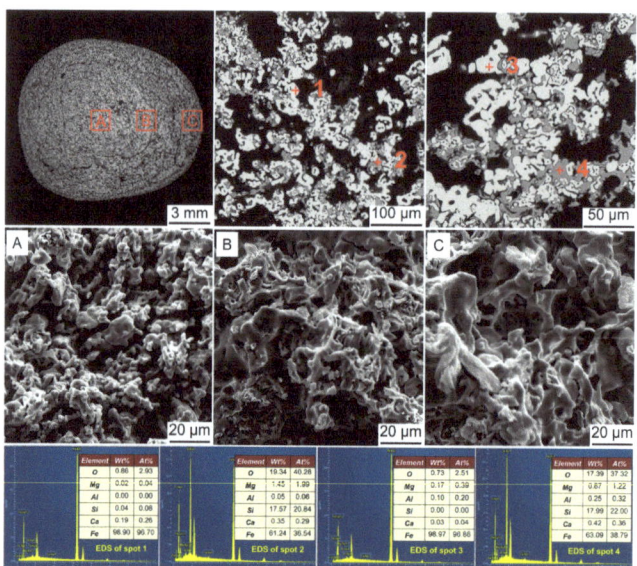

Figure 10. SEM-EDS analysis of the metalized pellets obtained by reduction with the H_2 flow rate of 0.6 L/min in case 3: (**A**)–central area, (**B**)–1/4-diameter area, and (**C**)–edge area.

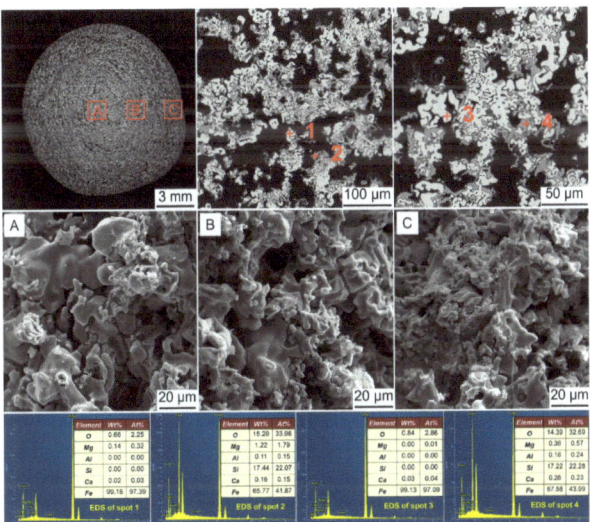

Figure 11. SEM-EDS analysis of the metalized pellets obtained by reduction with the H$_2$ flow rate of 0.8 L/min: (**A**)–central area, (**B**)–1/4-diameter area, and (**C**)–edge area.

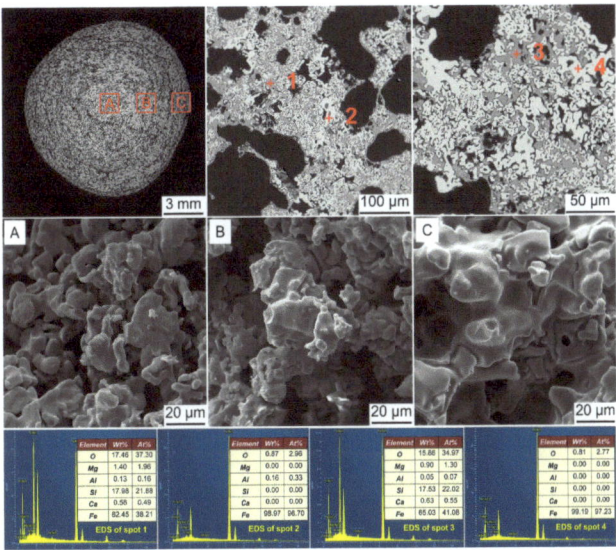

Figure 12. SEM-EDS analysis of the metalized pellets obtained by reduction with the H$_2$ flow rate of 1.0 L/min: (**A**)–central area, (**B**)–1/4-diameter area, and (**C**)–edge area.

Figure 9 shows the SEM-EDS analysis of the metalized pellets obtained by reduction with the H$_2$ flow rate of 0.4 L/min. According to EDS analysis, at spots 1 and 3, Fe was identified, while at spot 2, Fe and Fe$_2$SiO$_4$ were observed. At spot 4, Fe and FeO were detected. Compared with the metalized pellets obtained at the H$_2$ flow rate of 0.2 L/min, i.e., case 1, the metallic iron particles in the inner layer grew significantly. It suggested that increasing the flow rate facilitated the diffusion of H$_2$ into the central area of the pellets, thereby achieving a better reduction performance. Additionally, a small number of metallic iron whiskers were also found in the central and 1/4-diameter areas, which partially accounted for the increase in RSI of the metalized pellets [12].

Figure 10 shows the SEM-EDS analysis of the metalized pellets obtained by reduction with the H_2 flow rate of 0.6 L/min. According to the EDS analysis, at spots 1 and 3, Fe was observed, while at spots 2 and 4, both Fe and Fe_2SiO_4 were present. The absence of FeO indicated that FeO was fully reduced to Fe, which was consistent with the aforementioned XRD analysis. In the central area, there were a large number of metallic iron whiskers, leading to an increase in RSI of the metalized pellets. In the 1/4-diameter and edge areas, there were many small pores, probably resulting from the release of water vapor generated by the reduction of iron oxides inside the pellets. The internal structure of the pellets became highly porous, causing a considerable decrease in CS.

Figure 11 shows the SEM-EDS analysis of the metalized pellets obtained by reduction with the H_2 flow rate of 0.8 L/min. At spots 1 and 3, Fe was observed, while at spots 2 and 4, Fe and Fe_2SiO_4 co-existed. In the central area, there existed only a small quantity of iron whiskers, contributing to the low RSI of the metalized pellets.

Figure 12 shows the SEM-EDS analysis of the metalized pellets obtained by reduction with the H_2 flow rate of 1.0 L/min. Evidently, there were no significant structural changes in the metalized pellets. The metallic iron particles in the pellets continued to grow and no evident metallic iron whiskers were detected. It suggested that increasing the H_2 flow rate could suppress the growth of metallic iron whiskers, reducing the RSI of the metalized pellets [32].

The aforementioned SEM analysis revealed notable size changes of metallic iron particles in the metalized pellets as the H_2 flow rate changed. Figure 13 shows the size distributions of metallic iron particles. With increasing H_2 flow rate from 0.2 L/min (case 1) to 1.0 L/min (case 5), the particle sizes corresponding to the cumulative distribution percentages reaching 10%, 50%, and 90% (D_{10}, D_{50}, and D_{90}) increased from 1.13 μm, 2.01 μm and 3.25 μm to 2.19 μm, 4.01 μm, and 7.21 μm, respectively. The average metallic iron particle size of the metalized pellets increased from 2.16 μm to 4.48 μm, attributed to the faster reduction of iron oxides within the pellets when the H_2 flow rate increased.

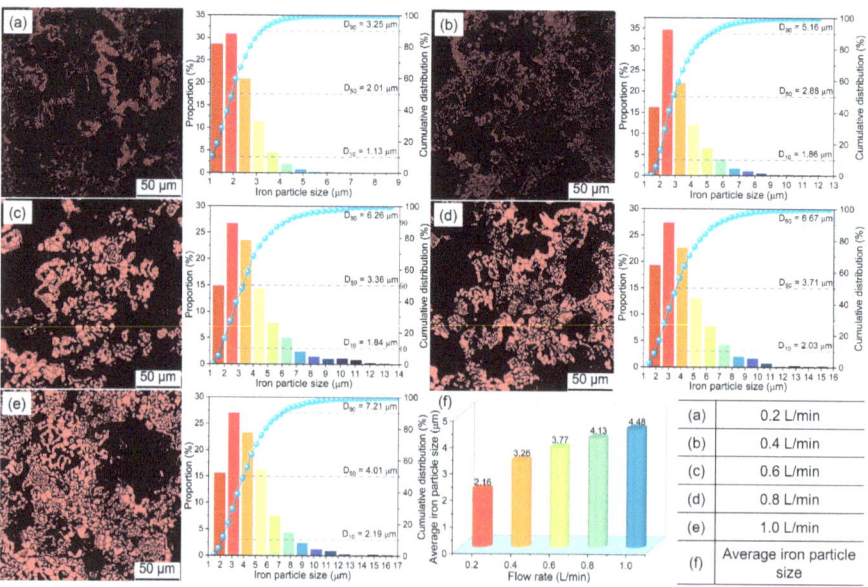

Figure 13. Distributions of metallic iron particles in the metalized pellets obtained by reduction with different H_2 flow rates.

The H_2 flow rate also affected the pore structure in the metalized pellets. To clarify its effect, the variations of pore size distribution, porosity, and average pore size of the metalized pellets with H_2 flow rate were measured. The results are shown in Figures 14 and 15. As shown in Figure 14a, in cases 1–5, the cumulative pore volume in the metalized pellets initially increased and then decreased with increasing H_2 flow rate. With the H_2 flow rate of 0.6 L/min (case 3), the metalized pellets exhibited the largest cumulative pore volume. Figure 14b shows the log-differential intrusion curves of the metalized pellets. The pore size distribution in the metalized pellets primarily ranged from 5 μm to 60 μm. As shown in Figure 15, when the H_2 flow rate increased from 0.2 L/min (case 1) to 0.6 L/min (case 3), the porosity and average pore size of the metalized pellets increased from 39.65% and 3.08 μm to 59.39% and 8.40 μm, respectively. When the H_2 flow rate further increased to 1.0 L/min (case 5), the porosity and average pore size decreased to 54.33% and 3.88 μm, respectively. The variation of porosity and average pore size was basically consistent with the changing trend of RSI [12].

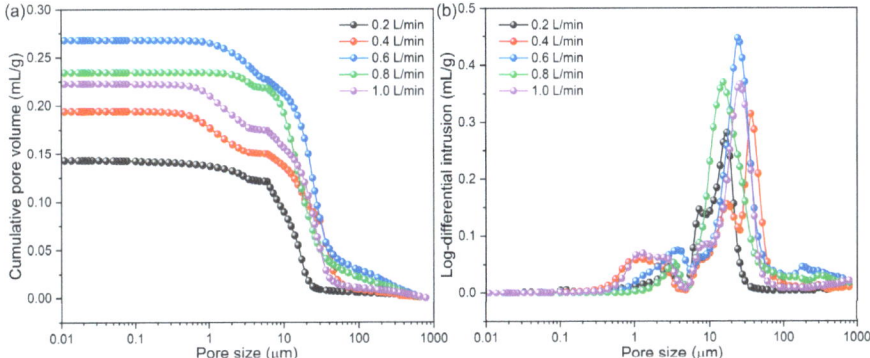

Figure 14. Pore size distributions of the metalized pellets obtained by reduction with different H_2 flow rates: (**a**) cumulative pore volume and (**b**) log-differential intrusion.

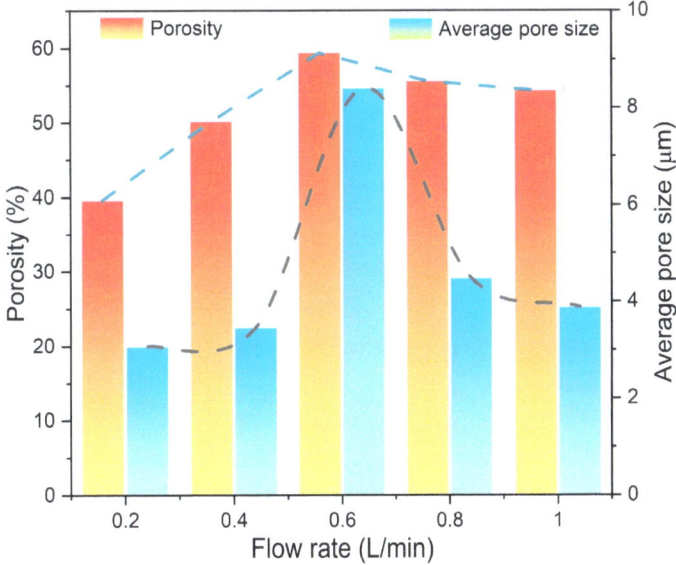

Figure 15. Effect of H_2 flow rate on the porosity and average pore size of the metalized pellets.

With the progress of inward reduction of the pellets, there would be variations in the porosities of different positions of the pellets, as shown in Figure 16. According to the observed variation pattern in porosity within the metalized pellets, there were two distinct stages for all areas. In stage I, i.e., in cases 1–3, the porosity increased with increasing H_2 flow rate. Conversely, in stage II, i.e., in cases 3–5, the porosity decreased. As shown in Figure 16d, when the H_2 flow rate reached 0.6 L/min, i.e., case 3, the porosities of all areas of the metalized pellets were the highest, in agreement with the aforementioned SEM analysis.

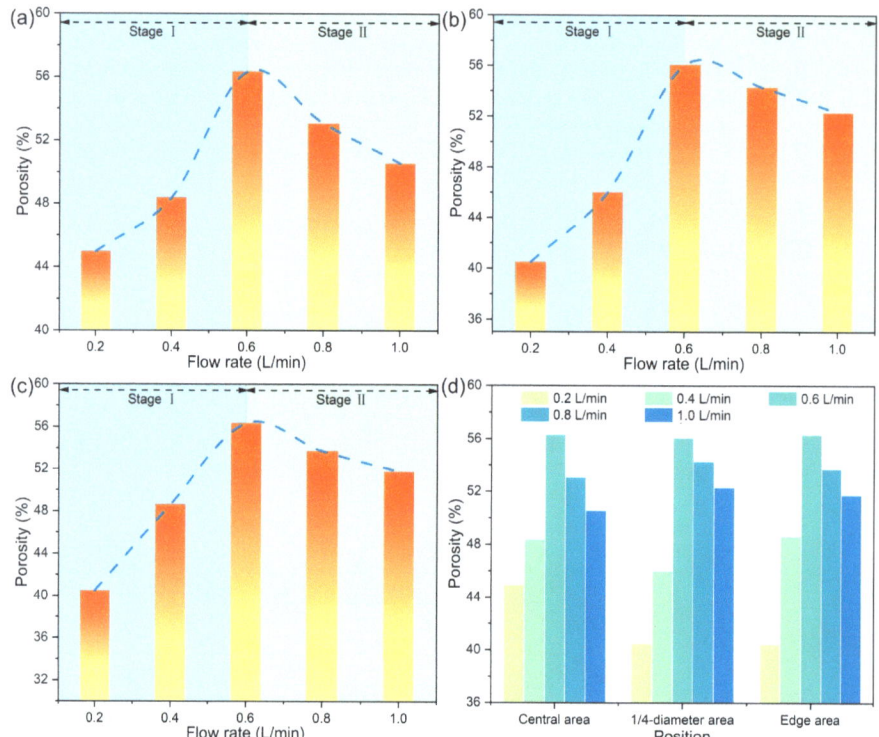

Figure 16. Variations of porosities of (**a**) central area, (**b**) 1/4-diameter area, and (**c**) edge area of the metalized pellets with H_2 flow rate and (**d**) comparison of porosities between different areas.

Considering the reduction performance of the pellets, the proper H_2 flow rate was found to be 0.8 L/min. Under this condition, the metalized pellets had the reduction degree of 91.45%, metallization degree of 84.07%, total iron content of 80.67 wt%, RSI of 4.66%, and CS of 1265 N/p. They were featured by high reduction and metallization degrees, low RSI, and modest CS.

Based on the above results, in cases 1–5, it was evident that increasing the H_2 flow rate effectively enhanced the reduction efficiency of iron ore pellets. However, when the counterdiffusion resistance of water vapor reached a critical value, further increasing the H_2 flow rate did not significantly enhance the reduction performance. In particular, under the condition of a high H_2 flow rate, such as 1.0 L/min, the structure of metalized pellets became more susceptible to damage, producing evident cracks. Therefore, selecting an appropriate H_2 flow rate is important for preparing high-quality metalized pellets with energy conservation.

4. Conclusions

In this study, the influence of H_2 flow rate on the HCR behavior of iron ore pellets was investigated. The results indicated that increasing the H_2 flow rate effectively enhanced the reduction efficiency of iron ore pellets. However, beyond a certain threshold, increasing the H_2 flow rate did not significantly elevate pellet reduction performance because of the countercurrent diffusion resistance of water vapor molecules and limited reaction sites in the pellets. The RSI initially increased and then decreased with increasing H_2 flow rate. This change was in association with the decrease in content of Fe_2SiO_4 in the metalized pellets and the changes in porosity and iron particle size. The CS decreased with increasing H_2 flow rate, showing a sharp decline when the H_2 flow rate reached 0.6 L/min, i.e., case 3. It was attributed to the significant increases in porosity and average pore size of the metallized pellets, with the formation of surface cracks. The content of Fe increased with increasing H_2 flow rate, while the contents of FeO and Fe_2SiO_4 decreased accordingly. Meanwhile, the metallic iron particles in the metalized pellets gradually increased in size, which was one of the significant factors lowering the RSI of the metalized pellets. The porosity and average pore size of the metalized pellets initially increased and then decreased. The initial increases were associated with the evaporation of water vapor generated during the reduction process. After iron oxides were reduced to metallic iron, they began to grow in a lamellar form, which filled the internal pores of the pellets. Therefore, when the H_2 flow rate increased to 0.6 L/min, both the porosity and average pore size decreased. When the H_2 flow rate was 0.8 L/min, i.e., case 4, the optimal reduction performance could be achieved. The reduction degree, metallization degree, total iron content, RSI, and CS of the metalized pellets were 91.45%, 84.07%, 80.67 wt%, 4.66%, and 1265 N/p, respectively. Overall, this study elucidated the importance of controlling the H_2 flow rate in the reduction of iron ore pellets during their cooling.

Author Contributions: Conceptualization, W.F.; data curation, W.F.; formal analysis, W.F., R.T. and G.L.; funding acquisition, Z.P.; methodology, W.F., R.T. and G.L.; investigation, W.F., L.Y. and M.R.; project administration, Z.P.; resources, Z.P.; software, W.F., R.T. and G.L.; supervision, Z.P.; validation, L.Y. and M.R.; writing—original draft, W.F.; writing—review & editing, Z.P. All authors have read and agreed to the published version of the manuscript.

Funding: This work was partially supported by the China Baowu Low Carbon Metallurgy Innovation Foundation under Grant BWLCF202103, the Hunan Provincial Natural Science Foundation of China under Grant 2023JJ10073, and the Hunan Provincial Key Research and Development Program under Grant 2023SK2079.

Institutional Review Board Statement: Not applicable.

Informed Consent Statement: Not applicable.

Data Availability Statement: The datasets presented in this article are not readily available because the data are part of an ongoing study. Requests to access the datasets should be directed to zwpeng@csu.edu.cn.

Conflicts of Interest: The authors declare no conflict of interest.

References

1. Lin, B.; Wu, R. Designing energy policy based on dynamic change in energy and carbon dioxide emission performance of China's iron and steel industry. *J. Clean. Prod.* **2020**, *256*, 120412. [CrossRef]
2. Zhang, S.; Yi, B.; Guo, F.; Zhu, P. Exploring selected pathways to low and zero CO_2 emissions in China's iron and steel industry and their impacts on resources and energy. *J. Clean. Prod.* **2022**, *340*, 130813. [CrossRef]
3. Lin, Y.; Yang, H.; Ma, L.; Li, Z.; Ni, W. Low-carbon development for the iron and steel industry in China and the world: Status quo, future vision, and key actions. *Sustainability* **2021**, *13*, 12548. [CrossRef]
4. Ren, M.; Lu, P.; Liu, X.; Hossain, M.S.; Fang, Y.; Hanaoka, T.; O'Gallachoir, B.; Glynn, J.; Dai, H. Decarbonizing China's iron and steel industry from the supply and demand sides for carbon neutrality. *Appl. Energy* **2021**, *298*, 117209. [CrossRef]
5. Wang, Y.; Liu, J.; Tang, X.; Wang, Y.; An, H.; Yi, H. Decarbonization pathways of China's iron and steel industry toward carbon neutrality, Resour. *Conserv. Recycl.* **2023**, *194*, 106994. [CrossRef]

6. Zang, G.; Sun, P.; Elgowainy, A.; Bobba, P.; McMillan, C.; Ma, O.; Podkaminer, K.; Rustagi, N.; Melaina, M.; Koleva, M. Cost and life cycle analysis for deep CO_2 emissions reduction of steelmaking: Blast furnace-basic oxygen furnace and electric arc furnace technologies. *Int. J. Greenhouse Gas Control* **2023**, *128*, 103958. [CrossRef]
7. Barik, K.; Prusti, P.; Soren, S.; Meikap, B.C.; Biswal, S.K. Mineralogical investigation on preheating studies of high LOI iron ore pellet. *Powder Technol.* **2023**, *418*, 118315. [CrossRef]
8. Palone, O.; Barberi, G.; Di Gruttola, F.; Gagliardi, G.G.; Cedola, L.; Borello, D. Assessment of a multistep revamping methodology for cleaner steel production. *J. Clean. Prod.* **2022**, *381*, 135146. [CrossRef]
9. Burnett, J.W.H.; Li, J.; McCue, A.J.; Kechagiopoulos, P.N.; Howe, R.F.; Wang, X. Directing the H_2-driven selective regeneration of NADH via Sn-doped Pt/SiO_2. *Green Chem.* **2022**, *24*, 1451–1455. [CrossRef]
10. Wang, R.R.; Zhao, Y.Q.; Babich, A.; Senk, D.; Fan, X.Y. Hydrogen direct reduction (H-DR) in steel industry—An overview of challenges and opportunities. *J. Clean. Prod.* **2021**, *329*, 129797. [CrossRef]
11. Tang, J.; Chu, M.; Li, F.; Feng, C.; Liu, Z.; Zhou, Y. Development and progress on hydrogen metallurgy. *Int. J. Miner. Metall. Mater.* **2020**, *27*, 713–723. [CrossRef]
12. Li, W.; Fu, G.; Chu, M.; Zhu, M. Effect of porosity of Hongge vanadium titanomagnetite-oxidized pellet on its reduction swelling behavior and mechanism with hydrogen-rich gases. *Powder Technol.* **2019**, *343*, 194–203. [CrossRef]
13. Yang, J.; Li, L.; Liang, Z.; Peng, X.; Deng, X.; Li, J.; Yi, L.; Huang, B.; Chen, J. Direct reduction of iron ore pellets by H_2-CO mixture: An in-situ investigation of the evolution and dynamics of swelling. *Mater. Today Commun.* **2023**, *36*, 106940. [CrossRef]
14. Liu, Z.; Lu, S.; Wang, Y.; Zhang, J.; Cheng, Q.; Ma, Y. Study on optimization of reduction temperature of hydrogen-based shaft furnace—Numerical simulation and multi-criteria evaluation. *Int. J. Hydrogen Energy* **2023**, *48*, 16132–16142. [CrossRef]
15. Loder, A.; Siebenhofer, M.; Böhm, A.; Lux, S. Clean iron production through direct reduction of mineral iron carbonate with low-grade hydrogen sources; the effect of reduction feed gas composition on product and exit gas composition. *Clean. Eng. Technol.* **2021**, *5*, 100345. [CrossRef]
16. Xu, R.; Zhang, J.; Zuo, H.; Jiao, K.; Hu, Z.; Xing, X. Mechanisms of swelling of iron ore oxidized pellets in high reduction potential atmosphere. *J. Iron Steel Res. Int.* **2015**, *22*, 1–8. [CrossRef]
17. Elsherbiny, A.A.; Qiu, D.; Wang, K.; Li, M.; Ahmed, M.; Hammam, A.; Zhu, Y.; Song, W.; Galal, A.M.; Chen, H.; et al. Parametric study on hematite pellet direct reduction by hydrogen. *Powder Technol.* **2024**, *435*, 119434. [CrossRef]
18. Wang, Z.; Chu, M.; Liu, Z.; Chen, S.; Xue, X. Effects of temperature and atmosphere on pellets reduction swelling index. *J. Iron Steel Res. Int.* **2012**, *19*, 7–19. [CrossRef]
19. Patenthub Home Page. Available online: https://www.patenthub.cn/patent/CN118186208A.html (accessed on 29 July 2024).
20. *GB/T 24236-2009*; Iron Ores for Shaft Direct-Reduction Feedstocks—Determination of the Reducibility Index, Final Degree of Reduction and Degree of Metallization. State General Administration of China for Quality Supervision and Inspection and Quarantine & Standardization Administration of China: Beijing, China, 2009.
21. *GB/T 6730.5-2022*; Iron Ores—Determination of Total Iron Content—Titrimetric Methods after Titanium (III) Chloride Reduction. State Administration for Market Regulation and Standardization Administration of China & Standardization Administration of China: Beijing, China, 2022.
22. *GB/T 14201-2018*; Iron Ore Pellets for Blast Furnace and Direct Reduction Feedstocks—Determination of the Crushing Strength. State Administration for Market Regulation and Standardization Administration of China & Standardization Administration of China: Beijing, China, 2018.
23. Wang, P.; Wang, C.; Wang, H.; Long, H.; Zhou, T. Effects of SiO_2, CaO and basicity on reduction behaviors and swelling properties of fluxed pellet at different stages. *Powder Technol.* **2022**, *396*, 477–489. [CrossRef]
24. Peng, Z.; Fan, W.; Tang, H.; Xiang, C.; Ye, L.; Yin, T.; Rao, M. Direct conversion of blast furnace ferronickel slag to thermal insulation materials. *Constr. Build. Mater.* **2024**, *412*, 134499. [CrossRef]
25. Feinman, J.; Smith, N.D.; Muskat, D.A. Effect of gas flow rate on kinetics of reduction of iron oxide pellets with hydrogen. *Ind. Eng. Chem. Process Des. Dev.* **1965**, *4*, 270–274. [CrossRef]
26. Singh, A.K.; Mishra, B.; Sinha, O.P. Reduction kinetics of fluxed iron ore pellets made of coarse iron ore particles. *Steel Res. Int.* **2024**, *in press*. [CrossRef]
27. Sakurai, S.; Namai, A.; Hashimoto, K.; Ohkoshi, S.-I. First observation of phase transformation of all four Fe_2O_3 phases ($\gamma \rightarrow \varepsilon \rightarrow \beta \rightarrow \alpha$-phase). *J. Am. Chem. Soc.* **2009**, *131*, 18299–18303. [CrossRef] [PubMed]
28. Han, G.; Zhang, D.; Huang, Y.; Jiang, T. Swelling behavior of hot preheated pellets in reduction roasting process. *J. Cent. South Univ.* **2016**, *23*, 2792–2799. [CrossRef]
29. Shi, Y.; Guo, Z.; Zhu, D.; Pan, J.; Lu, S. Isothermal reduction kinetics and microstructure evolution of various vanadium titanomagnetite pellets in direct reduction. *J. Alloys Compd.* **2023**, *953*, 170126. [CrossRef]
30. Wang, H.; Shen, L.; Bao, H.; Zhang, W.; Zhang, X.; Luo, L.; Song, S. Investigation of solid-state carbothermal reduction of fayalite with and without added metallic iron. *JOM* **2021**, *73*, 703–711. [CrossRef]
31. Ma, Y.; Filho, I.R.S.; Bai, Y.; Schenk, J.; Patisson, F.; Beck, A.; Van Bokhoven, J.A.; Willinger, M.G.; Li, K.; Xie, D.; et al. Hierarchical nature of hydrogen-based direct reduction of iron oxides. *Scr. Mater.* **2022**, *213*, 114571. [CrossRef]
32. Abu Tahari, M.N.; Salleh, F.; Tengku Saharuddin, T.S.; Samsuri, A.; Samidin, S.; Yarmo, M.A. Influence of hydrogen and carbon monoxide on reduction behavior of iron oxide at high temperature: Effect on reduction gas concentrations. *Int. J. Hydrogen Energy* **2021**, *46*, 24791–24805. [CrossRef]

33. Kazemi, M.; Glaser, B.; Sichen, D. Study on direct reduction of hematite pellets using a new TG setup. *Steel Res. Int.* **2014**, *85*, 718–728. [CrossRef]
34. Zhao, Z.; Tang, J.; Chu, M.; Wang, X.; Zheng, A.; Wang, X.; Li, Y. Direct reduction swelling behavior of pellets in hydrogen-based shaft furnaces under typical atmospheres. *Int. J. Miner. Metall. Mater.* **2022**, *29*, 1891–1900. [CrossRef]
35. Scharm, C.; Küster, F.; Laabs, M.; Huang, Q.; Volkova, O.; Reinmöller, M.; Guhl, S.; Meyer, B. Direct reduction of iron ore pellets by H_2 and CO: In-situ investigation of the structural transformation and reduction progression caused by atmosphere and temperature. *Miner. Eng.* **2022**, *180*, 107459. [CrossRef]

Disclaimer/Publisher's Note: The statements, opinions and data contained in all publications are solely those of the individual author(s) and contributor(s) and not of MDPI and/or the editor(s). MDPI and/or the editor(s) disclaim responsibility for any injury to people or property resulting from any ideas, methods, instructions or products referred to in the content.

Article

Study on the Migration Patterns of Oxygen Elements during the Refining Process of Ti-48Al Scrap under Electromagnetic Levitation

Xinchen Pang, Guifang Zhang *, Peng Yan *, Zhixiang Xiao and Xiaoliang Wang

Faculty of Metallurgical and Energy Engineering, Kunming University of Science and Technology, Kunming 650093, China; xinchen001118@163.com (X.P.); xiaozhixiang1998@163.com (Z.X.); wangxiaoliang@kust.edu.cn (X.W.)
* Correspondence: guifangzhang65@163.com (G.Z.); yanp_km@163.com (P.Y.)

Abstract: This study investigated the migration patterns of oxygen in the deoxidation process of Ti-48Al alloy scrap using electromagnetic levitation (EML) technology. Scanning electron microscopy (SEM), X-ray diffraction (XRD), and X-ray photoelectron spectroscopy (XPS) were employed to analyze the oxygen distribution patterns and migration path during EML. The refining process resulted in three types of oxygen migration: (1) escape from the lattice and evaporation in the form of AlO, Al_2O; (2) formation of metal oxides and remaining in the alloy melt; (3) attachment to the quartz tube wall in the form of metal oxides such as Al_2O_3 and Cr_2O_3. The oxygen content of the scrap was dropped with a deoxidation ratio of 62%. It indicated that EML can greatly promote the migration and removal of oxygen elements in Ti-Al alloy scrap.

Keywords: electromagnetic levitation; Ti-Al alloy scrap; refining deoxidation; elemental oxygen migration

Citation: Pang, X.; Zhang, G.; Yan, P.; Xiao, Z.; Wang, X. Study on the Migration Patterns of Oxygen Elements during the Refining Process of Ti-48Al Scrap under Electromagnetic Levitation. *Materials* **2024**, *17*, 3709. https://doi.org/10.3390/ma17153709

Academic Editor: Evgeny Levashov

Received: 17 June 2024
Revised: 21 July 2024
Accepted: 22 July 2024
Published: 26 July 2024

Copyright: © 2024 by the authors. Licensee MDPI, Basel, Switzerland. This article is an open access article distributed under the terms and conditions of the Creative Commons Attribution (CC BY) license (https://creativecommons.org/licenses/by/4.0/).

1. Introduction

Titanium alloys are widely utilized in the chemical and aerospace industries, marine engineering, and a variety of other areas owing to their excellent characteristics, such as low weight, high specific strength, high temperature resistance, corrosion resistance, and outstanding antioxidant properties [1–3]. Particularly, titanium–aluminum (Ti-Al) alloys are considered to have great application prospects in the development of high-temperature structural materials for aerospace, defense, and automotive industries [4]. They need to withstand harsh operating conditions, including high temperatures, severe air friction, and cyclic oxidation [5], leading to a large amount of oxidized scrap during melting, processing, and subsequent operations. Due to its high affinity for Ti [6], oxygen tends to dissolve preferentially in the α phase of Ti alloys, thereby strengthening this phase [7]. In particular, the solubility limit of oxygen can reach 14 wt.% in the hexagonal α-Ti phase [8], making it extremely challenging to deoxidize from titanium scrap. The ideal and most economical method of utilizing Ti scrap is to melt it back into ingots. However, it is essential to reduce the oxygen content of titanium scrap before returning to smelting. Thus, investigations about simple and effective deoxidation methods for Ti-Al alloys and Ti alloy scrap are highly focused. At present, the deoxygenation techniques for Ti scrap include molten-salt electrochemical deoxidation [9], hydrogenation–dehydrogenation (HDH) [10], and the smelting deoxidation process [11,12]. Suzuki et al. [13,14] have explored the molten-salt electrolytic deoxidation method. Metallic Ca is obtained by electrolyzing CaO in molten salt, followed by calciothermically reducing TiO_2 to form sponge Ti. This method has a lot of advantages, such as low process costs, minimal energy consumption, and high purity of the Ti sponge product. On the other hand, parasitic reactions result in a decrease in current efficiency, such as carbon precipitation during electrolysis. Then, the separation of the product from the electrolyte becomes challenging due to the structural issues in the electrolytic cell [15].

Su et al. [16] have used the HDH process to melt the Ti64 alloy, reducing its oxygen content from 0.12 to 0.028 wt.%. This process involves hydrogen atoms entering the lattice to form metal hydrides, followed by dehydrogenation under appropriate experimental conditions to yield Ti powder [17]. Although this method involves a short process flow and simple equipment, it presents difficulties in the complete removal of hydrogen during the dehydrogenation process, particularly for the scrap with a high oxygen content [18]. Bartosinski et al. [19,20] utilized a vacuum induction melting furnace to investigate the preparation of Ti-6Al-4V by aluminothermic reduction. This method resulted in a decrease in the oxygen content by 1500–3500 ppm and significantly lowered the production cost of Ti alloys. Despite these advancements, these challenges have to be overcome for the utilization of Ti scrap; for example, high oxygen content and the presence of numerous inclusions persist. Electromagnetic levitation (EML) technology, characterized by rapid surface renewal and vigorous internal stirring [21,22], can offer favorable kinetic conditions for volatilization and impurity removal, which can lead to an ideal metallurgical reaction process.

This study proposes the utilization of EML melting technology for deoxidation from titanium scrap, capitalizing on its advantages in strengthening metallurgy. To address the current limitations of conventional melting and deoxidation processes, such as sluggish diffusion of internal solute elements and restricted surface renewal frequency, the elemental migration process during the intensified deoxidation process was focused on in this study. Finally, the results from this work should aid in future adoptions of shorter and cost-effective processes in the utilization of oxygen-containing Ti scrap.

2. Materials and Methods

2.1. Materials

The raw material for this study is titanium scrap produced during the production processes of a factory, with Al scrap, Cr scrap, and TiO_2 powder added. The prepared alloy was tested for its composition as Ti-48Al-2Cr (Ti: Al: Cr = 50: 48: 2 at.%; O 0.5 at.%). The Ti-48Al scrap was cut into samples weighing approximately 1.2 g each for the EML experiments. The argon gas used in the experiments was supplied by Guangruida Gas Co., Ltd., Kunming, China, with a purity of 99.999%.

2.2. Method

The procedure and experimental setup for the EML experiment are depicted in Figure 1. The equipment mainly consists of an electromagnetic levitation melting system, atmosphere control system, and camera detection system. Before activating the device, a stream of pure argon gas was supplied at a rate of 1.5 L/min (liters per minute) to remove air presented in the glass tube. The deoxygenation method performed on the Ti-Al alloy scrap using EML has three stages: preheating, levitation melting, and levitation refining. At the end of the refining process, the power current was gradually reduced to below 0.7 A and the melt precipitated and fell into the copper crucible and solidified. Afterwards, the power supply was turned off, and bubbled Ar gas was used as a cooling atmosphere until the scrap approached room temperature. To ensure the reliability of the experimental data, three repetitions of the experiment were carried out, and the elemental oxygen content was tested.

X-ray Diffraction (X'pert 3, Malvern-Panalytical Powder) was used to analyze the phase composition of the samples before and after EML. The variation in oxygen content was identified using an oxygen/nitrogen analyzer (HORIBA EMGA-830, Kyoto, Japan). The microstructure and element composition of the samples was detected using a scanning electron microscope (HITACHI SU8010, Tokyo, Japan) and via energy-dispersive X-ray spectroscopy (OXFORD ULTIM MAX400, Oxford, UK). Additionally, X-ray photoelectron spectroscopy (pHI5000 VERSAPROB-II, Tokyo, Japan) was used to study the changes in the valence states of oxygen and other elements during the removal process under EML.

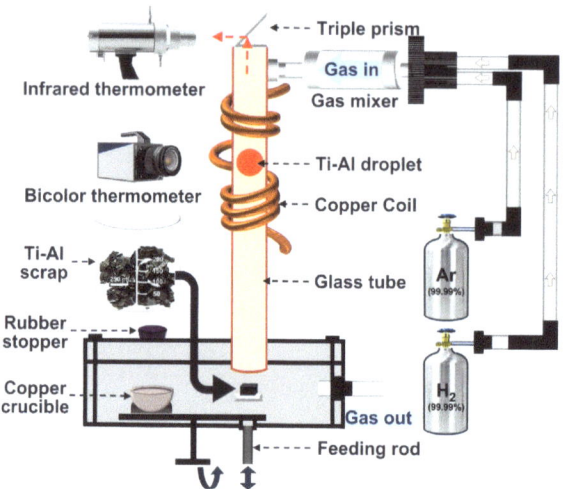

Figure 1. Schematic depiction of the EML system.

3. Results and Discussion

3.1. Thermodynamic Analysis of Deoxygenation

The volatilization behavior of the Ti-Al alloys was investigated through the calculation of saturated vapor pressures of Ti, Al, and Cr at different temperatures using the Clausius–Clapeyron equation (C-C equation) [23]:

$$logP_i^* = AT^{-1} + BlogT + CT + D \quad (1)$$

where T (K) represents thermodynamic temperature, P_i^* denotes the saturated vapor pressure (Pa) of ideal metal at temperature T. The fixed values in the C-C equation at 2173 K are shown in Table 1.

Table 1. The C-C equation parameters for Ti, Al, and Cr in Ti-Al alloys at 2173 K [24,25].

Element	A	B	C	D
Ti	−24,914	−2.52	—	20.832
Al	−16,450	−1.023	—	14.48
Cr	−20,680	−1.31	—	16.68

Figure 2 depicts the variation in saturation vapor pressure with temperature for the elements Ti, Al, and Cr in the Ti-Al alloys. At temperatures ranging from 0 to 1600 K, all the curves are approximately the same and converge to 0, suggesting that the Al, Ti, and Cr components are not easily volatilized under melting conditions of lower temperature. When the temperature exceeds 1600 K, the saturated vapor pressure of Al and Cr starts to increase exponentially, while Ti remains relatively constant; when it rises to 2173 K, the saturated vapor pressure of Al reaches 3313 Pa at this time, and Cr also reaches 619 Pa, while the Ti element remains at around 9 Pa. Ono et al. [26] categorized refractory metals based on the difficulty of deoxygenation, grouping them from I to IV according to the level of difficulty. The saturation vapor pressures of Ti and Ti oxides in group IV at a temperature of 1800 K are 5.2×10^{-3} Pa and 1.3×10^{-2} Pa, respectively. Therefore, it is highly unlikely for elemental oxygen to be removed through evaporation by combining it with Ti. Under identical thermal conditions, Cr and its oxides in group II have saturation vapor pressures of 9.6 Pa and 4.0×10^{-3} Pa, respectively. Metals in this group have low oxygen solubility and activity; deoxygenation is effective when Cr content is above a certain amount. However, the Cr content in the alloy matrix is much lower than that of the Ti and Al in this study; so, the impact of its volatilization on the oxygen content is currently not being taken into account.

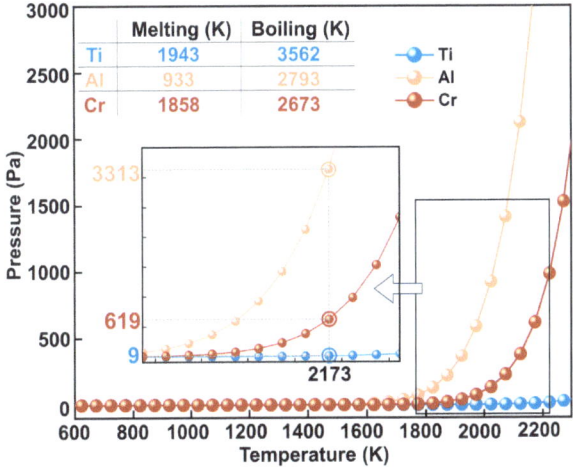

Figure 2. Temperature variation in saturation vapor pressure of Ti, Al, and Cr in Ti-Al alloys.

Based on the relevant literature [27], two main forms in which oxygen can be removed through evaporation reactions in Ti-Al system are as follows:

$$\text{Al (l) (in molten TiAl)} + \text{O(in molten TiAl)} \rightarrow \text{AlO(g)} \uparrow \tag{2}$$

$$\text{2Al (l) (in molten TiAl)} + \text{O(in molten TiAl)} \rightarrow \text{Al}_2\text{O(g)} \uparrow \tag{3}$$

This study covers a temperature range from 1973 to 2273 K [28,29]; the Gibbs free energy remains consistently negative and increases in magnitude with rising temperature. The result suggests that oxygen can be removed through the volatilization of Al in Ti-Al system, with reactions occurring more easily at higher temperatures.

3.2. Results of Deoxidation Experiments under Electromagnetic Levitation Conditions

The transfer behavior of impurity elements in the electromagnetic levitation refining process is significantly influenced by various EML conditions. Consequently, this study aimed to examine the effect of different melting time, temperature, and initial oxygen content on the removal efficiency of oxygen in Ti-Al alloys; the results are depicted in Figure 3. As depicted in Figure 3a, under the five melting times of 10 min, 20 min, 30 min, 40 min, and 50 min, the oxygen removal rate increases gradually from 10 min to 40 min and then decreases with longer melting times. The oxygen removal ratio at 40 min is 61.1%, significantly higher than at other melting times. Therefore, a levitation time of 40 min is suitable for the further experiments. Subsequently, as shown in Figure 3b, the experiments were conducted at five temperatures in the range of 1973–2373 K, with a levitation melting temperature of 2173 K exhibiting a remarkably high oxygen removal ratio of 61.3%. Ultimately, the experiments with varying initial oxygen contents were conducted at a temperature of 2173 K and levitation time of 40 min. The oxygen contents ranged from 0.25 at.% to 0.75 at.%, as illustrated in Figure 3c, which revealed that the highest oxygen removal ratio occurred at an initial oxygen content of 0.5 at.%. Consequently, it was determined that the optimal levitation parameters for a 61.3% deoxidation ratio are a temperature of 2173 K, levitation time of 40 min, and initial oxygen content of 0.5 at.%.

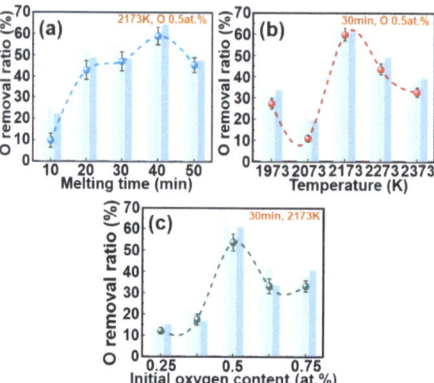

Figure 3. Results of deoxidation experiments under different electromagnetic levitation conditions: (**a**) melting time, (**b**) temperature, and (**c**) initial oxygen content.

3.3. Results of the Changes in Elemental Distribution and Phase Composition during EML Experiments

The results of the elemental distribution of the original alloy samples and the specimen obtained under the optimum conditions are depicted in Figure 4. As can be seen in Figure 4, a uniform distribution of Ti, Al, and Cr elements is observed both before and after EML, while oxygen exhibits a radically different distribution. Figure 4a reveals that the elemental composition of the original alloy matrix predominantly comprises Ti and Al, with the presence of aggregated black striped substances on the surface. As seen in the SEM–EDS result for point S2 in Figure 4a, the Al and O element contents are determined to be 41.8 at.% and 57.6 at.%, respectively. Consequently, the inference drawn is that the black substance corresponds to Al_2O_3 [30]. As seen in Figure 4b, element surface scanning from the SEM image of the levitated sample was conducted, and a small amount of oxygen content was detected. The black striped substances were determined to be an oxide by the surface scanning results. Quantitative test by the oxygen/nitrogen analyzer revealed that the refined sample contained 0.258 at.% of oxygen elements, compared to 0.5 at.% in the initial sample. This indicates a significant reduction in oxygen content in the alloy after electromagnetic levitation melting. The EDS surface scan images show that the black striped substances are dissipated and become aggregated flakes and interconnections. EML resulted in the disappearance of microcracks on the sample's surface, and the size and morphology of the black substance significantly improved. Combined with the thermodynamic calculations in the previous work, the impure oxygen notably decreased in levels and aggregated towards voids, and this may be the migration path of the oxygen removal process [31].

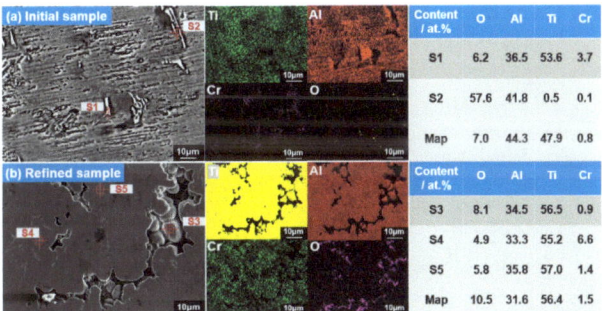

Figure 4. The EDS detection images of the Ti-Al alloy (**a**) before EML and (**b**) under conditions of 2173 K for 40 min.

The SEM–EDS analysis verifies significant alterations in the existence form, distribution, and composition of the oxygen element in the alloy after EML levitation. Meanwhile, the samples were subjected to an X-ray diffraction (XRD) analysis to investigate changes in their phase composition; the results are depicted in Figure 5. The initial sample exhibited crystalline peaks at diffraction angles (2θ) at approximately 31.7°, 38.7°, 44.5°, 45.4°, and 65.5°, corresponding to characteristic peaks of the tetragonal crystal system in TiAl (space group P4/mmm (No.#123)). The crystalline peaks observed at approximately 36.1°, 41.2°, and 54.1° are attributed to the hexagonal crystal structure of Ti$_3$Al (space group P6$_3$/mmc (No.#194)). The main diffraction peaks were observed at angles of 38.7° for the γ-TiAl phase and 41.2° for the α$_2$-Ti$_3$Al phase; it can be indicated the phase composition of the initial sample is the TiAl and Ti phase [32]. The refined sample exhibited crystalline peaks at diffraction angles (2θ) at approximately 26.4°, 36.1°, 38.9°, 41.2°, and 54.1°, corresponding to hexagonal crystal structure of Ti$_3$Al. This confirms the sample is composed solely of Ti$_3$Al phase. As shown in Table 2, the lattice parameters and axial ratio of the γ-TiAl phase decrease after EML, along with a reduction in the plane spacing of the main peak. The c/a ratio of the alloy sample decreased from 1.4416 to 0.8034 after levitation; when the oxygen content is less than 0.5 at.%, the c/a ratio decreases with decreasing oxygen content.

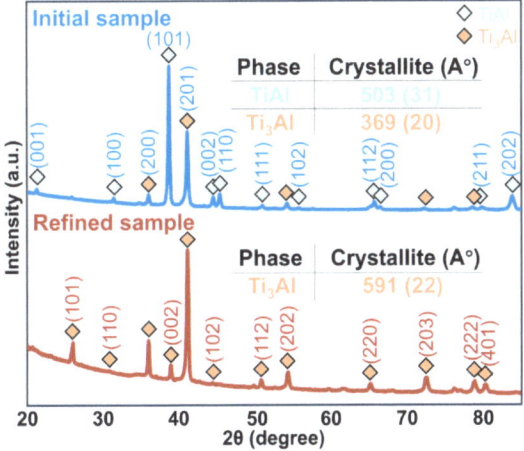

Figure 5. XRD images of the sample before and after EML.

Table 2. The lattice parameters a, c, axial ratio (c/a), and interplanar spacing d of the sample before and after EML.

Alloy	a (Å)	b (Å)	c/a	d (Å)		
				(110)	(201)	(110)
Initial sample: TiAl	2.8228	4.0693	1.4416	2.3172	2.3253	2.0031
Initial sample: Ti$_3$Al	5.7347	4.6327	0.8078			
Refined sample	5.7381	4.6101	0.8034	2.3070	2.1874	1.6906

3.4. Analysis Results of Element Migration during EML Deoxidation Process

X-ray photoelectron spectroscopy (XPS) was used to examine the surface characteristics of the alloy elements in the initial sample, the refined sample, and the residual volatile substance on the tube wall. The results were used to determine the binding states of oxygen with different alloy elements and to explore the migration pattern of oxygen elements. Figure 6 illustrates the XPS survey spectra of the initial, refined, and tube wall-adhered samples. In the XPS survey spectra under three different conditions, distinct O1s, Ti2p, and Al2p characteristic peaks are observed in the initial sample, indicating the presence of O, Ti, and Al as the main elements in the initial sample. In the levitated sample's survey spectrum, besides the presence of O1s, Ti2p, C1s, and Al2p, characteristic peaks of Cr2p are also evident. Based on the XPS survey spectra of the three samples, the elements Cr, Al, Ti, and O are detected at around 576 eV, 73 eV, 459 eV, and 530 eV, respectively. By analyzing the fine spectra of C1s and performing charge correction using a binding energy of 284.8 eV, the binding states of each element in the levitated alloy are further clarified.

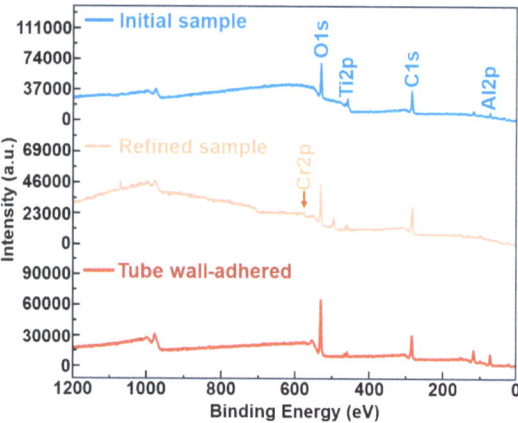

Figure 6. The XPS survey spectra of the substances under three different conditions.

Figure 7 presents the spectral analysis of Ti, Al, Cr, and O elements, and it reveals that in the XPS fine spectrum of Ti2p in the initial sample, two distinct peaks appear at 458.98 eV and 464.68 eV, corresponding to the spin–orbit split photoelectrons Ti2p3/2 and Ti2p1/2 of the Ti^{4+} chemical phase, respectively. This observation indicates the bonding of titanium with oxygen, attributed to Ti^{4+} within the TiO$_2$ lattice. Similarly, in both the levitated and tube wall volatile substances, peaks corresponding to the spin–orbit split photoelectrons Ti2p3/2 and Ti2p1/2 of the Ti^{4+} chemical phase within the TiO$_2$ lattice are observed at 458.88 eV, 464.58 eV and 458.86 eV, 464.56 eV, respectively. This indicates that the chemical state of titanium remains Ti^{4+} before and after the material processing. In the XPS fine spectrum of Al2p in the initial sample, two peaks appear at 72.28 eV and 74.58 eV, representing the characteristic XPS signals of Al (around 72 eV) and the binding peak of Al$_2$O$_3$. The proportions of these two peaks are 25.73% and 74.27%, respectively, indicating that Al mainly exists as Al_2O_3 in the initial sample, with some Al present as well. Similarly, peaks corresponding to Al and Al_2O_3 are detected in the levitated sample, with proportions of 37.79% and 62.21%, respectively. Compared to the initial sample, the relative content of Al in the levitated sample, in terms of Al, O, Ti, and Cr elements, is lower, decreasing from 32.88% to 12.24%. Furthermore, the overall spectrum confirms the lower total Al content in the refined sample, indicating a decrease in the content of both Al and Al_2O_3 after EML, which is due to the high vapor pressure of Al promoting the evaporation of Al_2O_3. In the tube wall volatile substances, peaks corresponding to Al and Al_2O_3 are also observed, with relative contents of 8.27% and 91.73% in the Al2p spectrum, indicating a higher content of Al_2O_3 adhering to the tube wall during the volatile process. Considering the low Cr content of only 1.0 at.% in the materials, the detected Cr content in both the initial sample and tube wall volatile substances is relatively low, while the peak intensity of Cr2p in the refined sample is higher, possibly due to the evaporation of other elements. In combination with the fine spectra, the relative content of Cr in terms of Al, O, Ti, and Cr elements is 3.04%, 10.47%, and 2.08%, respectively, and Cr exists in the form of Cr_2O_3 in all three materials.

In the fine spectra of O1s, oxygen elements are detected in the form of lattice oxygen (O Latt) and adsorbed oxygen (O ads) in both the initial, refined, and tube wall volatile samples. The relative elemental oxygen contents of each are 45.42%, 50.68%, and 42.8%, respectively. Furthermore, the relative contents of lattice oxygen are 32.58%, 25.99%, and 22.33%, respectively. The decrease in lattice oxygen corresponds to a decrease in the content of metal oxides on the alloy surface, while adsorbed oxygen can accelerate oxygen diffusion, thereby effectively improving the material's electronic structure, geometric structure, and magnetic properties. To confirm the departure of lattice oxygen, this study conducted oxygen content detection before and after EML, revealing a decrease in oxygen content from 0.5 at.% to 0.29 at.%, indicating a reduction of 62% compared to before melting. This suggests that the high-temperature treatment of Ti-Al alloys using EML melting technology greatly promotes the migration of oxygen atoms.

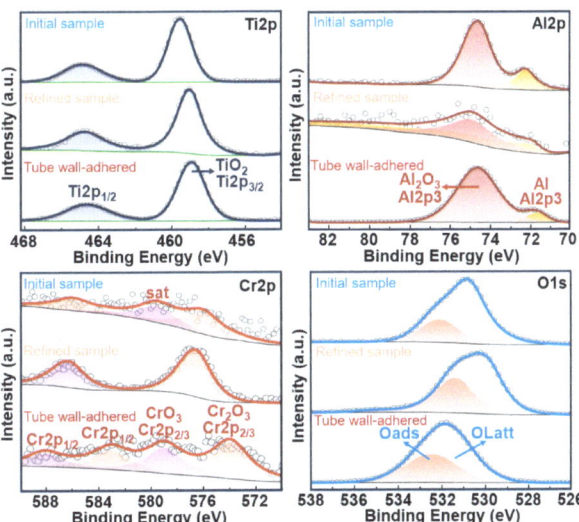

Figure 7. The XPS detection results for Ti, Al, Cr, and O elements.

As depicted in Figure 8, during the electromagnetic refining process of the Ti-Al alloy scrap, the migration modes of oxygen elements include the following three forms: (1) detachment from the crystal lattice to form AlO and Al_2O and evaporation from the alloy melt; (2) formation of metal oxides remaining in the alloy melt; and (3) evaporation and adherence to the quartz tube wall as metal oxides.

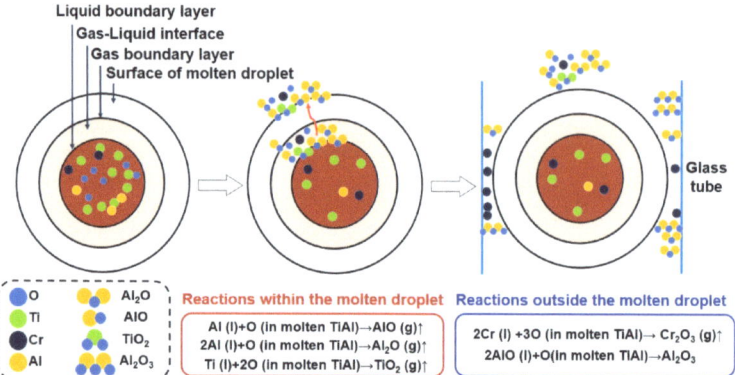

Figure 8. Deoxidation process schematic of Ti-Al alloys under EML.

The removal process of oxygen elements in Ti-Al alloys can be simplified into the following five steps:

(1) Oxygen atoms move into the melt boundary layer from the Ti-Al droplet interior.
(2) Oxygen atoms move through the melt boundary layer to the surface of Ti-Al droplets.
(3) Volatilization of oxygen atoms occurs at the gas–liquid interface.
(4) Oxygen atoms move through the gaseous boundary layer to the reaction chamber.
(5) Oxygen atoms condense on the glass wall or are extracted by a flowing gas system or vacuum system.

4. Conclusions

(1) The thermodynamic analysis of deoxidation indicated that oxygen in the Ti-Al system can be removed with the volatilization of aluminum, and the reaction is facilitated by increasing temperature. The deoxidation condition experiments revealed an optimal levitation temperature, time, and initial oxygen content of 2173 K, 40 min, and 0.5 at.%, resulting in an oxygen removal ratio of 62%.
(2) The impure oxygen in the alloy were significantly removed by EML, with a transition from a dispersed to a clustered distribution. The phase composition of the alloy changed from the TiAl and Ti_3Al phases to a single Ti_3Al phase. Furthermore, the lattice parameter c/a was reduced after EML.
(3) The refinement process of the Ti-48Al alloy scrap under EML can be divided into three forms of oxygen migration: (a) detachment from the crystal lattice to form AlO and Al_2O, (b) formation of metal oxides remaining in the alloy melt, and (c) evaporation and adherence to the quartz tube wall as metal oxides.

Author Contributions: Conceptualization, G.Z. and P.Y.; methodology, X.P.; software, X.P.; validation, P.Y. and X.P.; formal analysis, P.Y. and X.W.; investigation, G.Z. and P.Y.; resources, P.Y. and X.P.; data curation, X.P. and Z.X.; writing—original draft preparation, P.Y. and G.Z.; writing—review and editing, G.Z.; visualization, X.P.; supervision, G.Z. and P.Y.; project administration, G.Z.; funding acquisition, G.Z. All authors have read and agreed to the published version of the manuscript.

Funding: This research was funded by the Yunnan Provincial Department of Education Scientific Research Fund Project (grant number 2023J0130); National Natural Science Foundation of China (grant number 52074140); and Hunan Zhongke Electric Co., Ltd., Yueyang, China.

Institutional Review Board Statement: Not applicable.

Informed Consent Statement: Not applicable.

Data Availability Statement: Data are contained within this article.

Conflicts of Interest: The authors declare no conflict of interest.

References

1. Güther, V.; Allen, M.; Klose, J.; Clemens, H. Metallurgical processing of titanium aluminides on industrial scale. *Intermetallics* **2018**, *103*, 12–22. [CrossRef]
2. Pere, B.V.; Joachim, G.; Andreas, J.; Norbert, S.; Jan, H.; Guillermo, R. Peritectic titanium alloys for 3D printing. *Nat. Commun.* **2018**, *9*, 3426.
3. Mermer, E.; Çinici, H.; Uğur, G.; Ünal, R. Development of an Al rich Ti-Al alloy with better ductility. *Vacuum* **2023**, *214*, 112224. [CrossRef]
4. Kim, T.; Oh, J.-M.; Cho, G.-H.; Chang, H.; Jang, H.D.; Lim, J.-W. Surface and internal deoxidation behavior of titanium alloy powder deoxidized by Ca vapor: Comparison of the deoxidation capability of solid solution and intermetallic titanium alloys. *Appl. Surf. Sci.* **2020**, *534*, 147623. [CrossRef]
5. Uwanyuze, R.S.; Kanyo, J.E.; Myrick, S.F.; Schafföner, S. A review on alpha case formation and modeling of mass transfer during investment casting of titanium alloys. *J. Alloys Compd.* **2021**, *865*, 158558. [CrossRef]
6. Chong, Y.; Gholizadeh, R.; Tsuru, T.; Zhang, R.; Inoue, K.; Gao, W.; Godfrey, A.; Mitsuhara, M.; Morris, J., Jr.; Minor, A.M. Grain refinement in titanium prevents low temperature oxygen embrittlement. *Nat. Commun.* **2023**, *14*, 404. [CrossRef]
7. Reitz, J.; Lochbichler, C.; Friedrich, B. Recycling of gamma titanium aluminide scrap from investment casting operations. *Intermetallics* **2011**, *19*, 762–768. [CrossRef]
8. Kim, T.; Oh, J.-M.; Cho, G.-H.; Park, J.; Lim, J.-W. Comparison of deoxidation capability of solid solution and intermetallic titanium alloy powders deoxidized by calcium vapor. *J. Alloys Compd.* **2020**, *828*, 154220. [CrossRef]
9. Iizuka, A.; Ouchi, T.; Okabe, T.H. New Deoxidation Method of Titanium Using Metal Filter in Molten Salt. *Metall. Mater. Trans. B* **2022**, *53*, 1371–1382. [CrossRef]
10. Kim, T.; Lim, J.W. Synthesis of low–oxygen Ti_3AlC_2 powders by hydrogenation–dehydrogenation and deoxidation from titanium scraps. *J. Am. Ceram. Soc.* **2023**, *106*, 7311–7321. [CrossRef]
11. Takeda, O.; Ouchi, T.; Okabe, T.H. Recent progress in titanium extraction and recycling. *Metall. Mater. Trans. B* **2020**, *51*, 1315–1328. [CrossRef]
12. Zhang, Y.; Fang, Z.Z.; Xia, Y.; Sun, P.; Van Devener, B.; Free, M.; Lefler, H.; Zheng, S. Hydrogen assisted magnesiothermic reduction of TiO_2. *Chem. Eng. J.* **2017**, *308*, 299–310. [CrossRef]
13. Suzuki, R.O.; Fukui, S. Reduction of TiO_2 in molten $CaCl_2$ by Ca deposited during CaO electrolysis. *Mater. Trans.* **2004**, *45*, 1665–1671. [CrossRef]
14. Reddy, R.G.; Shinde, P.S.; Liu, A. Review—The Emerging Technologies for Producing Low-Cost Titanium. *J. Electrochem. Soc.* **2021**, *168*, 042502. [CrossRef]
15. Dring, K.; Dashwood, R.; Inman, D. Predominance diagrams for electrochemical reduction of titanium oxides in molten $CaCl_2$. *J. Electrochem. Soc.* **2005**, *152*, 184. [CrossRef]
16. Su, Y.; Wang, L.; Luo, L.; Jiang, X.; Guo, J.; Fu, H. Deoxidation of titanium alloy using hydrogen. *Int. J. Hydrogen Energy* **2009**, *34*, 8958–8963. [CrossRef]
17. Gökelma, M.; Celik, D.; Tazegul, O.; Cimenoglu, H.; Friedrich, B. Characteristics of Ti6Al4V powders recycled from turnings via the HDH technique. *Metals* **2018**, *8*, 336. [CrossRef]
18. Wang, L.; Su, Y.; Wang, S.; Luo, L.; Fu, H. Effect of melt hydrogenation on structure and hardness of TC_{21} alloy. *Rare. Metal Mat. Eng.* **2011**, *40*, 321–324.
19. Bartosinski, M.; Hassan-Pour, S.; Friedrich, B.; Ratiev, S.; Ryabtsev, A. Deoxidation Limits of Titanium Alloys during Pressure Electro Slag Remelting. *Mater. Sci. Eng. Conf. Ser.* **2016**, *143*, 12009. [CrossRef]
20. Cheng, C.; Dou, Z.H.; Zhang, T.A.; Zhang, H.J.; Su, J.M. Synthesis of As-Cast Ti-Al-V Alloy from Titanium-Rich Material by Thermite Reduction. *JOM* **2017**, *69*, 1818–1823. [CrossRef]
21. Brillo, J.; Egry, I.; Novakovic, R. Surface tension of liquid Cu-Ti binary alloys measured by electromagnetic levitation and thermodynamic modelling. *Appl. Surf. Sci.* **2011**, *257*, 7739–7745.
22. Yan, P.; Zhang, G.; Yi, B.; Mclean, A. Kinetic characterization of phosphorus removal from the surface of metallurgical grade (MG) silicon droplets during electromagnetic levitation. *Int. J. Heat Mass Trans.* **2023**, *211*, 124200. [CrossRef]
23. Vache, N.; Cadoret, Y.; Dod, B.; Monceau, D. Modeling the oxidation kinetics of titanium alloys: Review, method and application to Ti-64 and Ti-6242s alloys. *Corros. Sci.* **2021**, *178*, 109041. [CrossRef]
24. Brandes, E.A.; Brook, G.B. *Smithells Metals Reference Book*; Butterworth-Heinemann: Oxford, UK, 1983.
25. Baehr, H.D. *Thermochemical Properties of Inorganic Substances*; Springer: Berlin/Heidelberg, Germany, 1992; Volume 58, p. 103.
26. Ono, K.; Moriyama, J. Deoxidation of high-melting-point metals and alloys in vacuum. *Metall. Mater. Trans. B* **1982**, *13*, 241–249. [CrossRef]
27. Oh, J.-M.; Seo, J.-H.; Lim, J.-W. Refining effect of TiAl intermetallic compounds prepared by hydrogen plasma arc melting from scraps of Ti–Al mixture. *Jpn. J. Appl. Phys.* **2019**, *59*, SAAB07.
28. Chase, M.W. NIST-JANAF thermochemical tables. *J. Phys. Chem. Ref. Data* **1998**, *9*, 1529–1564.
29. Sabat, K.C.; Murphy, A.B. Hydrogen plasma processing of iron ore. *Metall. Mater. Trans. B* **2017**, *48*, 1561–1594. [CrossRef]
30. Song, Y.; Dou, Z.; Zhang, T.; Liu, Y. A novel continuous and controllable method for fabrication of as-cast TiAl alloy. *J. Alloys Compd.* **2019**, *789*, 266–275. [CrossRef]

31. Dong, G.; You, X.; Xu, Z.; Wang, Y.; Tan, Y. A new model for studing the evaporation behavior of alloy elements in DD98M alloy during electron beam smelting. *Vacuum* **2022**, *195*, 110641. [CrossRef]
32. Schuster, J.C.; Palm, M. Reassessment of the binary Aluminum-Titanium phase diagram. *J. Phase Equilib. Diff.* **2006**, *27*, 255–277. [CrossRef]

Disclaimer/Publisher's Note: The statements, opinions and data contained in all publications are solely those of the individual author(s) and contributor(s) and not of MDPI and/or the editor(s). MDPI and/or the editor(s) disclaim responsibility for any injury to people or property resulting from any ideas, methods, instructions or products referred to in the content.

Article

Effects of Mean Normal Stress and Microstructural Properties on Deformation Properties of Ultrahigh-Strength TRIP-Aided Steels with Bainitic Ferrite and/or Martensite Matrix Structure

Koh-ichi Sugimoto [1,*], Shoya Shioiri [1] and Junya Kobayashi [2]

1. Graduate School of Science and Technology, Shinshu University, Nagano 380-8553, Japan; mellow.pretty8330@gmail.com
2. Graduate School of Science and Engineering, Ibaraki University, Hitachi 316-8511, Japan; junya.kobayashi.jkoba@vc.ibaraki.ac.jp
* Correspondence: btxpr049@ymail.ne.jp

Abstract: The effects of mean normal stress on the deformation properties such as the strain-hardening, strain-induced martensite transformation, and micro-void initiation behaviors of low-carbon ultrahigh-strength TRIP-aided bainitic ferrite (TBF), bainitic ferrite/martensite (TBM), and martensite (TM) steels were investigated to evaluate the various cold formabilities. In addition, the deformation properties were related to the microstructural properties such as the matrix structure, retained austenite characteristics, and second-phase properties. Positive mean normal stress considerably promoted strain-induced martensite transformation and micro-void initiation, with an increased strain-hardening rate in an early strain range in all steels. In TM steel, the primary martensite matrix structure suppressed the micro-void initiation through high uniformity of a primary martensite matrix structure and a low strength ratio, although the strain-induced transformation was promoted, and a large amount of martensite/austenite constituent or phase was contained. A mixed matrix structure of bainitic ferrite/primary martensite in TBM steel also suppressed the micro-void initiation because of the refined microstructure and relatively stable retained austenite. Promoted micro-void initiation of TBF steel was mainly promoted by a high strength ratio.

Keywords: advanced ultrahigh-strength steel; TRIP-aided steel; deformation property; mean normal stress; microstructural property

Citation: Sugimoto, K.-i.; Shioiri, S.; Kobayashi, J. Effects of Mean Normal Stress and Microstructural Properties on Deformation Properties of Ultrahigh-Strength TRIP-Aided Steels with Bainitic Ferrite and/or Martensite Matrix Structure. *Materials* **2024**, *17*, 3554. https://doi.org/10.3390/ma17143554

Academic Editors: Chih-Chun Hsieh, Seong-Ho Ha, Young-Ok Yoon, Dong-Earn Kim and Young-Chul Shin

Received: 28 May 2024
Revised: 12 July 2024
Accepted: 16 July 2024
Published: 18 July 2024

Copyright: © 2024 by the authors. Licensee MDPI, Basel, Switzerland. This article is an open access article distributed under the terms and conditions of the Creative Commons Attribution (CC BY) license (https://creativecommons.org/licenses/by/4.0/).

1. Introduction

The third-generation advanced ultrahigh- and high-strength steels (AHSSs) have been developed for lightening automobiles and improving crash safety [1–3]. The third-generation AHSSs are classified into the following three groups.

"Group L": TRIP-aided bainitic ferrite (TBF) steel [4,5], carbide-free bainitic (CFB) steel [6,7], and duplex-type medium Mn (D-MMn) steel [8,9].

"Group M": TRIP-aided bainitic ferrite/martensite (TBM) steel [10] and quenching and partitioning (Q&P) steel [11,12].

"Group H": TRIP-aided martensite (TM) steel [10] and martensite-type medium Mn (M-MMn) steel [13].

Any group of AHSSs contains a certain amount of metastable-retained austenite. The AHSSs of "Group L" have a tensile strength of less than 1.0 GPa and a matrix structure of bainitic ferrite. Only D-MMn steel has an annealed martensite matrix structure, in the same way as the first-generation AHSS such as TRIP-aided annealed martensite (TAM) steel [14]. On the other hand, the AHSSs of "Group H" have a tensile strength of higher than 1.5 GPa and a matrix structure of primary martensite. The AHSSs of "Group M" have a tensile strength between "Group L" and "Group H" steels and a mixed matrix microstructure of bainitic ferrite and primary martensite. The TBF, TBM, and TM steels possess extremely high cold stretch formability (maximum stretch height: H_{max}) and stretch-flangeability (hole expansion ratio: HER), as shown in Figure 1 [8,10,13,14], in the same way as CFB [6,7] and Q&P steels [11,12]. The excellent formabilities are mainly associated with deformation properties such as high strain hardening, suppressed strain-induced martensite transformation, and suppressed micro-void initiation behaviors. As the formalities are measured under a different stress state or mean normal stress, it is essential to understand the effect of the mean normal stress on the deformation properties in the TBF, TBM, and TM steels, as well as the effect of microstructural properties.

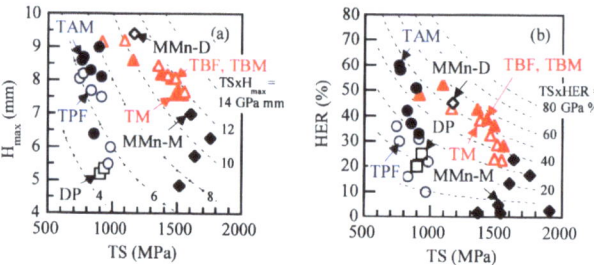

Figure 1. (a) Maximum stretch height (H_{max}), tensile strength (TS), and (b) hole expansion ratio (HER); TS relations at room temperature in low-carbon 0.15C-0.25Si-1.70Mn ferrite-martensite dual-phase (DP) steel [14], 0.2C-(1.0-2.5)Si-(1.0-2.0)Mn TRIP-aided polygonal ferrite (TPF) and TRIP-aided annealed martensite (TAM) steels [14], 0.20C-1.5Si-1.5Mn-0.05Nb TRIP-aided bainitic ferrite (TBF), bainitic ferrite/martensite (TBM), and martensite (TM) steels [10], and 0.21C-1.50Si-4.94Mn duplex type (D-MMn) and martensite-type medium Mn (M-MMn) steels [8,13]. This figure is redrawn by using the results of Refs. [8,10,13,14]. The DP, TPF, and TAM steels belong to the first-generation AHSSs.

This research investigates the effect of the mean normal stress on the deformation properties at room temperature to evaluate the different cold formabilities in the low-carbon TBF, TBM, and TM steels. In addition, the deformation properties were related to the microstructural properties such as matrix structure, retained austenite characteristics, and martensite/austenite constitute (MA phase) properties.

2. Material and Methods

A steel slab of 100 kg with the chemical composition listed in Table 1 was manufactured using laboratory-based vacuum melting and then air cooling. Then, the slab was hot rolled to a 13 mm diameter at a finish temperature of 850 °C. Tensile specimens, torsional specimens, and compressive specimens, shown in Figure 2, were machined from the hot-rolled bars. To produce TBF, TBM, and TM steels with bainitic ferrite, bainitic ferrite/primary martensite mixture, and primary martensite matrix structures, respectively, three kinds of isothermal transformation (IT) treatments were carried out after austenitizing (Figure 3). These IT temperatures (T_{IT}) are higher than M_s, between M_s and M_f, and lower than M_f, respectively, at which the maximum retained austenite fractions are achieved.

Table 1. Chemical composition (mass%) and measured martensite-start (M_S, °C) and -finish temperatures (M_f, °C) of a steel slab.

C	Si	Mn	P	S	Al	Nb	Cr	Mo	N	Fe	M_s	M_f
0.18	1.48	1.49	0.004	0.003	0.043	0.05	1.02	0.20	0.001	bal.	407	292

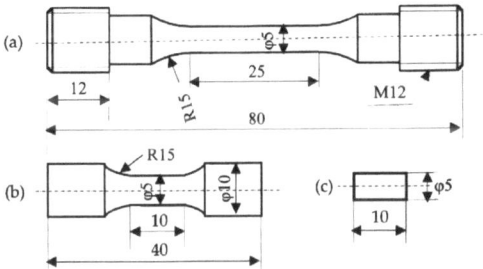

Figure 2. Dimensions (unit: mm) of (**a**) tensile, (**b**) torsional, and (**c**) compressive specimens.

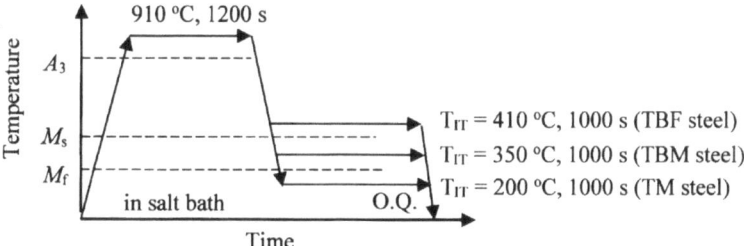

Figure 3. Heat treatment diagrams to produce TBF, TBM, and TM steels. The heat treatment was carried out in salt and oil baths. T_{IT}: isothermal transformation temperature; O.Q: quenching in oil at 50 °C.

The microstructure of the steels was observed at a 1/2 radius and a 1/2 height of the specimens before and after deformation using field-emission scanning electron microscopy (FE-SEM; JSM-6500F, JEOL Ltd., Akishima, Tokyo, Japan), which was performed using an instrument equipped with an electron backscatter diffraction system (EBSD; OIM system, TexSEM Laboratories, Inc., Prova, UT, USA). The beam area, beam diameter, beam step size, and acceleration voltage of the EBSD analysis were 40×40 μm^2, 1.0 μm, 0.15 μm, and 25 kV, respectively. The specimens for the FE-SEM–EBSD analysis were first ground with emery paper (#1200), alumina powder, and colloidal silica. Finally, ion thinning was carried out.

The retained austenite characteristics of the steels were quantified using an X-ray diffractometer (RINT2000, Rigaku Co., Akishima, Tokyo, Japan). The surfaces of the specimens were electropolished after being ground with emery paper (#1200). The volume fraction of the retained austenite phase (f_γ, vol.%) was calculated from the integrated intensity of the (200)α, (211)α, (200)γ, (220)γ, and (311)γ peaks obtained with X-ray diffractometry using Mo-Kα radiation [15]. The carbon concentration in the retained austenite (C_γ, mass%) was estimated from the lattice constant of the (200)γ, (220)γ, and (311)γ peaks of the Cu-Kα radiation and the empirical equation proposed by Dyson and Holmes [16]. The X-ray half-width (HW) of (211)α peak of the Cu-Kα radiation was measured to relate to the equivalent plastic strain ($\bar{\varepsilon}_p$) [17]. The above X-ray characteristics were measured using three or more samples.

The Vickers hardness (HV0.1) was measured using a Vickers microhardness tester (Shimadzu Co., DUH-201H, Kyoto, Japan) with a load of 0.98 N. Micro-void initiation behavior was observed by FE-SEM.

To develop various mean normal stress states, uniaxial tensile and compressive tests were conducted on a universal testing instrument (AD-10TD, Shimadzu Co., Kyoto, Japan) at 25 °C and a crosshead speed of 10 mm/min. Torsional tests were carried out on a torsion testing machine (AG-300kNXplus, Shimadzu Co., Kyoto, Japan) at 25 °C and a torsion rate of 10 deg./min. Three or more specimens were prepared to measure the mechanical properties.

Mean normal stress was defined by

$$\sigma_m = (\sigma_1 + \sigma_2 + \sigma_3)/3 \quad (1)$$

where σ_1, σ_2, and σ_3 are principal stresses, respectively. An equivalent stress $\bar{\sigma}$ and equivalent strain $\bar{\varepsilon}$ were calculated using von Mises criterion [18], as follows,

$$\bar{\sigma} = 1/\sqrt{2} \cdot \left[(\sigma_x - \sigma_y)^2 + (\sigma_y - \sigma_z)^2 + (\sigma_z - \sigma_x)^2 + 6\left(\tau_{xy}^2 + \tau_{yz}^2 + \tau_{xz}^2\right) \right]^{1/2} \quad (2)$$

$$\bar{\varepsilon} = \sqrt{2/3} \times \left[(\varepsilon_x - \varepsilon_y)^2 + (\varepsilon_y - \varepsilon_z)^2 + (\varepsilon_z - \varepsilon_x)^2 + 3/2 \times \left(\gamma_{xy}^2 + \gamma_{yz}^2 + \gamma_{zx}^2\right) \right]^{1/2} \quad (3)$$

where σ_i, ε_i and τ_{ij}, and γ_{ij} (i, j = x, y, z) represent normal stress, shear stress, normal strain, and shear strain in the X–Y–Z coordinate system, respectively. For tension tests, $\bar{\sigma}$ and $\bar{\varepsilon}$ of the necking region were calculated by

$$\bar{\sigma} = \frac{P}{\pi d^2 \left(1 + \frac{2R}{d}\right) ln\left(1 + \frac{d}{2R}\right)} \quad (4)$$

$$\bar{\varepsilon} = 2 \ln \frac{d_0}{d} \quad (5)$$

where P, d, d_0, and R are an applied load, a minimum diameter of the neck cross-section, an initial diameter of the specimen, and a radius of curvature of the neck profile, respectively [19].

3. Results

3.1. Microstructural Properties

Figure 4 shows the microstructures of TBF, TBM, and TM steels observed in terms of FE-SEM-EBSD. In the same way as Ref. [10], the matrix microstructures of TBF, TBM, and TM steels are bainitic ferrite, bainitic ferrite/primary martensite, and primary martensite, respectively. As the second phase, carbon-enriched soft-retained austenite and a hard MA phase are contained in these steels. The martensite in the MA phase is carbon-enriched secondary martensite. TBM steel's primary martensite fraction ($f\alpha_m$) can be estimated to be about 30 vol% using the following equation [20]:

$$f\alpha_m = 1 - \exp[-1.1 \times 10^{-2}(M_s - T_{IT})]. \quad (6)$$

The initial-retained austenite fractions of TBF, TBM, and TM steels are $f\gamma_0$ = 11.4, 7.2, and 5.5 vol.%, and the initial carbon concentrations are $C\gamma_0$ = 0.65, 1.08 and 0.45 mass%, respectively (Table 2). The products of $f\gamma_0$ and $C\gamma_0$ of TBF, TBM, and TM steels are 0.074, 0.078, and 0.024, respectively. The product of TM steel is considerably low. Most of the retained austenite is located along the lath boundaries of bainitic ferrite and primary martensite and in the MA phase, in the same way as Ref. [10]. The volume fractions of the MA phase are f_{MA} = 2.0, 10.8, and 15.8 vol.% in TBF, TBM, and TM steels, respectively. Most of the MA phase mainly exists along these steels' prior austenitic grain boundary and the lath boundaries of bainitic ferrite and primary martensite. The largest size of the MA phase is produced in TM steel. It seems that the prior austenitic grain size is the same for all three steels because of the same austenitizing temperature.

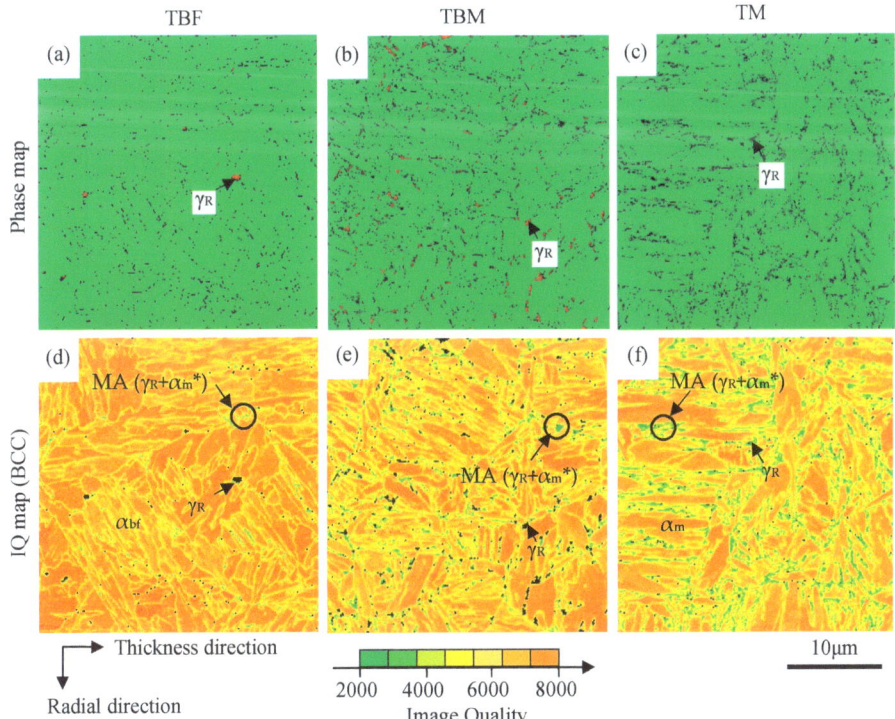

Figure 4. (a–c) Phase maps of BCC and FCC and (d–f) image quality (IQ) maps of BCC in TBF, TBM, and TM steels. α_{bf}, α_m, α_m^*, γ_R, and MA are bainitic ferrite, primary martensite, secondary martensite, retained austenite, and martensite/austenite (MA) phase, respectively.

Table 2. Retained austenite characteristics and martensite/austenite constituent (MA phase) properties of TBF, TBM, and TM steels.

Steel	$f\gamma_0$ (vol.%)	$C\gamma_0$ (mass%)	k Tension	k Torsion	k Comp.	$\Delta f\alpha_m$ Tension	$\Delta f\alpha_m$ Torsion	$\Delta f\alpha_m$ Comp.	f_{MA} (vol.%)	HV0.1
TBF	11.4 ± 1.2	0.65 ± 0.14	1.64	1.44	1.24	9.2	6.6	6.0	2.0 ± 0.3	350
TBM	7.2 ± 1.4	1.08 ± 0.22	2.05	2.41	0.68	5.1	5.5	2.4	10.8 ± 1.2	405
TM	5.5 ± 1.5	0.45 ± 0.20	2.84	5.08	0.80	4.3	4.5	2.1	15.8 ± 1.8	422

$f\gamma_0$: retained austenite fraction, $C\gamma_0$: carbon concentration of retained austenite, k: strain-induced transformation factor, $\Delta f\alpha_m$: strain-induced martensite fraction, f_{MA}: volume fraction of MA phase. The k-values were calculated in an equivalent plastic strain range of $\bar{\varepsilon}_p$ = 0 to 0.6. The k-value and $\Delta f\alpha_m$ of TM steel deformed in torsion were decided between $\bar{\varepsilon}_p$ = 0 and 0.3, HV0.1: Vickers hardness.

Vickers hardnesses of TBF, TBM, and TM steels are HV0.1 = 350, 405, and 422, respectively (Table 2). The HV0.1 of TBM steel is slightly lower than that of TM steel.

3.2. Strain-Hardening Behavior
3.2.1. Flow Stress, Mechanical Properties, and Strain-Hardening

Figure 5 shows the flow curves of TBF, TBM, and TM steels deformed in tension, torsion, and compression. The mechanical properties are shown in Table 3. In tensile deformation, the tensile yield stress (YS) and tensile strength (TS) of TM steel are the highest. TBF steel possesses the largest uniform (UEl) and total (TEl) elongations. This is associated with high strain hardening in a large strain range. TBM and TM steels also have relatively large total elongations, although uniform elongations are considerably lower than that of TBF steel. It is noteworthy that the reductions of area (RAs) of TBM and TM steels are higher than that of TBF steel.

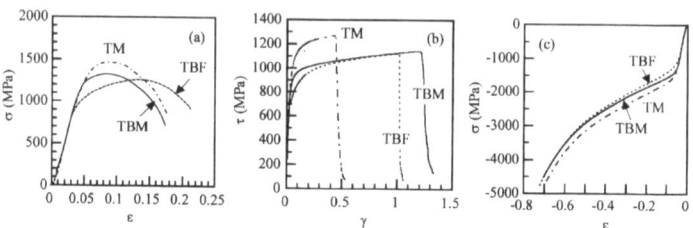

Figure 5. (a) Engineering tensile stress–strain (σ–ε) curves, (b) shear stress–strain (τ–γ) curves, and (c) compressive stress–strain curves (σ–ε) in TBF, TBM, and TM steels.

Table 3. Mechanical properties of TBF, TBM, and TM steels.

Steel	YS	TS	UEl	TEl	RA	τ_0	τ_{max}	σ_0
	(MPa)	(MPa)	(%)	(%)	(%)	(MPa)	(MPa)	(MPa)
TBF	709 ± 15	1276 ± 18	9.0 ± 0.8	17.7 ± 2.3	49.5 ± 2.8	932 ± 24	1981 ± 41	937 ± 14
TBM	1058 ± 35	1310 ± 38	3.8 ± 0.5	14.7 ± 3.4	69.9 ± 3.7	1206 ± 37	2021 ± 57	1125 ± 28
TM	1073 ± 46	1463 ± 52	4.5 ± 1.0	14.6 ± 3.8	63.5 ± 4.2	1251 ± 45	2174 ± 63	1227 ± 37

YS: tensile yield stress, TS: tensile strength, UEl: uniform elongation, TEl: total elongation, RA: reduction of area, τ_0: torsional shear yield stress, τ_{max}: torsional maximum shear stress, σ_0: compressive yield stress.

In torsional deformation, TM steel has the highest torsional shear yield stress (τ_0) and maximum shear stress (τ_{max}), with the lowest total shear strain. Differing from tensile deformation, TBM steel has a higher total shear strain than TBF steel, with just higher shear stress than TBF steel.

In compressive deformation, TM steel has higher compressive yield stress (σ_0) and flow stress than those of TBF and TBM steels, in the same way as the tensile and torsional deformations. The compressive flow stress of TBM steel is slightly higher than that of TBF steel.

Figure 6 shows the equivalent stress–equivalent plastic strain ($\bar{\sigma}$ – $\bar{\varepsilon}_p$) curves of TBF, TBM, and TM steels. Notably, the equivalent strain-hardening rate in an early strain range is higher in tension, compared to in torsion and compression. The $\bar{\sigma}$ – $\bar{\varepsilon}_p$ curves in torsion tend to be higher than those in tension and compression, in all steels, in the same way as previously presented TRIP-aided polygonal ferrite (TPF) and TAM steels [21]. This is because the von Mises criterion was applied to calculate the $\bar{\sigma}$ – $\bar{\varepsilon}_p$ curves.

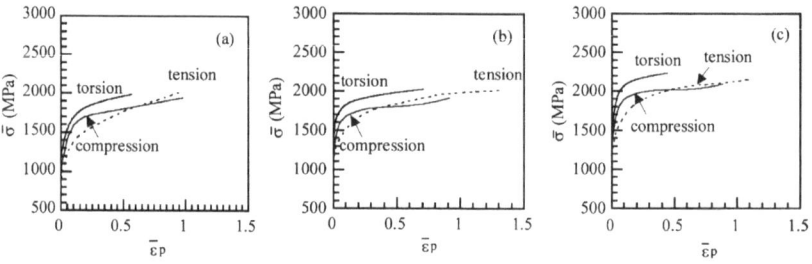

Figure 6. Equivalent stress–equivalent plastic strain ($\bar{\sigma}$–$\bar{\varepsilon}_p$) curves of (a) TBF, (b) TBM, and (c) TM steels plastically deformed in tension, torsion, and compression.

3.2.2. X-ray Half-Width and Equivalent Plastic Strain Relation

Figure 7 shows the X-ray half-width and equivalent plastic strain (HW–$\bar{\varepsilon}_p$) relations in TBF, TBM, and TM steels plastically deformed in tension, torsion, and compression. The half-width linearly increases with increasing equivalent plastic strain in all steels. When the half-width characteristics are quantified by the half-width at $\bar{\varepsilon}_p = 0$ (HW$_0$) and the slope of the straight line (n-value), TM steel has the highest HW$_0$ (0.68 deg) and the smallest n-value (0.04), as shown in Figure 8. On the other hand, TBF steel exhibits the lowest HW$_0$ (0.58 deg.) and the largest n-value (0.133). In this case, the n-value of TBF steel was measured in an equivalent strain range below $\bar{\varepsilon}_p = 0.6$. The HW$_0$ and n-values of TBM steel are between those of TBF and TM steels. These HW$_0$s and n-values are higher and lower than those of TPF and TAM steels, respectively [21]. Linear relationships in Figure 7 agree with a modified Williamson–Hall equation [22,23]. The HW$_0$ and n-value may be correlated with the yield stress and strain-hardening rate in all steels, respectively. The relation between these HW$_0$ and n-value and valuables in the modified Williamson–Hall equation will be investigated in the future.

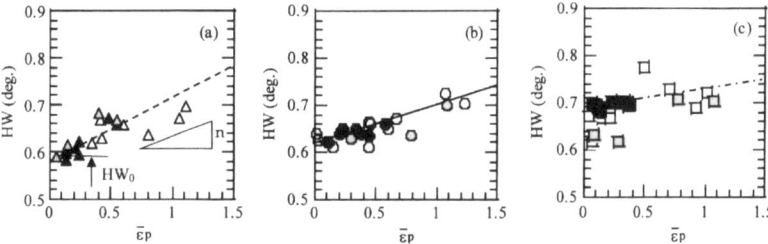

Figure 7. Relationship between X-ray half-width (HW) and equivalent plastic strain ($\bar{\varepsilon}_p$) in (**a**) TBF (triangle marks), (**b**) TBM (circle marks), and (**c**) TM (square marks) steels plastically deformed in tension (open marks), torsion (black solid marks), and compression (gray solid marks).

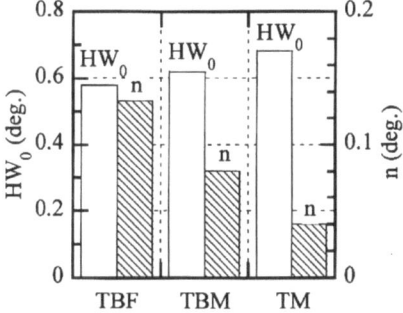

Figure 8. Comparison of HW$_0$ and n-value in TBF, TBM, and TM steels. HW$_0$ and n-value of TBF steel are decided in an equivalent plastic strain range of $\bar{\varepsilon}_p = 0$ to 0.6.

3.3. Strain-Induced Martensite Transformation Behavior

Figure 9 shows the variations in the volume fraction of untransformed retained austenite as a function of equivalent plastic strain in TBF, TBM, and TM steels. The strain-induced martensite transformation is most suppressed in compressive deformation in all steels. Tensile deformation promotes the strain-induced martensite transformation, especially in TBF steel. In TBM and TM steels, the strain-induced martensite transformation in tension is slightly promoted compared to in torsion (or zero mean normal stress).

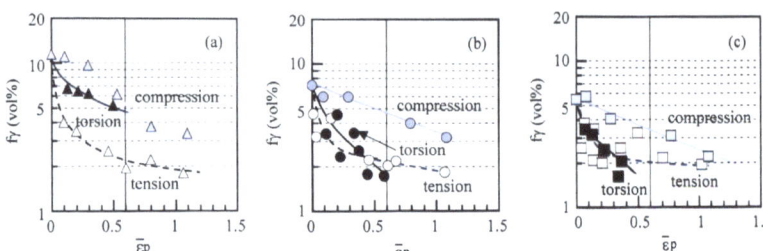

Figure 9. Variations in volume fraction of untransformed retained austenite (*f*γ) as a function of equivalent plastic strain ($\bar{\varepsilon}_p$) in (**a**) TBF (triangle marks), (**b**) TBM (circle marks), and (**c**) TM (square marks) steels. Open marks: tension; black solid marks: torsion; gray solid marks: compression.

Figure 10 shows the relationships between the *k*-value (the strain-induced transformation factor [10]) and mean normal stress and between strain-induced martensite fraction and mean normal stress in TBF, TBM, and TM steels. The *k*-value means the mechanical stability of the retained austenite and is defined by

$$k = (\ln f\gamma_0 - \ln f\gamma)/\bar{\varepsilon}_P \qquad (7)$$

where *f*γ is the volume fraction of retained austenite in the steels subjected to an equivalent plastic strain of $\bar{\varepsilon}_p$. When the *k*-values were measured in a range of $\bar{\varepsilon}_P = 0$ and 0.6 (Figure 9), the *k*-value approximately increases with increasing mean normal stress in all steels (Figure 10a). In this case, the *k*-value of TM steel deformed in torsion is decided between $\bar{\varepsilon}_P = 0$ and 0.3 (Table 2), because the TM steel fractured at the equivalent plastic strain below $\bar{\varepsilon}_P = 0.6$ (Figure 6c). The *k*-values in tension and torsion in TM steel are higher than those in TBF and TBM steels. On the other hand, when the strain-induced martensite fraction ($\Delta f\alpha_m$) is defined by a difference between $f\gamma_0$ and $f\gamma$ at $\bar{\varepsilon}_P = 0.6$ (Figure 9), the strain-induced martensite fraction increases with increasing mean normal stress (Figure 10b). In addition, TBF steel has the highest strain-induced martensite fraction. On the other hand, TM steel exhibits the smallest strain-induced martensite fraction. The strain-induced martensite fraction of TBM steel is slightly higher than that of TM steel.

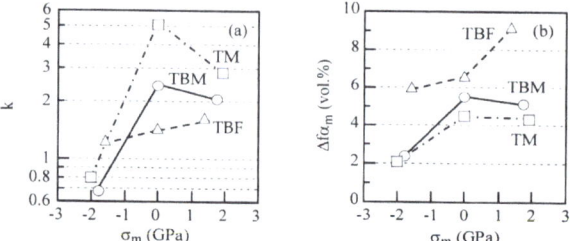

Figure 10. Variations in (**a**) *k*-value and (**b**) strain-induced martensite fraction ($\Delta f\alpha_m$) as a function of mean normal stress (σ_m) in TBF (triangle marks), TBM (circle marks), and TM (square marks) steels. The *k*-value was calculated in an equivalent plastic strain range of $\bar{\varepsilon}_p = 0$ to 0.6. The *k*-value and $\Delta f\alpha_m$ of TM steel deformed in torsion are decided between $\bar{\varepsilon}_p = 0$ and 0.3.

3.4. Micro-Void Initiation Behavior

Figure 11 shows FE-SEM images of micro-voids in TBF, TBM, and TM steels plastically deformed to an equivalent plastic strain of $\bar{\varepsilon}_p = 0.6$. In this case, micro-voids larger than 0.5 μm are counted. It is found that no micro-voids are formed on compressive deformation. The most frequent micro-void initiation occurs by plastic deformation in tension. Many micro-voids initiate at the lath boundaries of bainitic ferrite and primary martensite, at the interfaces of the MA phase/matrix structure, and at the strain-induced martensite/matrix structure in all steels.

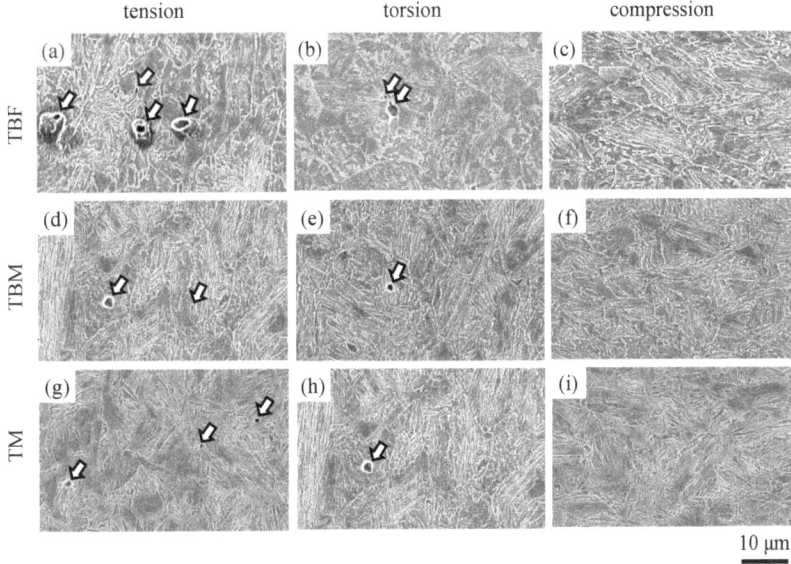

Figure 11. FE-SEM images of micro-voids initiated in (**a**–**c**) TBF, (**d**–**f**) TBM, and (**g**–**i**) TM steels plastically deformed to $\bar{\varepsilon}_p = 0.6$ in (**a**,**d**,**g**) tension, (**b**,**e**,**h**) torsion, and (**c**,**f**,**i**) compression. Arrows denote the micro-voids.

Figure 12 shows the variations in the mean size (D_v) and mean number per unit area (N_v) of micro-voids as a function of mean normal stress in TBF, TBM, and TM steels plastically deformed to $\bar{\varepsilon}_p = 0.6$. The mean size and mean number of micro-voids increase with increasing mean normal stress. TBF steel has the maximum mean size and mean number of micro-voids. TBM steel shows the minimum mean size of micro-voids, and TM steel shows the minimum mean number of micro-voids, although the differences in the mean size and mean number between TBM and TM steels are small. Toji et al. [12] also showed a similar result using 0.19C-1.5Si-2.9Mn Q&P and TBF steels. Namely, in the Q&P steel with a mixed-matrix structure of bainitic ferrite and primary martensite, the micro-void initiation in tension was suppressed in comparison with that of TBF steel.

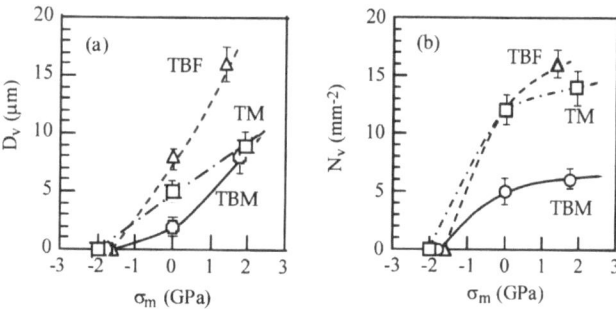

Figure 12. Variations in (**a**) the mean size (D_v) and (**b**) mean number of per unit area (N_v) of micro-voids are a function of mean normal stress (σ_m) in TBF, TBM, and TM steels subjected to an equivalent plastic strain of $\bar{\varepsilon}_p = 0.6$.

4. Discussion

In general, the flow stress of TRIP-aided steel can be decided by the sum of the following items (i) to (iv) [21]:

(i) "Flow stress of matrix structure", including strain hardening.
(ii) "Long-range internal stress hardening", which results from the difference in plastic strain between the matrix structure and second phase (retained austenite, strain-induced martensite, MA phase, etc.) [24].
(iii) "Strain-induced transformation hardening", which results from an increase in strain-induced martensite fraction. The transformation also relaxes the localized stress concentration through an expansion strain [25]. In an early stage, the expansion strain brings on an initial yielding or continuous yielding.
(iv) "Forest dislocation hardening", which is estimated by the Ashby equation [26].

On the other hand, the micro-void initiation behavior is controlled by [10,14].

(v) "Matrix structure": acicular or lath-type structure suppresses the void formation, compared to granular structure, by refining the structure size [10,12,14].
(vi) "Retained austenite characteristics": a large amount of mechanically stable retained austenite suppresses the micro-void initiation due to the plastic relaxation by expansion strain on the strain-induced martensite transformation [10,12,14].
(vii) "A strength ratio" or a ratio of the second phase strength to the matrix structure strength: a high strength ratio increases the localized stress concentration and promotes void initiation at the matrix/second phase interface. Carbon-enriched strain-induced martensite enhances the strength ratio [12,27,28].

In the following, the effects of mean normal stress and microstructural properties on the deformation properties such as the strain hardening, strain-induced martensite transformation, and micro-void initiation behaviors in TBF, TBM, and TM steels are discussed considering items (i) to (vii).

4.1. Effect of Mean Normal Stress on Deformation Properties

In Figure 10, the k-values and $\Delta f \alpha_m$ increased with increasing mean normal stress in all steels, although these levels differed in each steel. According to Hiwatashi et al. [29], stretch forming (an equi-biaxial tension or positive mean normal stress) significantly enhanced the strain-induced martensite transformation of the retained austenite in 0.11C-1.18Si-1.55Mn TPF steel. On the other hand, shrink flanging (a compression or negative mean normal stress) suppressed the strain-induced martensite transformation, and uniaxial tension slightly suppressed the strain-induced martensite transformation, compared to the stretch forming. They explained these results as follows. A positive mean normal stress assists the strain-induced martensite transformation because of the expansion stress or strain. Kawata et al. [19] also reported that compression stress suppressed the strain-induced transformation compared to tensile stress in 0.1C-1.2Si-1.5Mn and 0.2C-1.2Si-2.0Mn TPF steels. Therefore, high k-values and $\Delta f \alpha_m$ under positive mean normal stress (Figure 10) are considered to be caused by high expansion stress or strain as proposed by Hiwatashi et al. [29].

As shown in Figure 12, micro-void initiation was promoted by a positive mean normal stress in all steels, particularly in TBF steel. This is because the positive mean stress originates the expansion stress which facilitates the micro-void initiation at the matrix/second phase interface, in the same way as the above strain-induced martensite transformation.

In Figure 6, the equivalent strain-hardening rate in an early strain range of tensile deformation was much higher than those of torsional and compressive deformation in all steel. The k-value in tension was higher than those in torsion and in compression in all steels (Figure 10a). Considering items (i) to (iv), the high equivalent strain-hardening behavior in tension may be associated with easy strain-induced martensite transformation, which promotes initial or continuous yielding [25], although the strain-induced martensite fraction differed in each steel (Figure 10b).

4.2. Effects of Microstructural Properties on Deformation Properties

In Figure 5a, TBF steel had a high strain-hardening rate in a large strain range, compared to TBM and TM steels. Considering items (i) to (iv) and the results of Figure 10, the high strain-hardening rate may be mainly associated with the large strain-induced martensite transformation hardening due to a large amount of mechanically stable retained austenite, with a contribution of the high strain hardening of its bainitic ferrite matrix. It is considered that a large uniform elongation of TBF steel is associated with large strain hardening in a large strain range. On the other hand, TM steel exhibited a high flow stress and high strain-hardening rate in an early strain range (Figure 5a). This is considered to be mainly associated with the high initial strain-hardening rate due to (1) the high dislocation density of primary martensite matrix structure and (2) the high long-range internal stress hardening due to high MA phase fraction with a contribution of early strain-induced martensite transformation hardening. TBM steel had the intermediate flow stress and strain-hardening rate between TBF and TM steels. Notably, the refined matrix structure of TBM steel contributes to an increase in the flow stress, total elongation, and reduction of area.

In Figure 10, the k-values in tension and torsion of TM steel were higher than those of TBF and TBM steels. Generally, retained austenite stability is controlled by the carbon concentration, size, morphology, and matrix structure surrounding the retained austenite. So the high k-values of TM steel may be caused by the low carbon concentration of retained austenite (Table 2) and the high flow stress of the primary martensite matrix structure, although the size was relatively small, and the morphology was filmy. In this case, retained austenite stability in the MA phase is relatively high, because most of the retained austenite is surrounded by harder secondary martensite and is highly carbon-enriched compared to the retained austenite at the primary martensite lath boundary.

The mean size and mean number per unit area of micro-voids were the largest in TBF steel (Figure 12). On the other hand, TBM showed the minimum mean size of micro-voids, and TM steel showed the minimum mean number of micro-voids, although the differences between TBM and TM steels were small. Considering items (v) to (vii), the easy micro-void initiation behavior of TBF steel may be associated with a relatively coarse soft matrix structure and a high strength ratio resulting from a large quantity of strain-induced martensite, although the strain-induced martensite transformation plays a role in lowering the localized stress concentration at the matrix structure/second phase interface. The suppressed micro-void initiation behavior of TBM steel is considered to be related to the refined mixed-matrix structure of bainitic ferrite/primary martensite and relatively stable retained austenite, despite a large MA phase fraction. Meanwhile, the suppressed micro-void initiation behavior of TM steel may be caused by a high uniformity of the primary lath-martensite matrix structure and the low strength ratio, despite the low volume fraction and mechanical stability of retained austenite and a large amount of MA phase.

5. Conclusions

The effect of mean normal stress on the deformation properties such as the strain-hardening, strain-induced martensite transformation, and micro-void initiation behaviors of TBF, TBM, and TM steels was investigated to evaluate the various cold formabilities. In addition, the deformation properties were related to the microstructural properties such as the matrix structure, retained austenite characteristics, and second-phase properties. The main results are summarized as follows:

(1) The positive mean normal stress increased the strain-hardening rate in an early strain range in all steels. This was mainly caused by facilitated strain-induced martensite transformation in an early strain range, resulting in an initial yielding or a continuous yielding.

(2) The equivalent plastic strain was linearly related to the X-ray half-width in all mean normal stress, which enabled the estimation of the equivalent stress in press-formed products. In this case, TBF steel had the lowest Vickers hardness and the highest

n-value. On the other hand, TM steel exhibited the highest Vickers hardness and the lowest *n*-value.

(3) The positive mean normal stresses promoted the strain-induced martensitic transformation because of expansion strain. The strain-induced martensite transformation behavior of TM steel was promoted compared to TBF and TBM steels, although the transformation fraction was the smallest.

(4) The positive mean normal stress promoted the micro-void initiation by developing the expansion stress/strain, especially in TBF steel. The effect of the mean normal stress on the micro-void initiation behavior was small in TBM and TM steels. This was associated with (1) the mixed-matrix structure of bainitic ferrite and primary martensite structure and a relatively stable retained austenite and (2) the high uniformity of primary martensite matrix structure and a low strength ratio for TBM and TM steels, respectively.

Author Contributions: Conceptualization, K.-i.S.; methodology, K.-i.S. and S.S.; formal analysis, K.-i.S. and S.S.; investigation, S.S. and J.K.; resources, K.-i.S.; data curation, K.-i.S., S.S. and J.K.; writing—original draft preparation, S.S. and J.K.; writing—review and editing, K.-i.S. and J.K.; visualization, S.S. and J.K.; supervision, K.-i.S. and J.K. All authors have read and agreed to the published version of the manuscript.

Funding: This research received no external funding.

Institutional Review Board Statement: Not applicable.

Informed Consent Statement: Not applicable.

Data Availability Statement: The original contributions presented in the study are included in the article; further inquiries can be directed to the corresponding author.

Conflicts of Interest: The author declares no conflicts of interest.

References

1. Rana, R.; Singh, S.B. *Automotive Steels—Design, Metallurgy, Processing and Applications*; Woodhead Publishing: Cambridge, UK, 2016; pp. 1–469.
2. Soleimani, M.; Kalhor, A.; Mirzadeh, H. Transformation-induced plasticity (TRIP) in advanced steels: A review. *Mater. Sci. Eng. A* **2020**, *795*, 140023. [CrossRef]
3. Frómeta, D.; Lara, A.; Grifé, L.; Dieudonné, T.; Dietsch, P.; Rehrl, J.; Suppan, C.; Casellas, D.; Calvo, J. Fracture resistance of advanced high-strength steel sheets for automotive applications. *Metall. Mater. Trans. A* **2021**, *52A*, 840–856. [CrossRef]
4. Polatids, E.; Haidemenopoulos, G.N.; Krizan, D.; Aravas, N.; Panzner, T.; Šmíd, M.; Papadioti, I.; Casati, N.; Van Petegem, S.; Van Swygenhoven, H. The effect of stress triaxiality on the phase transformation in transformation induced plasticity steels: Experimental investigation and modeling the transformation kinetics. *Mater. Sci. Eng. A* **2021**, *800*, 140321. [CrossRef]
5. Wang, Y.; Xu, Y.; Wang, Y.; Zhang, J.; Guo, C.; Wang, X.; Zhao, W.; Liu, H. Enhanced stretch flangeability and crack propagation behavior of an 1100 MPa grade TRIP-aided bainitic ferrite steel. *J. Mater. Res. Technol.* **2023**, *26*, 5503–5517. [CrossRef]
6. Weißensteiner, I.; Suppan, C.; Hebesberger, T.; Winkelhofer, F.; Clemens, H.; Maier-Kiener, V. Effect of morphological differences on the cold formability of an isothermally heat-treated advance high-strength steel. *JOM* **2018**, *70*, 1567–1575. [CrossRef]
7. Tang, S.; Lan, H.; Liu, Z.; Wang, G. Enhancement of balance in strength, ductility and stretch flangeability by two-step austempering in a 1000 MPa grade cold rolled bainitic steel. *Metals* **2021**, *11*, 96. [CrossRef]
8. Sugimoto, K.; Hidaka, S.; Tanino, H.; Kobayashi, J. Warm formabiliy of 0.2 Pct C-1.5 Pct Si-5 Pct Mn transformation-induced plasticity-aided steel. *Metall. Mater. Trans. A* **2017**, *48*, 2237–2246. [CrossRef]
9. Kim, J.; Seo, E.; Kwon, M.; Kang, S.; De Cooman, B.C. Effect of quenching temperature on stretch flangeability of a medium Mn steel processed by quenching and partitioning. *Mater. Sci. Eng. A* **2018**, *729*, 276–284. [CrossRef]
10. Kobayashi, J.; Pham, D.V.; Sugimoto, K. Stretch-flangeability of 1.5 GPa grade TRIP-aided martensitic cold rolled sheet steels. In Proceedings of the 10th International Conference on Technology of Plasticity (ICTP 2011), Aachen, Germany, 25–30 September 2011; pp. 598–603.
11. Im, Y.; Kim, E.; Song, T.; Lee, J.; Suh, D. Tensile properties and stretch-flangeability of TRIP steels produced by quenching and partitioning (Q&P) process with different fractions of constituent phases. *ISIJ Int.* **2021**, *61*, 572–581.
12. Toji, Y.; Nakagaito, T.; Matsuda, H.; Hasegawa, K.; Kaneko, S. Effect of microstructure on mechanical properties of quenching and partitioning steel. *ISIJ Int.* **2023**, *63*, 758–765. [CrossRef]
13. Sugimoto, K.; Tanino, H.; Kobayashi, J. Cold formabilities of martensite-type medium Mn steel. *Metals* **2021**, *11*, 1371. [CrossRef]

14. Sugimoto, K.; Hojo, T.; Nagasaka, A.; Hashimoto, S.; Ikeda, S. The effects of Nb and Mo additions on the microstructure and formability of C-Mn-Si-Al TRIP-aided ferrous sheet steels with an annealed martensite matrix. *Steel Grips* **2004**, *2*, 483–487.
15. Maruyama, H. X-ray measurement of retained austenite. *Jpn. Soc. Heat Treat.* **1977**, *17*, 198–204.
16. Dyson, D.J.; Holmes, B. Effect of alloying additions on the lattice parameter of austenite. *J. Iron Steel Inst.* **1970**, *208*, 469–474.
17. Cullity, B.D. *Elements of X-Ray Diffraction*, 2nd ed.; Addison-Wesley Publishing Company, Inc.: Boston, MA, USA, 1978; p. 287.
18. Hill, R. *The Mathematical Theory of Plasticity*; Oxford University Press: New York, NY, USA, 1985; p. 20.
19. Kawata, H.; Yasutomi, T.; Shirakami, S.; Nakamura, K.; Sakurada, E. Deformation-induced martensite transformation behavior during tensile and compressive deformation in low-alloy TRIP steel sheets. *ISIJ Int.* **2021**, *61*, 527–536. [CrossRef]
20. Koistinen, D.P.; Marburger, R.E. A general equation prescribing the extent of the austenite-martensite transformation in pure iron-carbon alloys and plain carbon steels. *Acta Metall.* **1959**, *7*, 59–60. [CrossRef]
21. Sugimoto, K.; Shioiri, S.; Kobayashi, J. Effects of mean normal stress on strain-hardening, strain-induced martensite transformation, and void-formation behaviors in high-strength TRIP-aided steels. *Metals* **2024**, *14*, 61. [CrossRef]
22. Williamson, G.K.; Hall, W.H. X-ray line broadening from filed aluminum and wolfram. *Acta Metall.* **1953**, *1*, 22–31. [CrossRef]
23. Takebayashi, S.; Kunieda, T.; Yoshinaga, N.; Ushioda, K.; Ogata, S. Comparison of the dislocation density in martensitic steels evaluated by some X-ray diffraction methods. *ISIJ Int.* **2010**, *50*, 875–882. [CrossRef]
24. Mura, T.; Mori, T. *Micromechanics*; Baifukan Co., Ltd.: Tokyo, Japan, 1976; p. 23. (In Japanese)
25. Sakaki, T.; Sugimoto, K.; Fukuzato, T. Role of internal stress for continuous yielding of dual-phase steels. *Acta Metall.* **1983**, *31*, 1737–1746. [CrossRef]
26. Ashby, M.F. Work hardening of dispersion-hardened crystals. *Philo. Mag.* **1966**, *14*, 1157–1178. [CrossRef]
27. Azuma, M.; Goutianos, S.; Hansen, N.; Winther, G.; Huang, X. Effect of hardness of martensite and ferrite on void formation in dual phase steel. *Mater. Sci. Technol.* **2012**, *28*, 1092–1100. [CrossRef]
28. Shoji, H.; Hino, K.; Ohta, M.; Shinohara, Y.; Minami, F. Ductile fracture mechanism for dual phase steel with high strength second phase. *Trans. Jpn. Weld. Soc.* **2015**, *33*, 341–348. (In Japanese) [CrossRef]
29. Hiwatashi, S.; Takahashi, M.; Katayama, T.; Usuda, M. Effect of deformation-induced transformation on deep drawability—Forming mechanism of TRIP type high-strength steel sheet. *J. Jpn. Soc. Technol. Plast.* **1994**, *35*, 1109–1114. (In Japanese)

Disclaimer/Publisher's Note: The statements, opinions and data contained in all publications are solely those of the individual author(s) and contributor(s) and not of MDPI and/or the editor(s). MDPI and/or the editor(s) disclaim responsibility for any injury to people or property resulting from any ideas, methods, instructions or products referred to in the content.

Article

On Strain-Hardening Behavior and Ductility of Laser Powder Bed-Fused Ti6Al4V Alloy Heat-Treated above and below the β-Transus

Emanuela Cerri and Emanuele Ghio *

Department of Engineering for Industrial Systems and Technologies, University of Parma, Via G. Usberti, 181/A, 43124 Parma, Italy; emanuela.cerri@unipr.it
* Correspondence: emanuele.ghio@unipr.it

Abstract: Laser powder bed-fused Ti6Al4V alloy has numerous applications in biomedical and aerospace industries due to its high strength-to-weight ratio. The brittle α′-martensite laths confer both the highest yield and ultimate tensile strengths; however, they result in low elongation. Several post-process heat treatments must be considered to improve both the ductility behavior and the work-hardening of as-built Ti6Al4V alloy, especially for aerospace applications. The present paper aims to evaluate the work-hardening behavior and the ductility of laser powder bed-fused Ti6Al4V alloy heat-treated below (704 and 740 °C) and above (1050 °C) the β-transus temperature. Microstructural analysis was carried out using an optical microscope, while the work-hardening investigations were based on the fundamentals of mechanical metallurgy. The work-hardening rate of annealed Ti6Al4V samples is higher than that observed in the solution-heat-treated alloy. The recrystallized microstructure indeed shows higher work-hardening capacity and lower dynamic recovery. The Considère criterion demonstrates that all analyzed samples reached necking instability conditions, and uniform elongations (>7.8%) increased with heat-treatment temperatures.

Keywords: work-hardening rate; strain-hardening; titanium alloy; laser powder bed fusion; mechanical properties; uniform elongation; work-hardening capacity

Citation: Cerri, E.; Ghio, E. On Strain-Hardening Behavior and Ductility of Laser Powder Bed-Fused Ti6Al4V Alloy Heat-Treated above and below the β-Transus. *Materials* 2024, 17, 3401. https://doi.org/10.3390/ma17143401

Academic Editors: Seong-Ho Ha, Young-Ok Yoon, Young-Chul Shin and Dong-Earn Kim

Received: 19 June 2024
Revised: 3 July 2024
Accepted: 7 July 2024
Published: 9 July 2024

Copyright: © 2024 by the authors. Licensee MDPI, Basel, Switzerland. This article is an open access article distributed under the terms and conditions of the Creative Commons Attribution (CC BY) license (https://creativecommons.org/licenses/by/4.0/).

1. Introduction

Ti6Al4V is a titanium alloy increasingly used in aerospace and biomedical applications due to its excellent combination of high strength, corrosion resistance, high fatigue life, and toughness [1–4]. Due to its high strength-to-weight ratio, Ti6Al4V is undoubtedly used to manufacture space capsule components, compressor blades, helicopter rotor hubs, and orthopedic and cranial implants. In this scenario, in which customization and flexibility are among the main design requirements, the laser powder bed fusion (LPBF) process finds a wide application area. In fact, the melt and fusion process of the metallic powder in a layer-by-layer methodology makes it possible to manufacture components with complex and topology-optimized geometry [5,6]. The LPBF process is also motivated by the absence of tools and minimal post-processing machining requirements [7].

Ti6Al4V is an α + β alloy where the α-stabilizers (Al, O, N, C) and β-stabilizers (V, Fe) stabilize the hcp (hexagonal close-packed) α-phase and the bcc (body-centered cubic) β-phase, respectively, at room temperature. Due to the nonequilibrium solidification process that occurs during the LPBF process, the as-built microstructure of the Ti6Al4V samples is composed of a hierarchical structure of needle-like α′-martensite laths arranged within columnar prior-β grains. In detail, β-grains nucleate directly on the build platform and grow from the bottom region to the top one, following the several solidified powder layers (i.e., towards the molten pools). The diffusionless β → α′-martensite transformation takes place within each molten pool [6,8–10]. Thanks to these microstructural features, laser powder bed-fused Ti6Al4V alloy shows higher strengths but lower work-hardening and ductility than wrought Ti6Al4V alloys, thus limiting its applications [6,11,12]. At the same time, the fast and localized cycles of heating and cooling caused by the laser–powder interactions trigger differential expansion and contraction of localized zones of the manufactured component. This generates stresses and strains that remain within the components as residual stresses and strains [6,13].

Post-processing heat treatments can improve the ductility of the as-built Ti6Al4V due to the following reasons [3,6]:

1. Decomposition of the brittle α'-martensite laths into $\alpha + \beta$ phase during the exposure at temperatures below the β-transus, i.e., during stress relief, annealing, and sub-transus solution heat treatment (SHT). Ductility improves with increasing heat-treatment temperatures.
2. Recrystallization of the columnar β-grains during the exposure at solution temperatures (above the β-transus) and the formation of a desired $\alpha + \beta$ microstructure by controlling the cooling pathway from the β-region to the room temperature.

For example, Vracken et al. [14] showed that the strain of as-built LPBFed Ti6Al4V alloy (e = 7 ± 1%) increases by 25% after annealing at 705 °C per 3 h, and by 85% in slowly cooled samples (furnace) from 1020 °C. When the solution-heat-treatment temperature increases up to 1150–1200 °C, no ductility enhancement is observed, as summarized in [6].

Focusing on the mechanical strength, both the as-built YS (YS, 980 ÷ 1200 MPa) and UTS (UTS, 1100 ÷ 1300 MPa) decrease with an increase in the heat-treatment temperature. For the solution-heat-treated samples, both UTS and YS can be recovered due to the $\beta \rightarrow \alpha'$-martensite transformation that occurs during a rapid cooling [15].

Work-hardening (i.e., strain-hardening) is closely related to the characterization of the plastic deformation of metallic materials, as well as strength, deformability, toughness, and ductility [16–18]. It is fundamentally based on an intricate interaction between several microstructural features such as grain boundaries, misorientation, dislocation, and crystal lattice [19,20]. In cubic structures, the strain-hardening behavior is well understood, with the main hardening mechanism primarily based on the accumulation of a dislocation forest. On the contrary, the significant plastic anisotropy and the low symmetry characterizing the hcp (hexagonal close-packed) lattice complicate the characterization of the strain-hardening behavior [21,22]. For these reasons, the various microstructural features characterizing the Ti6Al4V alloy and the different lattice structures of both hcp α-phase and bcc (body-centered cubic) β-phase make the work-hardening analysis meaningful. Several studies have investigated the work-hardening behavior of cast Ti6Al4V at different strain rates and high testing temperatures [23–26]. In recent years, few studies have focused on the work-hardening analysis of additively manufactured Ti6Al4V alloys [7,16,27–29]. De Formanoir et al. [29] investigated electron beam powder bed-fused Ti6Al4V in annealed (850 °C), sub-transus SHTed (920 °C), and hot-isostatic pressing conditions. Jankowski [28] briefly analyzed the mechanical behavior of additively manufactured Ti6Al4V alloys, considering only the morphology of the work-hardening curves and the softening coefficient. Muiruri et al. [27] investigated direct metal laser-sintered Ti6Al4V after a cycle of different heat treatments. Lastly, Song et al. [7] briefly analyzed the work-hardening behavior of an LPBFed Ti6Al4V alloy in as-built conditions by comparing it to Ti44 and Ti84 alloys.

To improve the literature review and considering the importance of the work-hardening behavior in various aerospace applications, the present study aims to evaluate the work-hardening behavior of a Ti6Al4V alloy laser powder bed-fused in different orientations and heat-treated below and above the β-transus temperature. Specially, work-hardening exponent, plastic instability, and work-hardening capacity are investigated, taking into account the microstructural variations observed after:

1. Annealing heat treatments at 704 and 740 °C;
2. Super-transus SHT at 1050 °C.

The results presented and discussed in this study build upon the microstructural investigations previously performed by the authors in [30,31].

2. Materials and Methods

Gas-atomized Ti6Al4V powder, whose chemical composition is listed in Table 1, was used to additively manufacture dog-bone test samples.

Table 1. Chemical composition (wt.%) of the gas-atomized Ti64 powder.

Ti	Al	V	Fe	C	N	O
Bal.	6.5	4.1	0.21	0.01	0.01	0.1

Tensile samples were laser powder bed-fused using an SLM®280 machine (SLM: selective laser melting, SLM Solution, Lübeck, Germany) with different orientations relative to the build platform (i.e., 0, 45, and 90 °C), utilizing the process parameters reported in [30,31]. Before their removal from the build platform, the samples were heat-treated in a vacuum furnace at temperatures below (704 °C, 740 °C) and above (1050 °C) the β-transus (Figure 1) to prevent the possible formation of cracks considering the brittleness of the as-built microstructure. As shown in Figure 1, samples heat-treated at 704 °C and 1050 °C were directly cooled in argon gas (dotted lines) for 60 min. Samples exposed to 740 °C were first furnace-cooled down to 520 °C over 90 min and then cooled with argon gas to room temperature. The annealing heat treatment at 704 °C was carried out in accordance with ASM 2801b standard [32], while that performed at 740 °C was developed to further enhance the anisotropic mechanical behavior of the as-built Ti64 alloy.

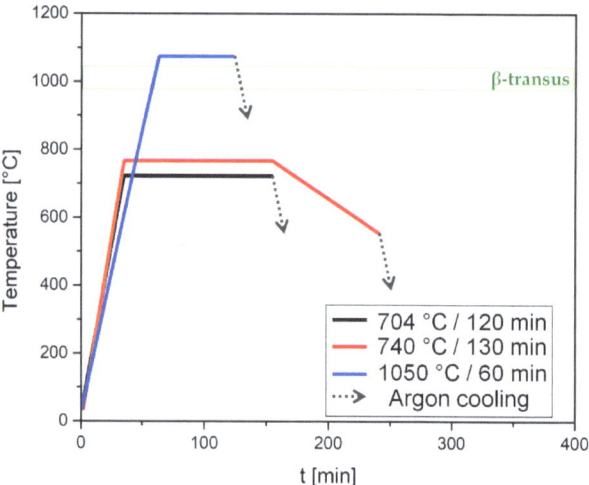

Figure 1. Temperature–time curves of the heat treatments performed at 704 °C × 120 min (black line), 740 °C × 130 min (red line), and 1050 °C × 60 min (blue line).

Microhardness was measured using a Vickers tester machine (VMHT Leica, Wetzlar, Germany) with a load of 100 gf and a dwell time of 15 s. The microhardness of heat-treated samples was obtained as the average of 9 indentations arranged in a 3 × 3 matrix, as discussed in [30].

The microstructure and fracture profiles of the tensile samples were observed through an inverted microscope (DMi8 Leica, Wetzlar, Germany). The investigated surfaces were grinded (P80-P4000), polished with silica colloidal suspension, and then chemically etched with Keller's reagent. In-depth microstructural analysis was previously performed and discussed in our earlier works [30,31].

Dog-bone samples, with a gauge length of 30 mm and a cross-sectional area of 36 mm², were tensile tested at room temperature using a Zwick Z100 machine (Zwick/Roell, Ulm,

Germany) at a constant strain rate of 1.6×10^{-3} s^{-1}. Tensile tests were repeated three times for each heat-treated condition to ensure the reliability of the results. The as-built mechanical properties were obtained by the literature review. To characterize the cross-section of each tested sample, two parallel hardness profiles (red dotted lines in Figure 2) were performed. These profiles extended from the fracture surface to the zone where both profiles converged to the microhardness of the undeformed sample. The microhardness profiles were spaced at 100 µm apart.

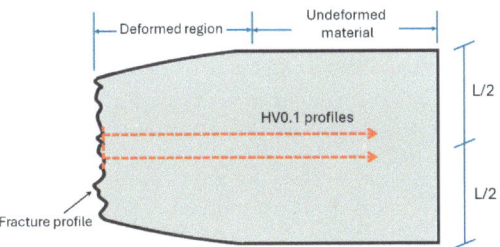

Figure 2. Graphical representation of the microhardness profiles performed on the cross-section of the tested samples. L represents the diameter of the dog-bone samples (L = 6.0 ± 0.1 mm).

To investigate the work-hardening behavior of the heat-treated Ti6Al4V alloy, the Ludwik–Hollomon equation was used and applied to the plastic flow region. This equation (Equation (1)) is also known as the power-law hardening equation:

$$\sigma = K\varepsilon^n \qquad (1)$$

where σ is the true stress (MPa), ε is the true strain (-), K is the strength coefficient (MPa), and n is the strain-hardening exponent (-). By differentiating Equation (1) with respect to the true elongation

$$\frac{d\sigma}{d\varepsilon} = n\frac{\sigma}{\varepsilon} = \theta, \qquad (2)$$

the obtained work-hardening rate (θ) can be used to investigate the ductile effects of the matrix. If the plastic region is well approximated by the power-law (Equation (1)), the work-hardening rate (Equation (2)) will intersect the true stress–strain curves at a point representing the true UTS (σ_{UTS}) and the respective strain value (ε_{UTS}) [33]. This point also defines the instability condition that may occur during a tensile test, namely, the onset of necking. This condition meets the following equivalence [34]:

$$\varepsilon_{UTS} = \varepsilon_u = n \qquad (3)$$

For these observations, the power-law equation (Equation (1)) at the UTS point can lastly be rewritten as follows:

$$\sigma_{UTS} = K\varepsilon_u^n = Kn^n \qquad (4)$$

Through this brief contextualization, the work-hardening rate was also used to investigate the point at which the necking occurred. Newer studies [35,36] supported these statements.

To obtain both the strain-hardening exponent and the strength coefficient for each heat-treated condition, Equation (1) was considered in a logarithmic form:

$$\ln(\sigma) = n\ln(\varepsilon) + \ln(K) \qquad (5)$$

where n represents the slope and K is the y-intercept (i.e., when $\varepsilon = 1$) of the linear fit (Figure 3a) carried out in the $\varepsilon_{YS} - \varepsilon_{UTS}$ range. Due to the morphology of the true stress–strain curve and to better evaluate the correlation between the plastic region of the true stress–strain curve and the power-law, both the slope (n) and the y-intercept

(K) were considered first as constant values, as shown in the single linear fit in Figure 3a (top row), and second as variable values, as highlighted by the four different linear fits in Figure 3b (bottom row). Another study [37] considered n and K as variable values due to the nonlinear morphology of the true stress–strain curve. Figure 3b shows how the true stress–strain curve (black line) can be approximated to the curves obtained by Equation (1) (i.e., power-law) so that the instability Considère criterion (Equation (2)) can be applied. For the discussions made on Equations (2)–(5), first, the true stress–strain curve is considered from YS to UTS points (namely, in the uniform elongation region). Second, the blue and red dotted lines in Figure 3b indicate the stress–strain curves obtained using the power-law equation (Equation (1)) in which the values n and K are considered first as constant (blue line in the top row of Figure 3a), and then as variable (red line in bottom row of Figure 3b). By comparing the power-law functions, the best approximation of the plastic region was evidently provided by the red dotted line, namely, when the n and K values are considered variable.

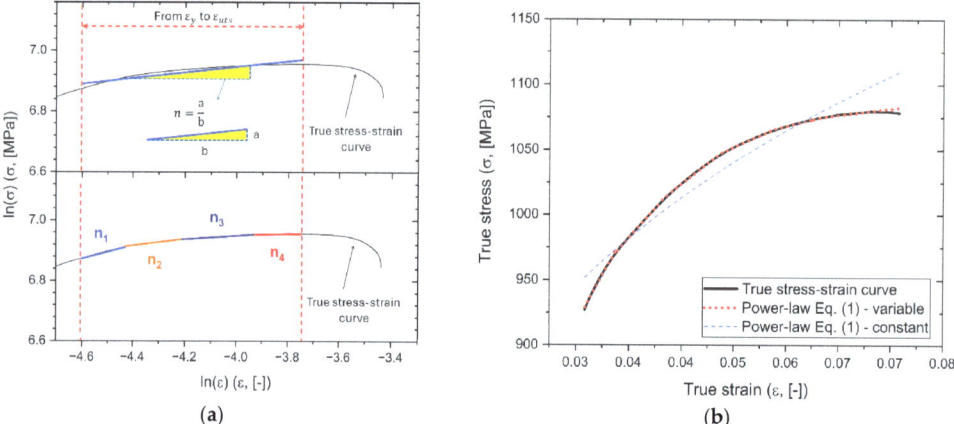

Figure 3. (a) True stress–strain curve plotted into a double-logarithmic graph to obtain the strain-hardening exponent (n) and the strength coefficient (K) as constant (first row) or variable (second row) values. (b) Representative true stress–strain curves, limited to the YS to the UTS, obtained by tensile test (black line), power-law (Equation (1)) where n and K are variable from YS to UTS (red dotted line), and power-law (Equation (1)) where n and K are constant.

The ratio (R) between the true yield strength (σ_{YS}) and ultimate tensile strength (σ_{UTS}),

$$R = \frac{\sigma_{YS}}{\sigma_{UTS}}, \tag{6}$$

was used to investigate the brittleness conferred by the different microstructures. The true YS was obtained by the intersection between the true stress–strain curve and the offset line at 0.002 positive true strain from the linear portion. The UTS was obtained by applying the Considère criterion to the stress–strain curve (Equation (2)).

As previously mentioned, the Considère criterion defines the point in which the necking instability occurs and the respective UTS (σ_{UTS}) value. To obtain the true plastic strain (ε_p (-)) characterizing each Ti64 sample, the following integral, evaluated from σ_{YS} to σ_{UTS}, was considered:

$$\varepsilon_p = \int_{\sigma_{YS}}^{\sigma_{UTS}} \frac{d\varepsilon}{d\sigma} \cdot d\sigma \tag{7}$$

Observing that the derivative $\left(\frac{d\varepsilon}{d\sigma}\right)$ is the inverse of the work-hardening rate (Equation (2)), Equation (7) can be rewritten as follows:

$$\varepsilon_p = \int_{\sigma_{YS}}^{\sigma_{UTS}} [\theta(\sigma)]^{-1} \cdot d\sigma \qquad (8)$$

By considering and integrating the Kocks–Mecking linear relationship [22,38], Equation (8) confers the following result:

$$\varepsilon_p = \frac{\ln[1 + C_b(1-R)]}{C_b} \qquad (9)$$

where C_b is the softening coefficient, which is related to the dynamic recovery occurring in Stage III of the work-hardening curve (Figure 4), and R is as expressed in Equation (6). In detail, the softening coefficient (C_b) represents the slope of the Kocks–Mecking linear relationship (red dotted line) between the work-hardening and the flow stress [22,38,39].

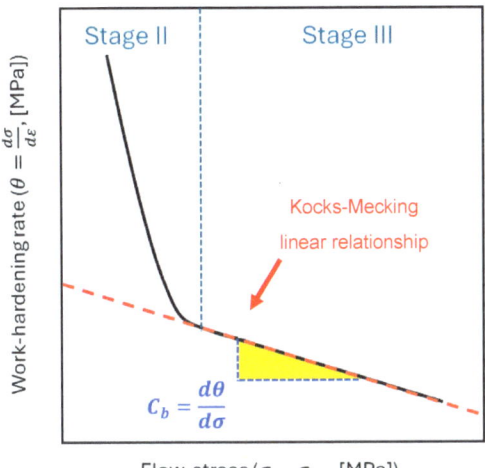

Figure 4. Schematic representation of the work-hardening curve (black line) elucidating the Kocks–Mecking analysis, in which the red dotted line represents the linear relationship between the work-hardening and the flow stress and Cb is its slope.

3. Results

Figure 5 shows the as-built (Figure 5a) and heat-treated (Figure 5b) microstructures acquired on the xz-plane of the Ti6Al4V dog-bone samples. The as-built microstructure (Figure 5a) is characterized by columnar β-grains containing hcp α'-martensite laths (white arrows in Figure 5a). As investigated in several studies [10,31,40], the LPBF process promotes, first, the nucleation of the β-grain directly on the build platform and, secondly, its growth along the build direction, namely, across the several molten pools. For this reason, columnar β-grains are disposed parallel to the build direction. The high cooling rate controlling the solidification process of the molten pools induces the diffusionless β → α'-martensite transformation and forms a cross-hatched structure of α'-martensite laths (dotted circle in Figure 5a). As deeply studied by Yang et al. [8], the cross-hatched structure is composed of a twine of several laths disposed perpendicular and/or parallel to each other. In detail, primary α'-martensite laths always extend across the whole columnar grain. Secondary, tertiary, and quartic α'-martensite laths, characterized by finer dimensions, are instead disposed of parallel and/or perpendicular to the primary ones. As widely investigated in our previous works [30,31], the heat treatments performed at temperatures below the β-transus (704 °C and 740 °C, Figure 1) induced the α′-martensite → α + β decomposition due to the diffusion of the β-stabilizer alloying elements (i.e., V, Fe) from the supersaturated hcp lattice to the α-phase grain boundary. The EBSD observations in [31] showed the presence of about 3% of the newly formed β-phase and that the newly formed α-phase retains the same orientation as its progenitor α′-martensite lath. The same EBSD maps in [31] indicated that the α-phase has an orientation relation with the retained columnar β-grain (yellow dotted line in Figure 5b).

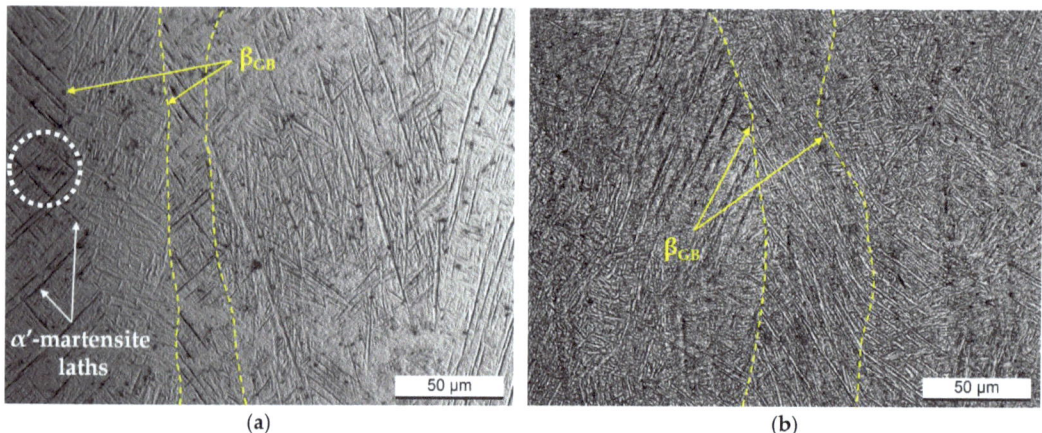

Figure 5. Optical micrographs acquired on the xz-plane (parallel to the build direction) of the Ti64 Z-samples in (**a**) as-built conditions and after the heat treatments at 740 °C/120 min (**b**).

Exposure at 1050 °C (Figure 6) recrystallizes the columnar β-grains into equiaxed grains whose boundaries are formed by the $α_{GB}$ (GB: grain boundary), thereby removing the microstructural anisotropy conferred by the LPBF process. On the other hand, the argon cooling from the β-region (Figure 1) conferred another degree of microstructural anisotropy. In fact, some newly β-equiaxed grains are formed by α + β laths distributed into a Widmanstätten structure and α + β colonies (Figure 6a), while others consist of globular or coarsened α-phase (Figure 6a). This microstructural variation is generally conferred by dissimilar cooling rates affecting the β → α + β transformation as also supported by the $α_{GB}$ and β-phase distribution along the grain boundaries (orange dotted lines). Specially, the β-phase precipitates along the $α_{GB}$-phase and separates it from the α + β laths or colonies (Figure 6a) during fast cooling. Conversely, lower cooling rates promote the interconnection between the $α_{GB}$-phase and the α laths (Figure 6b). These statements are supported and well documented in [41,42].

Figure 7 summarizes the mechanical properties trend of the XZ-, Z-, and 45-samples heat-treated at 704 °C (red columns), 740 °C (blue columns), and 1050 °C (green columns). Generally, the heat treatments at both 704 °C and 740 °C reduced the as-built UTS ($σ_{UTS}$ > 1.1 GPa, [6,9,43–45]) and YS ($σ_{YS}$ > 950 MPa, [6,9,43–45]) values due to the α′-martensite → α + β decomposition. Considering that the α′-martensite decomposition typically occurs above 400 °C, the furnace cooling from 740 °C to 520 °C continues to increase the amount of the decomposed martensite [6,46]. For this reason, and because of the coarsening phenomena affecting the α-phase (according to the Hall–Petch law), both the YS and UTS decrease from the samples annealed at 704 °C to those at 740 °C (Figure 7, Table 2). Our previous work [30] showed that increasing the heat-treatment temperatures from 704 to 740°C, and varying the cooling pathway, resulted in an increase in grain size from (540 ± 60) nm to (799 ± 10) nm. For the same findings, UTS and YS values slightly decreased (Figure 7, Table 2) when the heat-treatment temperature increased from 704 °C to 740 °C. The highest strength reduction and the anisotropy removal were conferred by the

SHT at 1050 °C per 60 min because of the recrystallized microstructure shown in Figure 6. Contrary to the expectation, the elongation values did not exhibit an adequate increment relative to the strength reduction, likely due to the presence of:

1. The fine α + β laths and colonies;
2. The α-case layer (as will later be discussed).

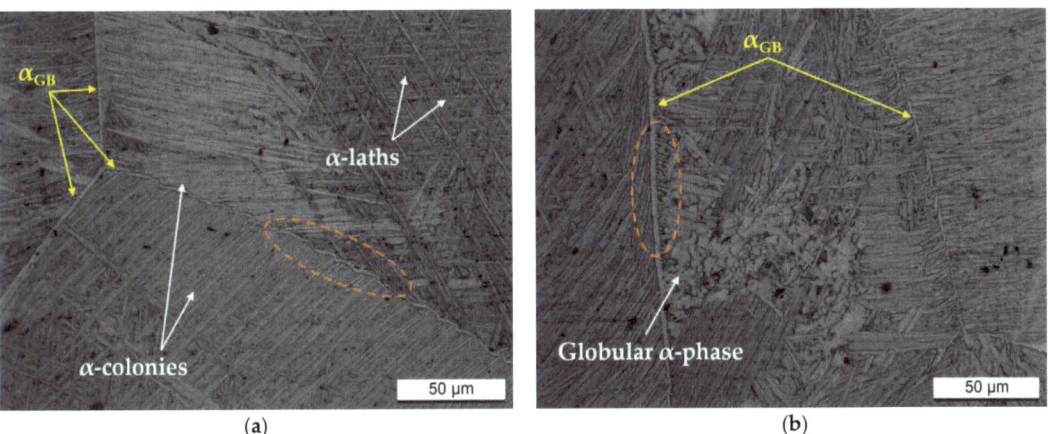

Figure 6. (a,b) Optical micrographs acquired on the xz-plane (parallel to the build direction) of the Ti64 Z-samples after the heat treatments at 1050 °C/60 min and representing two different zones of the same sample.

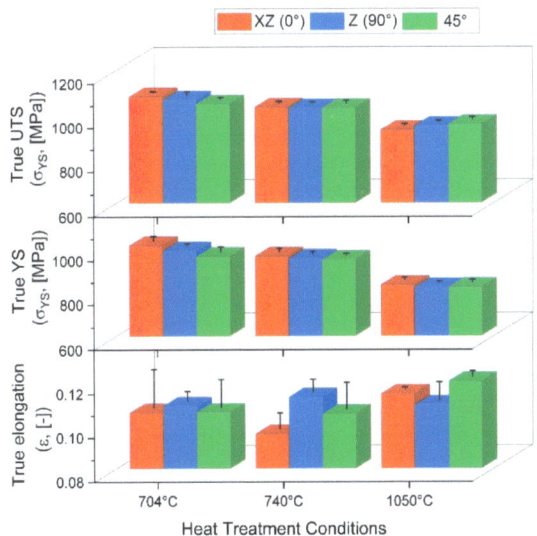

Figure 7. True UTS (σ_{UTS} (MPa)), YS (σ_{YS} (MPa)), and elongation (ε (-)) values of the Ti64 samples (XZ, Z, and 45°) heat-treated at 704 °C, 740 °C, and 1050 °C.

Table 2. Average of engineering UTS (s$_{UTS}$), YS (s$_{YS}$), and elongation (e (%)) values of the heat-treated XZ-, Z-, and 45-samples.

Samples	s$_{UTS}$ (MPa)			s$_{YS}$ (MPa)			e (%)		
	704 °C	740 °C	1050 °C	704 °C	740 °C	1050 °C	704 °C	740 °C	1050 °C
XZ	1059 ± 2	1008 ± 4	909 ± 6	997 ± 21	950 ± 14	820 ± 12	11 ± 2	10 ± 1	12 ± 1
Z	1043 ± 14	1000 ± 8	924 ± 3	974 ± 9	939 ± 10	809 ± 6	11 ± 1	12 ± 1	12 ± 1
45°	1022 ± 10	1002 ± 12	930 ± 11	949 ± 22	933 ± 6	810 ± 11	11 ± 1	11 ± 1	13 ± 1

Similar findings were observed in [6,47,48].

Figure 8 shows the work-hardening curves obtained by considering the variable strain-hardening exponents and applying Equation (2) to the Ti6Al4V samples heat-treated at 704 °C (Figure 8a), 740 °C (Figure 8b), and 1050 °C (Figure 8c). It is important to note that the work-hardening behavior of a polycrystalline metallic material is closely related not only to the grain size and distribution, dislocations, and misorientations, but also to dislocation annihilation, formation of local shear zones, and new sub-grains. These factors influence the stages of the work-hardening. Each curve presents the same three distinct stages (Stages I–III). Stage I describes the dislocation multiplication, leading to rapid decrease in the work-hardening rate, with an increase in plastic strain [49]. Immediately after this stage, towards the end of the elasto/plastic region, Stage II shows an increase in the work-hardening rate up to a relative maximum, possibly due to the possible presence of stacking faults or twins, as investigated in [50–53]. Stage III is dominated by the dynamic recovery (represented by C_b in Figure 4), where dislocation annihilation occurs. Finally, the sudden decrease in the work-hardening rate (Stage IV) is not well described in the literature yet. Muiruri et al. [27] suggested that intense localization of shear described the final part of the work-hardening curve. Considering the shape of the curves, it can be concluded that the microstructural variations (Figure 5) resulting from the heat treatments did not significantly affect the strain-hardening behaviour.

Despite the similarity between the work-hardening curves (Figure 8a,b vs. Figure 8c), the work-hardening rate, calculated as widely described in [54], decreases as the heat-treatment temperature increases from 704 °C to 1050 °C, as listed in Table 3. Simultaneously, the orange curve plotted in Figure 8d exhibits a slower work-hardening rate than those described by the red and black curves in the same Stage II. The smaller number of dislocations within the lattice structure of the SHTed Ti6Al4V sample may support these findings. As affirmed in [55], additively manufactured Ti6Al4V alloy heat-treated below the β-transus (<800 °C) showed a higher amount of dislocation compared to that observed in over-transus SHTed samples. Lastly, in accordance with the expectations, in samples annealed at 704 °C and 740 °C, the equality between the black and red work-hardening curves in the first stage indicates that there is no significant variation in dislocation multiplication.

Table 3. Maximum values of work-hardening rate ($\theta_{max} \times 10^3$ (MPa)) obtained by heat-treated Ti64 samples.

	704 °C	740 °C	1050 °C
XZ	3.7 ± 0.2	3.8 ± 0.2	2.5 ± 0.2
Z	3.8 ± 0.1	3.7 ± 0.2	2.6 ± 0.2
45°	3.6 ± 0.2	3.8 ± 0.2	2.5 ± 0.2

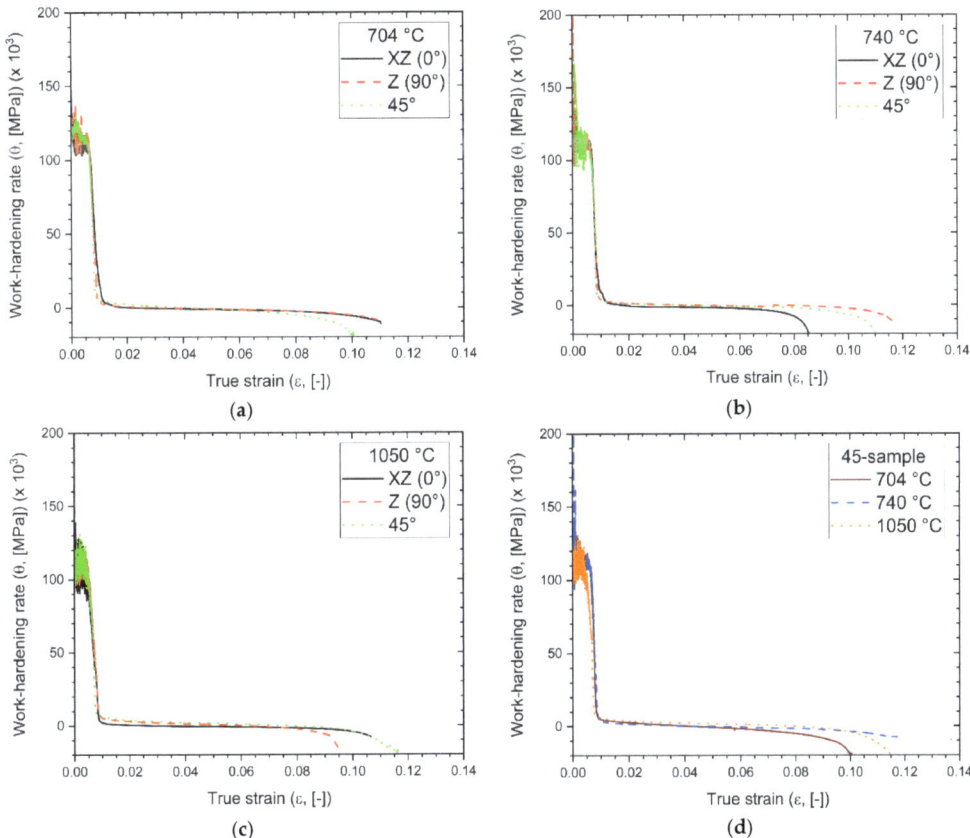

Figure 8. Work-hardening curves of the Ti64 samples heat-treated at 704 °C (**a**), 740 °C (**b**), and 1050 °C (**c**). The graph in (**d**) displays the work-hardening curves of the 45-sample heat-treated at 704 °C (red line), 740 °C (blue line), and 1050 °C (orange line).

Figure 9a shows the work-hardening capacity of the heat-treated Ti6Al4V samples as a function of their constant strain-hardening exponent. Work-hardening capacity (WHC) is defined as follows:

$$WHC = \frac{\sigma_{UTS} - \sigma_y}{\sigma_y} = \frac{\sigma_{UTS}}{\sigma_y} - 1 \tag{10}$$

and represents the ability to accommodate dislocations during a plastic deformation. For this reason, the SHTed Ti6Al4V samples show a better capacity to store dislocation compared to those heat-treated below the β-transus temperature due to a greater amount of the bcc β-phase, different grain sizes, and crystal orientations, as supported in [11,31,56,57]. In fact, the fully lamellar structure of the annealed Ti6Al4V samples limits strain-hardening ability and negative effects on uniform elongation [29,58,59]. Conversely, the SHTed Ti6Al4V alloy exhibits a lower degree of elasticity (R obtained from Equation (6) and shown in Figure 9a) and, thus, a higher capacity for plastic deformation before failure compared to the annealed alloy. Figure 9b shows the variation of the softening factor, which represents the capacity to recover dislocation during deformation (Figure 4), in relation to the true plastic strain (Equation (9)) [53,60–63]. From the plotted results, it can be concluded that all Ti6Al4V samples analyzed in the present work exhibit higher elasticity compared to those produced by electron beam powder bed fusion (square-shaped red symbols, [60,61]), as well as those produced by laser powder bed fusion both in as-built (circle-shaped symbols [53,61–63]) and heat-treated (hexagonal-shaped symbols [61,62]) conditions. Specifically, the columnar β-grains that characterize the annealed Ti6Al4V samples conferred similar true elongation strains to the as-built LPBFed Ti6Al4V samples, but the decomposed α′-martensite laths may increase both the softening factor and elasticity. The higher amount of β-phase and the presence of an α-β microstructure, arranged into colonies or Widmanstätten structure, enhance the true plastic strain at the expense of the softening factor. Considering that both the as-built EBM and heat-treated LPBF Ti6Al4V samples exhibit an α + β microstructure within the β-grains, it is possible to conclude that the mechanical behavior of SHTed Ti6Al4V is more influenced by the α + β structure (a,b) and the α-case (Figure 10c) than by the morphology of the β-grains.

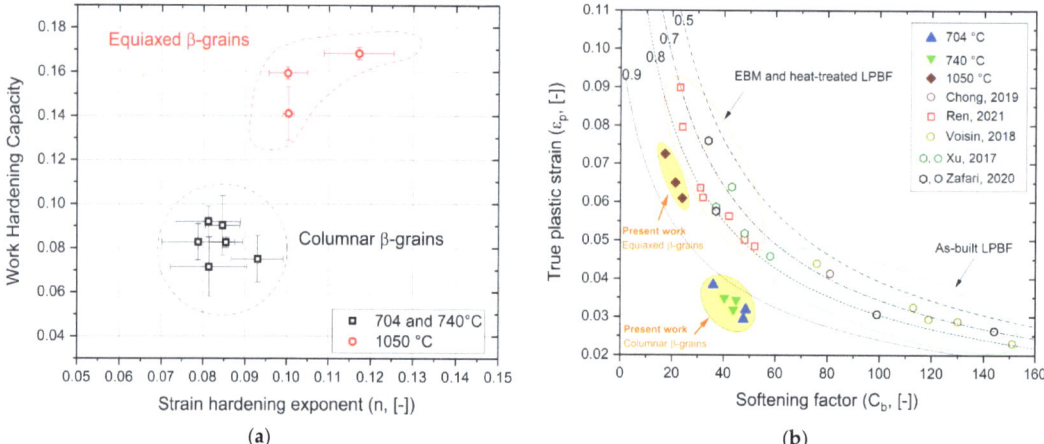

Figure 9. (a) WHC versus strain-hardening exponents of the Ti64 samples heat-treated at 704 °C, 740 °C, and 1050 °C. (b) Variation of the softening factor in relation to the true plastic strain of the samples analyzed in the present work and in [53,60–63].

As highlighted in Figure 10, the main crack appears to propagate indeed along the boundaries of both the α + β colonies (red arrows in Figure 10a) and the α + β laths (yellow arrows in Figure 10a), as well as along the boundaries of the coarsened α-phase (white arrows in Figure 10b). In summary, cracks propagate across the softer bcc β-phase, as extensively discussed in [6]. Furthermore, mechanical behavior can be also influenced by the presence of the irregular-shaped lack of fusion (Figure 10d), thoroughly argued in [6]. Lastly, Figure 10e summarizes Figure 10a–d and clearly highlights the cross-sectional area reduction due to the necking instability conditions that occurred during the tensile test.

Considering that the power-law equation (Equation (1)) effectively describes the plastic region (see Figure 3b), the Considère criterion can be applied to evaluate the uniform elongation and the potential neck formation in the Ti6Al4V samples. Each work-hardening curve was obtained by Equation (2), where the strain-hardening exponents are considered as constant value (see linear fit in Figure 3a), ensuring the satisfaction of Equations (2)–(4). Indeed, only work-hardening curves obtained by constant n and K values intersect the true stress–strain curves at the (σ_{UTS}, ε_{UTS}) point (Figure 11). Furthermore, these intersections between the work-hardening curve $\left(\theta = -\frac{d\sigma}{d\varepsilon}\right)$ and the true stress–strain curves confirm the onset of the necking instability in all analyzed Ti6Al4V samples, regardless of the post-processing heat treatments (Figure 11). Focusing on the Ti6Al4V samples annealed at 704 °C (Figure 11a), slight variations in uniform elongation values are observed due to the different build orientations. A clear increase in uniform elongation was obtained by increasing the heat-treatment temperatures while maintaining constant build orientation (red arrow in Figure 11b).

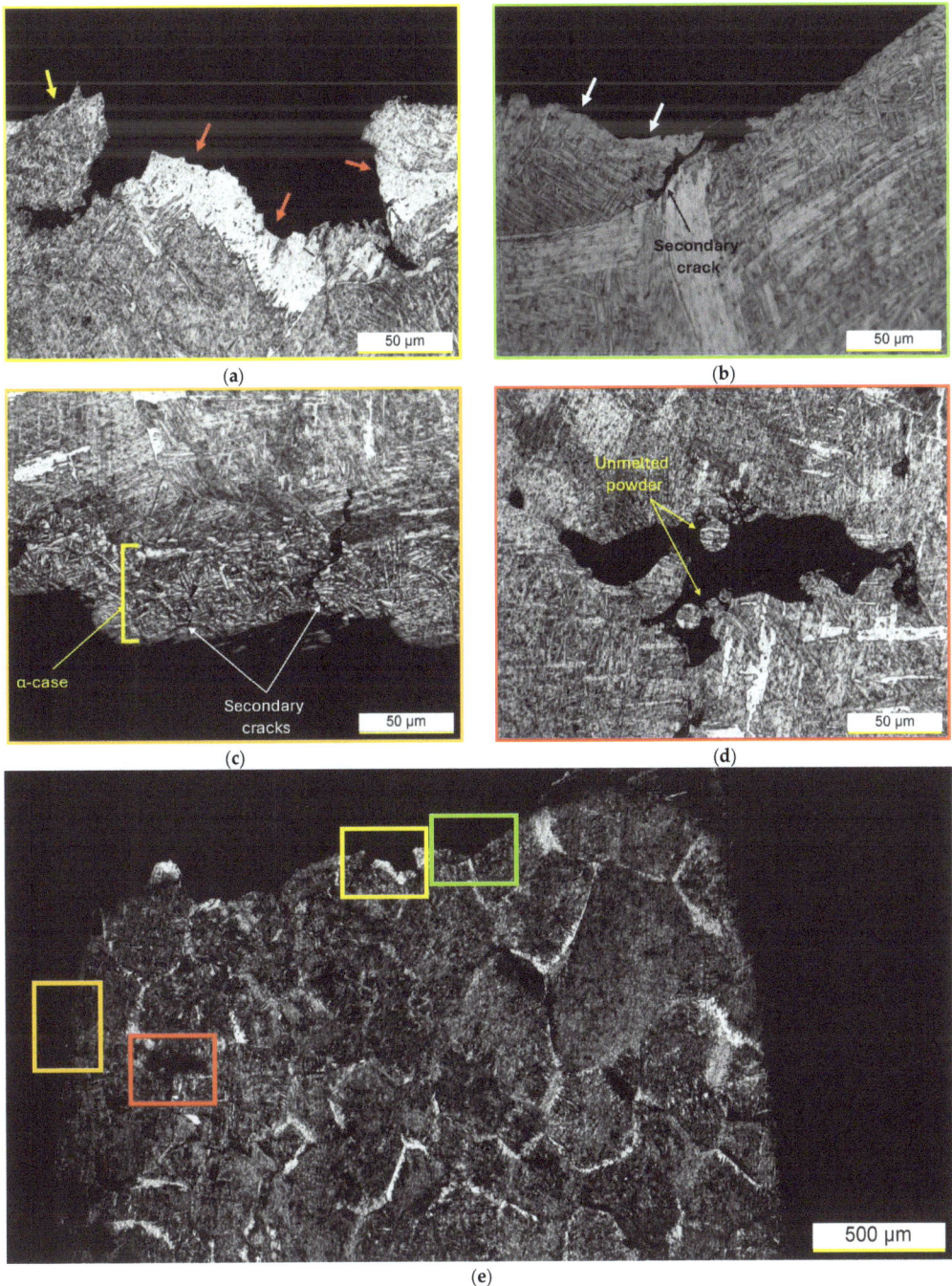

Figure 10. Fracture profile of a Ti64 samples heat-treated at 1050 °C per 60 min. Panels (**a**–**d**) belong to the sectioned tensile sample (**e**) and exhibit (**a**,**b**) fracture profile, (**c**) edge of the sample with α-case layer, and (**d**) lack-of-fusion pore. Yellow, red, and white arrows indicate the crack pathways along the boundaries of the α-phases.

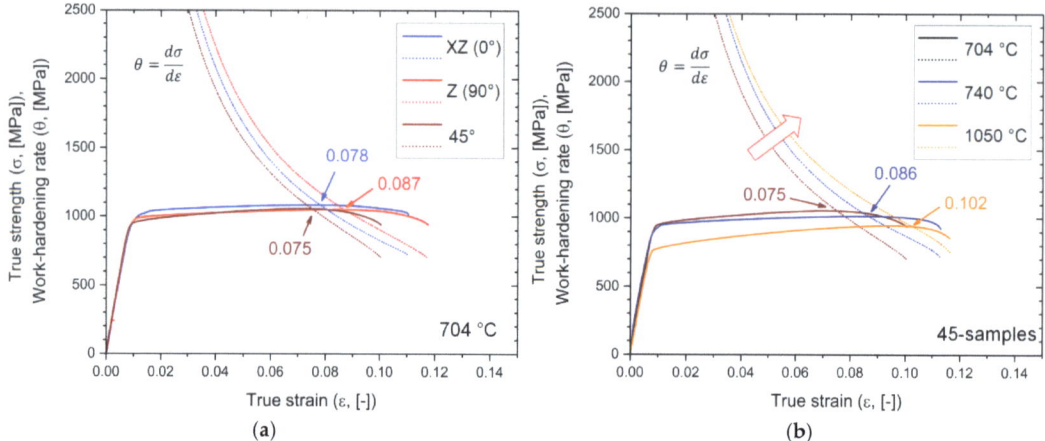

Figure 11. Representative true stress–strain curves and work-hardening rates of (**a**) XZ-, Z-, and 45-samples heat-treated at 704 °C, (**b**) 45-samples heat-treated at 704 °C (black lines), 740 °C (blue lines), and 1050 °C (orange lines).

Figure 12 displays the Vickers microhardness profiles (see) of the tested Ti64 samples after the exposure at 704 °C (Figure 12a), 740 °C (Figure 12b), and 1050 °C (Figure 12c). Starting from the farthest point of each profile, Vickers microhardness trends increase up to a maximum value due to the work-hardening in the necking region. From this point, hardness values decrease up to the closest zone to the fracture profile, probably due to the presence of several damages that are undetectable through optical microscopy. The high strain rates characteristic of the closest zone to the fracture profile can induce the formation of several secondary cracks and tears [27,64,65]. Considering the Vickers microhardness profiles plotted in Figure 12a,b, it appears that Ti6Al4V samples heat-treated at 740 °C exhibit a slightly larger damaged area compared to those at 740 °C (Figure 12a). These zones are located between the fracture profile and the maximum value of the Vickers profile. As discussed in Figures 8 and 9a, the highest work-hardening capacity observed in the SHTed Ti6Al4V samples is also reflected in the greatest variation in hardness between the maximum and the undeformed region (Figure 12c). In this context, it is important to note that the undeformed region may still contain a certain amount of strain that is not detectable through the Vickers profile [64,65].

Lastly, the average microhardness values of the undeformed zones (Figure 12) reflect the decline in mechanical properties (see Table 2) from the Ti6Al4V samples annealed at 704 °C to those solubilized at 1050 °C.

Figure 13 correlates the engineering yield strength (s_{YS}) to the engineering strain (e) of the heat-treated Ti6Al4V samples analyzed in this study (Table 2) and compares these values with the LPBF Ti6Al4V and CP-Ti alloys investigated in [6,66–69]. It is important to note that the YS of Ti6Al4V samples heat-treated at 704 and 740 °C is comparable to the LPBF Ti6Al4V samples in as-built conditions. Both recrystallization phenomena affecting the columnar β-grains and the argon cooling from the β-region significantly reduce the YS; thereby, the ASMT standard is not satisfied [67]. Therefore, aging heat treatments in the 450–600 °C range can improve mechanical performance by precipitation of the $TiAl_3$ and Ti_3Al phases, as demonstrated in [50]. In contrast, according to the results obtained by using the Considère criterion (Figure 11), all analyzed samples exhibited excellent uniform elongations (>7.8%). These values are higher than those exhibited by Ti6Al4V samples produced by LPBF and powder metallurgy in [7,11,53,70,71], and are comparable to those shown by wrought Ti64 alloys in [11] with the same strain rate range ($10^{-3} \div 10^{-4}$ s^{-1}).

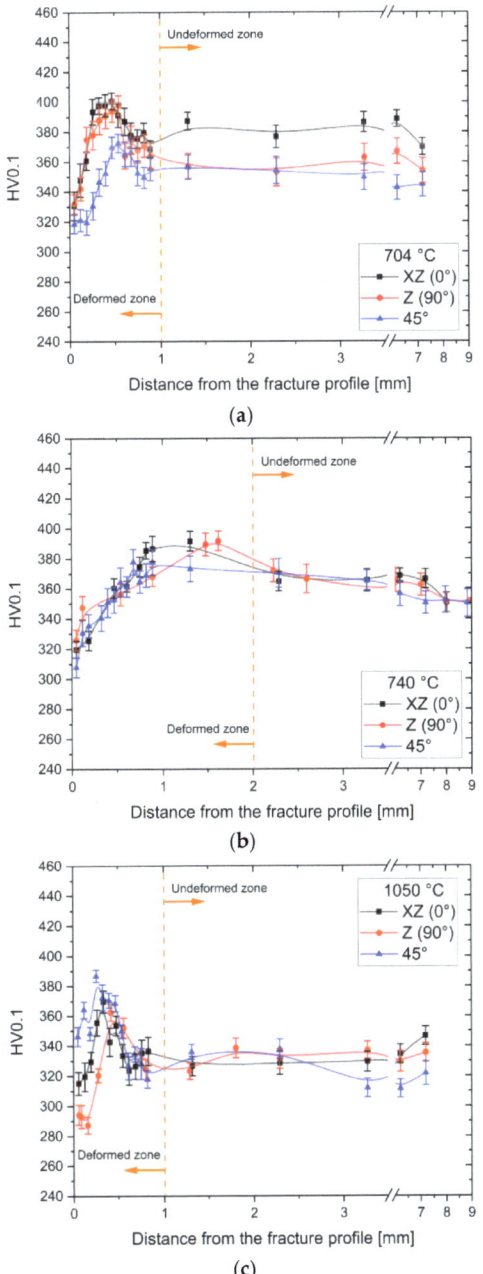

Figure 12. Vickers microhardness profiles (see) measured on the cross-section of the broken dog-bone samples after the heat treatments at 704 °C (**a**), 740 °C (**b**), and 1050 °C (**c**). Orange dotted lines divide the highly deformed zone and the undeformed one.

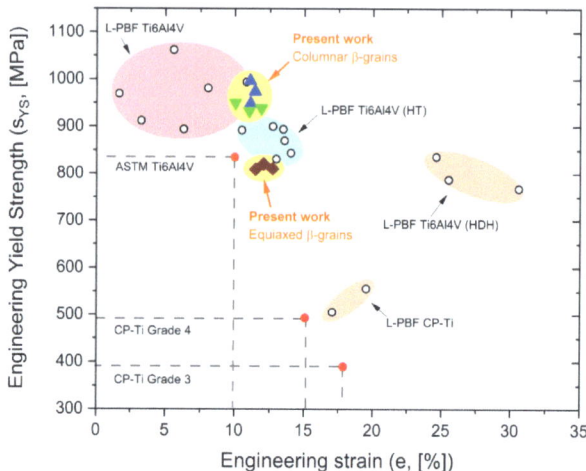

Figure 13. Engineering YS versus engineering strains of the Ti64 samples analyzed in the present work (colored symbols) and compared with Ti64 and CP-Ti (CP: commercially pure) samples studied in [6,66–69]. HDH-Ti means hydrogenated–dehydrogenated titanium alloy.

4. Conclusions

An investigation of the strain-hardening and ductility of laser powder bed-fused Ti6Al4V samples heat-treated below (704 and 740 °C) and above (1050 °C) the β-transus was presented in this paper. The following conclusions can be drawn from the presented results:

1. The microstructure of the annealed Ti6Al4V samples exhibits retained columnar β-grains containing α-phase laths arranged in a cross-hatched structure. When treatment temperature increases from 704 to 740 °C, α-phase width increases by about +30%.
2. The columnar β-grains recrystallize into equiaxed β-grains during the exposure at 1050 °C, and subsequent argon cooling forms an α + β microstructure with several morphologies: (i) colonies, (ii) Widmanstätten structure, and (iii) globular.
3. Due to the rise in heat-treatment temperature, both yield and ultimate tensile strengths decreased by about—16% and—12%, respectively. However, elongation values did not show significant improvement. All analyzed samples exhibited excellent uniform elongations (>7.8%), which increased from Ti64 samples annealed at 704 °C to those solution-heat-treated at 1050 °C, along with necking instability.
4. The equality between the uniform elongations obtained by the Considère criterion and the ε_{UTS} indicates that constant n and K values must be considered.
5. Annealed samples show higher σ_{YS}/σ_{UTS} ratios than those exhibited by the Ti6Al4V samples with recrystallized microstructure. The consequent lower work-hardening capacity values promote low work-hardening of the necking region. Thereby, the recrystallized microstructure shows both lower work-hardening rates in Stage II and lower softening values (i.e., dynamic recovery). This is reflected in the highest hardness increment from the undeformed region obtained during the plastic deformation.
6. Despite the anisotropy conferred by the build orientations, they did not significantly influence the work-hardening behavior.

Future works should focus on the work-hardening behavior of aged samples, given that the tensile strength of solution-heat-treated Ti6Al4V samples must be improved.

Author Contributions: Conceptualization, E.G. and E.C.; methodology, E.G.; software, E.G.; validation, E.C.; formal analysis, E.G. and E.C.; investigation, E.G.; resources, E.G.; data curation, E.G.; writing—original draft preparation, E.G. and E.C.; writing—review and editing, E.G. and E.C.; visualization, E.G.; supervision, E.C.; project administration, E.C.; funding acquisition, E.C. All authors have read and agreed to the published version of the manuscript.

Funding: This project was funded under the National Recovery and Resilience Plan (NRRP), Mission 4, Component 2, Investment 1.5.—Call for tender No. 3277 of 30/12/2021 of Italian Ministry of University and Research funded by European Union—NextGenerationEU.

Institutional Review Board Statement: Not applicable.

Informed Consent Statement: Not applicable.

Data Availability Statement: The original contributions presented in the study are included in the article, further inquiries can be directed to the corresponding author.

Conflicts of Interest: The authors declare no conflicts of interest.

References

1. Donachie, M.J. *Titanium: A Technical Guide*, 2nd ed.; ASM International Materials Park: Novelty, OH, USA, 2000.
2. Carson, C. Heat Treating of Titanium and Titanium Alloys. In *Heat Treating of Nonferrous Alloys*; ASM International: Novelty, OH, USA, 2016; pp. 511–534.
3. Haydar, H.J.; Al-Deen, J.; AbidAli, A.K.; Mahmoud, A.A. Improved Performance of Ti6Al4V Alloy in Biomedical Applications—Review. *J. Phys. Conf. Ser.* **2021**, *1973*, 012146. [CrossRef]
4. Hamza, H.M.; Deen, K.M.; Haider, W. Microstructural Examination and Corrosion Behavior of Selective Laser Melted and Conventionally Manufactured Ti$_6$Al$_4$V for Dental Applications. *Mater. Sci. Eng. C* **2020**, *113*, 110980. [CrossRef] [PubMed]
5. Ahmadi, M.; Tabary, S.A.A.B.; Rahmatabadi, D.; Ebrahimi, M.S.; Abrinia, K.; Hashemi, R. Review of Selective Laser Melting of Magnesium Alloys: Advantages, Microstructure and Mechanical Characterizations, Defects, Challenges, and Applications. *J. Mater. Res. Technol.* **2022**, *19*, 1537–1562. [CrossRef]
6. Ghio, E.; Cerri, E. Additive Manufacturing of AlSi$_{10}$Mg and Ti$_6$Al$_4$V Lightweight Alloys via Laser Powder Bed Fusion: A Review of Heat Treatments Effects. *Materials* **2022**, *15*, 2047. [CrossRef] [PubMed]
7. Song, Z.; Zeng, X.; Wang, L. Laser Additive Manufacturing of Titanium Alloys with Various Al Contents. *Mater. Res. Lett.* **2023**, *11*, 391–398. [CrossRef]
8. Yang, J.; Yu, H.; Yin, J.; Gao, M.; Wang, Z.; Zeng, X. Formation and Control of Martensite in Ti-$_6$Al-$_4$V Alloy Produced by Selective Laser Melting. *Mater. Des.* **2016**, *108*, 308–318. [CrossRef]
9. Buhairi, M.A.; Foudzi, F.M.; Jamhari, F.I.; Sulong, A.B.; Radzuan, N.A.M.; Muhamad, N.; Mohamed, I.F.; Azman, A.H.; Harun, W.S.W.; Al-Furjan, M.S.H. Review on Volumetric Energy Density: Influence on Morphology and Mechanical Properties of Ti6Al4V Manufactured via Laser Powder Bed Fusion. *Prog. Addit. Manuf.* **2023**, *8*, 265–283. [CrossRef]
10. Antonysamy, A.A.; Meyer, J.; Prangnell, P.B. Effect of Build Geometry on the β-Grain Structure and Texture in Additive Manufacture of Ti$_6$Al$_4$V by Selective Electron Beam Melting. *Mater. Charact.* **2013**, *84*, 153–168. [CrossRef]
11. Luo, Y.; Xie, Y.; Zhang, Z.; Liang, J.; Zhang, D. Improving Strain Hardening Capacity of High-Strength Ti-$_6$Al-$_4$V Alloy by a Dual Harmonic Structure. *J. Mater. Res. Technol.* **2023**, *26*, 1122–1135. [CrossRef]
12. Zhang, D.; Wang, L.; Zhang, H.; Maldar, A.; Zhu, G.; Chen, W.; Park, J.-S.; Wang, J.; Zeng, X. Effect of Heat Treatment on the Tensile Behavior of Selective Laser Melted Ti-$_6$Al-$_4$V by in Situ X-Ray Characterization. *Acta Mater.* **2020**, *189*, 93–104. [CrossRef]
13. Xiao, Z.; Chen, C.; Zhu, H.; Hu, Z.; Nagarajan, B.; Guo, L.; Zeng, X. Study of Residual Stress in Selective Laser Melting of Ti$_6$Al$_4$V. *Mater. Des.* **2020**, *193*, 108846. [CrossRef]
14. Vrancken, B.; Thijs, L.; Kruth, J.-P.; Van Humbeeck, J. Heat Treatment of Ti$_6$Al$_4$V Produced by Selective Laser Melting: Microstructure and Mechanical Properties. *J. Alloys Compd.* **2012**, *541*, 177–185. [CrossRef]
15. Jaber, H.; Kónya, J.; Kulcsár, K.; Kovács, T. Effects of Annealing and Solution Treatments on the Microstructure and Mechanical Properties of Ti$_6$Al$_4$V Manufactured by Selective Laser Melting. *Materials* **2022**, *15*, 1978. [CrossRef] [PubMed]
16. Afrin, N.; Chen, D.L.; Cao, X.; Jahazi, M. Strain Hardening Behavior of a Friction Stir Welded Magnesium Alloy. *Scr. Mater.* **2007**, *57*, 1004–1007. [CrossRef]
17. Meyers, M.A.; Chawla, K.K. *Mechanical Behavior of Materials*, 2nd ed.; Cambridge University Press: Cambridge, UK, 2009.
18. Chen, X.H.; Lu, L. Work Hardening of Ultrafine-Grained Copper with Nanoscale Twins. *Scr. Mater.* **2007**, *57*, 133–136. [CrossRef]
19. Nes, E. Modelling of Work Hardening and Stress Saturation in FCC Metals. *Prog. Mater. Sci.* **1997**, *41*, 129–193. [CrossRef]
20. Lavrentev, F.F. The Type of Dislocation Interaction as the Factor Determining Work Hardening. *Mater. Sci. Eng.* **1980**, *46*, 191–208. [CrossRef]
21. Fan, C.L.; Chen, D.L.; Luo, A.A. Dependence of the Distribution of Deformation Twins on Strain Amplitudes in an Extruded Magnesium Alloy after Cyclic Deformation. *Mater. Sci. Eng. A* **2009**, *519*, 38–45. [CrossRef]
22. Kocks, U.F.; Mecking, H. Physics and Phenomenology of Strain Hardening: The FCC Case. *Prog. Mater. Sci.* **2003**, *48*, 171–273. [CrossRef]
23. Akhonin, S.V.; Mishchenko, R.N.; Petrichenko, I.K. Investigation of the Weldability of Titanium Alloys Produced by Different Methods of Melting. *Mater. Sci.* **2006**, *42*, 323–329. [CrossRef]

24. Rao, P.P.; Tangri, K. Yielding and Work Hardening Behaviour of Titanium Aluminides at Different Temperatures. *Mater. Sci. Eng. A* **1991**, *132*, 49–59. [CrossRef]
25. Honarmandi, P.; Aghaie-Khafri, M. Hot Deformation Behavior of Ti$_{-6}$Al$_{-4}$V Alloy in β Phase Field and Low Strain Rate. *Metallogr. Microstruct. Anal.* **2013**, *2*, 13–20. [CrossRef]
26. Gupta, R.K.; Kumar, V.A.; Mathew, C.; Rao, G.S. Strain Hardening of Titanium Alloy Ti6Al4V Sheets with Prior Heat Treatment and Cold Working. *Mater. Sci. Eng. A* **2016**, *662*, 537–550. [CrossRef]
27. Muiruri, A.; Maringa, M.; du Preez, W. Effects of Quasi-Static Strain Rate and Temperature on the Microstructural Features of Post-Processed Microstructures of Laser Powder Bed Fusion Ti$_6$Al$_4$V Alloy. *Appl. Sci.* **2024**, *14*, 4261. [CrossRef]
28. Jankowski, A.F. A Constitutive Structural Parameter c_b for the Work Hardening Behavior of Additively Manufactured Ti$_{-6}$Al$_{-4}$V. *Mater. Des. Process. Commun.* **2021**, *3*, e262. [CrossRef]
29. de Formanoir, C.; Brulard, A.; Vivès, S.; Martin, G.; Prima, F.; Michotte, S.; Rivière, E.; Dolimont, A.; Godet, S. A Strategy to Improve the Work-Hardening Behavior of Ti$_{-6}$Al$_{-4}$V Parts Produced by Additive Manufacturing. *Mater. Res. Lett.* **2017**, *5*, 201–208. [CrossRef]
30. Cerri, E.; Ghio, E.; Bolelli, G. Ti6Al4V-ELI Alloy Manufactured via Laser Powder-Bed Fusion and Heat-Treated below and above the β-Transus: Effects of Sample Thickness and Sandblasting Post-Process. *Appl. Sci.* **2022**, *12*, 5359. [CrossRef]
31. Cerri, E.; Ghio, E.; Bolelli, G. Effect of Surface Roughness and Industrial Heat Treatments on the Microstructure and Mechanical Properties of Ti6Al4V Alloy Manufactured by Laser Powder Bed Fusion in Different Built Orientations. *Mater. Sci. Eng. A* **2022**, *851*, 143635. [CrossRef]
32. *AMS 2801b*; Heat Treatment of Titanium Alloy Parts. SAE International: Danvers, MA, USA, 2014.
33. Dieter, G.E. *Mechanical Metallurgy*; Mehl, R.F., Bever, M.B., Eds.; McGraw-Hill: London, UK, 1961; Volume 1.
34. Considère, A. Annales Des Ponts et Chaussées. *Comm. Ann. Ponts Chaussees Paris* **1885**, *9*, 574–575.
35. Zhu, C.; Xu, J.; Yu, H.; Shan, D.; Guo, B. Size Effect on the High Strain Rate Micro/Meso-Tensile Behaviors of Pure Titanium Foil. *J. Mater. Res. Technol.* **2021**, *11*, 2146–2159. [CrossRef]
36. Yasnikov, I.S.; Vinogradov, A.; Estrin, Y. Revisiting the Considère Criterion from the Viewpoint of Dislocation Theory Fundamentals. *Scr. Mater.* **2014**, *76*, 37–40. [CrossRef]
37. Liu, W.H.; Lu, Z.P.; He, J.Y.; Luan, J.H.; Wang, Z.J.; Liu, B.; Liu, Y.; Chen, M.W.; Liu, C.T. Ductile CoCrFeNiMox High Entropy Alloys Strengthened by Hard Intermetallic Phases. *Acta Mater.* **2016**, *116*, 332–342. [CrossRef]
38. Mecking, H.; Kocks, U.F. Kinetics of Flow and Strain-Hardening. *Acta Metall.* **1981**, *29*, 1865–1875. [CrossRef]
39. Mondal, C.; Singh, A.K.; Mukhopadhyay, A.K.; Chattopadhyay, K. Tensile Flow and Work Hardening Behavior of Hot Cross-Rolled AA7010 Aluminum Alloy Sheets. *Mater. Sci. Eng. A* **2013**, *577*, 87–100. [CrossRef]
40. Qi, M.; Huang, S.; Ma, Y.; Youssef, S.S.; Zhang, R.; Qiu, J.; Lei, J.; Yang, R. Columnar to Equiaxed Transition during β Heat Treatment in a near β Alloy by Laser Additive Manufacture. *J. Mater. Res. Technol.* **2021**, *13*, 1159–1168. [CrossRef]
41. Yi, H.-J.; Kim, J.-W.; Kim, Y.-L.; Shin, S. Effects of Cooling Rate on the Microstructure and Tensile Properties of Wire-Arc Additive Manufactured Ti–6Al–4V Alloy. *Met. Mater. Int.* **2020**, *26*, 1235–1246. [CrossRef]
42. Ahmed, T.; Rack, H.J. Phase Transformations during Cooling in A+β Titanium Alloys. *Mater. Sci. Eng. A* **1998**, *243*, 206–211. [CrossRef]
43. Xue, M.; Chen, X.; Ji, X.; Xie, X.; Chao, Q.; Fan, G. Effect of Particle Size Distribution on the Printing Quality and Tensile Properties of Ti-6Al-4V Alloy Produced by LPBF Process. *Metals* **2023**, *13*, 604. [CrossRef]
44. Etesami, S.A.; Fotovvati, B.; Asadi, E. Heat Treatment of Ti$_{-6}$Al$_{-4}$V Alloy Manufactured by Laser-Based Powder-Bed Fusion: Process, Microstructures, and Mechanical Properties Correlations. *J. Alloys Compd.* **2022**, *895*, 162618. [CrossRef]
45. Sabban, R.; Bahl, S.; Chatterjee, K.; Suwas, S. Globularization Using Heat Treatment in Additively Manufactured Ti$_{-6}$Al$_{-4}$V for High Strength and Toughness. *Acta Mater.* **2019**, *162*, 239–254. [CrossRef]
46. Boccardo, A.D.; Zou, Z.; Simonelli, M.; Tong, M.; Segurado, J.; Leen, S.B.; Tourret, D. Martensite Decomposition Kinetics in Additively Manufactured Ti-6Al-4V Alloy: In-Situ Characterisation and Phase-Field Modelling. *Mater. Des.* **2024**, *241*, 112949. [CrossRef]
47. Jha, J.S.; Toppo, S.P.; Singh, R.; Tewari, A.; Mishra, S.K. Deformation Behavior of Ti$_{-6}$Al$_{-4}$V Microstructures under Uniaxial Loading: Equiaxed vs. Transformed-β Microstructures. *Mater. Charact.* **2021**, *171*, 110780. [CrossRef]
48. Seth, P.; Jha, J.S.; Alankar, A.; Mishra, S.K. Alpha-Case Formation in Ti$_{-6}$Al$_{-4}$V in a Different Oxidizing Environment and Its Effect on Tensile and Fatigue Crack Growth Behavior. *Oxid. Met.* **2022**, *97*, 77–95. [CrossRef]
49. Dodd, B.; Bai, Y. (Eds.) *Adiabatic Shear Localization*; Elsevier: Amsterdam, The Netherlands, 2012; ISBN 9780080977812.
50. Cerri, E.; Ghio, E.; Spigarelli, S.; Cabibbo, M.; Bolelli, G. Static and Dynamic Precipitation Phenomena in Laser Powder Bed-Fused Ti6Al4V Alloy. *Mater. Sci. Eng. A* **2023**, *880*, 145315. [CrossRef]
51. Tian, Y.Z.; Zhao, L.J.; Chen, S.; Shibata, A.; Zhang, Z.F.; Tsuji, N. Significant Contribution of Stacking Faults to the Strain Hardening Behavior of Cu-15%Al Alloy with Different Grain Sizes. *Sci. Rep.* **2015**, *5*, 16707. [CrossRef] [PubMed]
52. Pierce, D.T.; Jiménez, J.A.; Bentley, J.; Raabe, D.; Wittig, J.E. The Influence of Stacking Fault Energy on the Microstructural and Strain-Hardening Evolution of Fe–Mn–Al–Si Steels during Tensile Deformation. *Acta Mater.* **2015**, *100*, 178–190. [CrossRef]
53. Voisin, T.; Calta, N.P.; Khairallah, S.A.; Forien, J.-B.; Balogh, L.; Cunningham, R.W.; Rollett, A.D.; Wang, Y.M. Defects-Dictated Tensile Properties of Selective Laser Melted Ti$_{-6}$Al$_{-4}$V. *Mater. Des.* **2018**, *158*, 113–126. [CrossRef]

54. Chen, X.; Chen, H.; Ma, S.; Chen, Y.; Dai, J.; Bréchet, Y.; Ji, G.; Zhong, S.; Wang, H.; Chen, Z. Insights into Flow Stress and Work Hardening Behaviors of a Precipitation Hardening AlMgScZr Alloy: Experiments and Modeling. *Int. J. Plast.* **2024**, *172*, 103852. [CrossRef]
55. Muiruri, A.; Maringa, M.; du Preez, W. Evaluation of Dislocation Densities in Various Microstructures of Additively Manufactured Ti$_6$Al$_4$V (Eli) by the Method of X-Ray Diffraction. *Materials* **2020**, *13*, 5355. [CrossRef] [PubMed]
56. Ren, C.X.; Wang, Q.; Hou, J.P.; Zhang, Z.J.; Zhang, Z.F. Effect of Work-Hardening Capacity on the Gradient Layer Properties of Metallic Materials Processed by Surface Spinning Strengthening. *Mater. Charact.* **2021**, *177*, 111179. [CrossRef]
57. Xu, W.; Lui, E.W.; Pateras, A.; Qian, M.; Brandt, M. In Situ Tailoring Microstructure in Additively Manufactured Ti-$_6$Al-$_4$V for Superior Mechanical Performance. *Acta Mater.* **2017**, *125*, 390–400. [CrossRef]
58. Zafari, A.; Lui, E.W.; Jin, S.; Li, M.; Molla, T.T.; Sha, G.; Xia, K. Hybridisation of Microstructures from Three Classes of Titanium Alloys. *Mater. Sci. Eng. A* **2020**, *788*, 139572. [CrossRef]
59. Chong, Y.; Deng, G.; Yi, J.; Shibata, A.; Tsuji, N. On the Strain Hardening Abilities of A+β Titanium Alloys: The Roles of Strain Partitioning and Interface Length Density. *J. Alloys Compd.* **2019**, *811*, 152040. [CrossRef]
60. Rafi, H.K.; Karthik, N.V.; Gong, H.; Starr, T.L.; Stucker, B.E. Microstructures and Mechanical Properties of Ti6Al4V Parts Fabricated by Selective Laser Melting and Electron Beam Melting. *J. Mater. Eng. Perform.* **2013**, *22*, 3872–3883. [CrossRef]
61. Leicht, A. Analyzing the Mechanical Behavior of Additive Manufactured Ti-$_6$Al-$_4$V Using Digital Image Correlation. Diploma Thesis, Master Programme Materials Engineering, Vancouver, BC, Canada, 2015.
62. Tao, P.; Li, H.; Huang, B.; Hu, Q.; Gong, S.; Xu, Q. Tensile Behavior of Ti-6Al-4V Alloy Fabricated by Selective Laser Melting: Effects of Microstructures and as-Built Surface Quality. *China Foundry* **2018**, *15*, 243–252. [CrossRef]
63. He, B.; Wu, W.; Zhang, L.; Lu, L.; Yang, Q.; Long, Q.; Chang, K. Microstructural Characteristic and Mechanical Property of Ti6Al4V Alloy Fabricated by Selective Laser Melting. *Vacuum* **2018**, *150*, 79–83. [CrossRef]
64. Centeno, G.; Martínez-Donaire, A.J.; Morales-Palma, D.; Vallellano, C.; Silva, M.B.; Martins, P.A.F. Novel Experimental Techniques for the Determination of the Forming Limits at Necking and Fracture. In *Materials Forming and Machining*; Elsevier: Amsterdam, The Netherlands, 2015; pp. 1–24.
65. Hwang, J.-K. Revealing the Small Post-Necking Elongation in Twinning-Induced Plasticity Steels. *J. Mater. Sci.* **2020**, *55*, 8285–8302. [CrossRef]
66. Popovich, A.; Sufiiarov, V.; Borisov, E.; Polozov, I.A. Microstructure and Mechanical Properties of Ti-$_6$Al-$_4$V Manufactured by SLM. *Key Eng. Mater.* **2015**, *651–653*, 677–682. [CrossRef]
67. ASTM B381-13; Standard Specification for Titanium and Titanium Alloys Forging. ASTM: West Conshohocken, PA, USA, 2013.
68. Song, B.; Zhao, X.; Li, S.; Han, C.; Wei, Q.; Wen, S.; Liu, J.; Shi, Y. Differences in Microstructure and Properties between Selective Laser Melting and Traditional Manufacturing for Fabrication of Metal Parts: A Review. *Front. Mech. Eng.* **2015**, *10*, 111–125. [CrossRef]
69. Dong, Y.P.; Tang, J.C.; Wang, D.W.; Wang, N.; He, Z.D.; Li, J.; Zhao, D.P.; Yan, M. Additive Manufacturing of Pure Ti with Superior Mechanical Performance, Low Cost, and Biocompatibility for Potential Replacement of Ti-$_6$Al-$_4$V. *Mater. Des.* **2020**, *196*, 109142. [CrossRef]
70. He, B.B.; Hu, B.; Yen, H.W.; Cheng, G.J.; Wang, Z.K.; Luo, H.W.; Huang, M.X. High Dislocation Density–Induced Large Ductility in Deformed and Partitioned Steels. *Science* **2017**, *357*, 1029–1032. [CrossRef] [PubMed]
71. Chen, M.; Van Petegem, S.; Zou, Z.; Simonelli, M.; Tse, Y.Y.; Chang, C.S.T.; Makowska, M.G.; Ferreira Sanchez, D.; Moens-Van Swygenhoven, H. Microstructural Engineering of a Dual-Phase Ti-Al-V-Fe Alloy via in Situ Alloying during Laser Powder Bed Fusion. *Addit. Manuf.* **2022**, *59*, 103173. [CrossRef]

Disclaimer/Publisher's Note: The statements, opinions and data contained in all publications are solely those of the individual author(s) and contributor(s) and not of MDPI and/or the editor(s). MDPI and/or the editor(s) disclaim responsibility for any injury to people or property resulting from any ideas, methods, instructions or products referred to in the content.

Article

Numerical Simulation of Electromagnetic–Thermal–Fluid Coupling for the Deformation Behavior of Titanium–Aluminum Alloy under Electromagnetic Levitation

Xiaoliang Wang [1], Guifang Zhang [1,2,*], Peng Yan [1,*], Xinchen Pang [1] and Zhixiang Xiao [1]

1. Faculty of Metallurgical and Energy Engineering, Kunming University of Science and Technology, Kunming 650093, China; wangxiaoliang@kust.edu.cn (X.W.); xinchen001118@163.com (X.P.); xiaozhixiang1998@163.com (Z.X.)
2. Key Laboratory of Clean Metallurgy for Complex Iron Resources in Colleges and Universities of Yunnan Province, Kunming University of Science and Technology, Kunming 650093, China
* Correspondence: guifangzhang65@163.com (G.Z.); yanp_km@163.com (P.Y.)

Citation: Wang, X.; Zhang, G.; Yan, P.; Pang, X.; Xiao, Z. Numerical Simulation of Electromagnetic–Thermal–Fluid Coupling for the Deformation Behavior of Titanium–Aluminum Alloy under Electromagnetic Levitation. *Materials* 2024, 17, 3338. https://doi.org/10.3390/ma17133338

Academic Editors: Seong-Ho Ha, Young-Ok Yoon, Dong-Earn Kim and Young-Chul Shin

Received: 2 May 2024
Revised: 30 June 2024
Accepted: 2 July 2024
Published: 5 July 2024

Copyright: © 2024 by the authors. Licensee MDPI, Basel, Switzerland. This article is an open access article distributed under the terms and conditions of the Creative Commons Attribution (CC BY) license (https://creativecommons.org/licenses/by/4.0/).

Abstract: Electromagnetic levitation (EML) is a good method for high-temperature processing of reactive materials such as titanium–aluminum (Ti–Al) alloys. In this study, the oscillation and deformation processes of Ti-48Al-2Cr alloy specimens at different high-frequency currents during the EML process were simulated using the Finite Element Method and Arbitrary Lagrangian–Eulerian (ALE) methods. The data of oscillation, stabilization time, deformation, and distribution of electromagnetic–thermal–fluid fields were finally obtained. The accuracy of the simulation results was verified by EML experiments. The results show the following: the strength and distribution of the induced magnetic field inside the molten droplet are determined by the high-frequency current; under the coupling effect of the electromagnetic field, thermal field, and fluid field, the temperature rise of electromagnetic heating is rapid, and accompanied by strong stirring, resulting in a uniform distribution of the internal temperature and a small temperature difference. Under the joint action of gravity and Lorentz force, the molten droplets are first within a damped oscillation and then tend to stabilize with time, and finally maintain the "near rhombus" shape.

Keywords: EML; Ti–Al alloy; ALE; electromagnetic–thermal–fluid field coupling; oscillatory deformation

1. Introduction

Titanium–aluminum (Ti–Al) alloys are widely used in aerospace applications as a high-temperature structural material due to their low density, high specific strength, and good flame-retardant properties [1–4]. However, despite these advantages, their low ductility and low-temperature fracture toughness limit their application [5]. Electromagnetic levitation (EML) technology is a type of non-contact melting technology that has many advantages, such as non-contact, fast heating speed, wide temperature range, and internal uniformity, among others [6–10]. By subjecting metal alloy materials to deep subcooling and rapid cooling treatment, EML can significantly alter the morphology and composition of the solidification organization of metal alloys, thereby improving the overall performance of alloy materials [11–14]. EML offers several advantages in the production of high-performance Ti–Al alloys, including significantly reducing material contamination to maintain purity, promoting composition homogenization, allowing finer control of the cooling process, thereby tuning the microstructure, reducing thermal stresses and microcracking, improving mechanical properties, and increasing energy efficiency and precision of heating control. Due to the levitation and oscillation behavior of liquid metals during EML is complex and affects alloy preparation, conventional experiments are unable to investigate the role of magnetic, thermal, and fluid fields inside the molten droplets. Many researchers have investigated the deformation of molten droplets during EML by simulation methods, taking into account the mutual coupling of electromagnetic, thermal, and fluid fields.

Kermanpur et al. [15] developed a simulation model that coupled electromagnetic and thermal fields. They verified it using an analytical solution and experimental results. Asakuma et al. [16] also performed a numerical simulation of the deformation behavior of silicon droplets under EML. Their simulation was based on a mathematical model that used mixed finite elements and boundary elements. The simulation results showed that the droplets gradually became flat with an increase in current. Liang et al. [17] studied the deformation mode and flow behavior of liquid Ti–Al–Nb alloys at high temperatures, over a wide current range of 700–1400 A, using numerical simulations based on the arbitrary Lagrangian–Eulerian method, along with corresponding EML experiments. The arbitrary Lagrangian–Eulerian (ALE) method combines the advantages of the Eulerian and Lagrangian methods and has been widely used to study fluid dynamics problems, such as those involving free surfaces, large deformations, and fluid–solid coupling [18,19]. To effectively control the solidification organization of Ti–Al-based alloys, it is necessary to study the effect of currents on the levitation behavior and internal transfer behavior of these alloys.

In this study, we have established a mathematical model using a combination of FEM and ALE methods to simulate and predict the behavior of Ti-48Al-2Cr alloy droplets at different currents. This model considers the coupling of the electromagnetic, thermal, and fluid fields, and analyzes the oscillation, deformation, and transfer of the internal electromagnetic, thermal, and fluid fields. The results provide valuable insights for controlling the levitation stability of titanium–aluminum alloys during the EML melting process and regulating the internal organization of materials.

2. Physical and Numerical Models

2.1. Physical Model

The EML system used in the experiment mainly includes a high-frequency power supply, pure copper coil, quartz tube, triangular prism, pure copper crucible, levitation specimen, gas inlet and outlet, thermodetector and high-speed camera, and so on. The Ti-48Al-2Cr alloy used in the EML experiment was obtained by melting high-purity Ti grains (Ti, 99.99 wt%), high-purity Al grains (Al, 99.9 wt%) and high-purity Cr grains (Cr, 99.95 wt%) in a vacuum-water-cooled copper crucible arc furnace after repeated melting times to ensure the homogeneity of the compositions, and the mass was 1.2 g. The schematic of the EML is shown in Figure 1a. The specimen is placed in a vacuum environment, where a high-frequency harmonic magnetic field is then applied to it. This causes electromagnetic induction, which generates eddy currents. The eddy currents are concentrated on the surface of the specimen due to the skin effect. The coil consists of two sections, namely the upper and lower parts, where the upper part of the coil is two turns, and the lower part of the coil is three turns.

(1) In the lower part, a specimen with a moderate mass made of a Ti–Al alloy is levitated by the Lorentz force. It is then quickly heated up and melted due to Joule heat.
(2) The upper part is designed to prevent oscillation of the specimen.

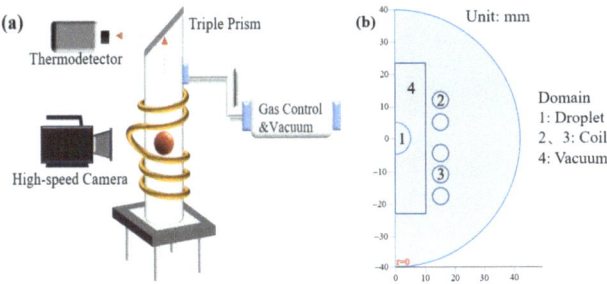

Figure 1. EML system: (**a**) EML schematic; (**b**) EML simulation computation domain.

The stabilizing coil is connected in series with the levitation coil but in the opposite direction. Figure 1b shows the simulation domain, where domain 1 is the Ti–Al alloy specimen, domains 2 and 3 are the coils, and domain 4 is vacuum domain.

2.2. Mathematical Model

In this study, the transient behavior of the alloy in the frequency domain was simulated using COMSOL 5.6 multiphysics field software, which considers three modes of physical field coupling, namely the electromagnetic, fluid, and thermal fields, according to the characteristics of electromagnetic levitation [17,20]. At each time step, the Joule heat and Lorentz force generated by the harmonic electromagnetic field are transferred to the thermal and flow fields, respectively, and the Lorentz force, gravity, and surface tension are applied to the alloy molten droplets to predict the oscillations, deformations, and flow fields of the melt. Dynamic meshing (ALE method) was used throughout the calculations, and the results were re-meshed at each time step to analyze the exact shape of the melt drop and the oscillation process. When the set time is reached, the simulation ends and the results for the different physical fields are released. The multiphysics field coupling considered by the calculation is specifically where the Lorentz force couples the electromagnetic field to the fluid field, while the non-isothermal flow couples the fluid field to the thermal field, simplifying the Navier–Stokes equations with the Boussinesq approximation. This multiphysics field coupling method considers the bidirectional influence of different fields and provides higher computational accuracy compared to traditional methods.

2.2.1. Assumptions

Based on the related literature [16], this study makes some assumptions about the model:

(1) Since the numerical model is a two-dimensional axisymmetric model, only the oscillations occurring in the vertical direction are considered;
(2) In the fluid model, both the molten droplet and the gas are set as incompressible fluids, and the pressure generated by the gas flow on the surface of the suspended droplet is ignored;
(3) The mass of the molten droplet is constant without considering the volatilization that occurs when the temperature of the droplet is above the liquid phase line.

2.2.2. Mathematical Equations

The electromagnetic field generated during EML melting is described by a system of Maxwell's equations as follows [21]:

$$\nabla \times \boldsymbol{H} = \boldsymbol{J} + \partial \boldsymbol{D}/\partial t \tag{1}$$

$$\nabla \times \boldsymbol{E} = -\partial \boldsymbol{B}/\partial t \tag{2}$$

$$\nabla \cdot \boldsymbol{B} = 0 \tag{3}$$

$$\nabla \cdot \boldsymbol{D} = \rho \tag{4}$$

where $\boldsymbol{H}$ is the magnetic field intensity, $\boldsymbol{E}$ is the electric field intensity, $\boldsymbol{B}$ is the magnetic induction intensity, $\boldsymbol{D}$ is the electric flux density, $\boldsymbol{J}$ is the current density, and ρ is the electric density.

In order to solve the Maxwell equations, a magnetic vector potential $\boldsymbol{A}$, and an electric scalar potential V, are defined as follows:

$$\boldsymbol{B} = \nabla \times \boldsymbol{A} \tag{5}$$

$$E = -\frac{\partial A}{\partial t} - \nabla V \tag{6}$$

The Lorentz Force and Joule heat generated would be used to analyze the fluid and thermal fields, respectively. The time-averaged Lorentz force F and Joule heat Q inside the specimen can be calculated using the following equations:

$$F = \frac{1}{2}\text{Re}(J \times B^*) \tag{7}$$

$$Q = \frac{1}{2}\text{Re}(J \cdot J^* \sigma^{-1}) \tag{8}$$

where σ is the conductivity, Re represents the real part of a complex quantity, and the asterisk designates the complex conjugate.

The governing equations for the thermal and flow fields (incompressible and Newtonian fluids) for the transient analysis are as follows:

$$\rho \nabla \cdot u = 0 \tag{9}$$

$$\rho \partial u / \partial t + \rho u \cdot (\nabla u) = \nabla \cdot \left\{ -pI + \mu \left[\nabla u + (\nabla u)^T \right] \right\} + F + \rho g_{\text{const}} + F_{st} \tag{10}$$

$$\rho C_p [\partial T / \partial t + u \cdot \nabla T] + \nabla \cdot (-k \nabla T) = Q - \nabla \cdot q_s \tag{11}$$

where ρ is the density of the alloy, μ is the viscosity, which varies with temperature, k is the thermal conductivity (which varies with temperature), u is the velocity vector, I is the unit-diagonal matrix, F_{st} is the surface tension, T is the temperature, p is the pressure, C_p is the corrected specific heat taking into account the latent heat, and q_s is the surface to ambient radiation heat.

The surface tension F_{st} is defined by the following:

$$F_{st} = \nabla \cdot [\gamma (I - nn^T) \delta] \tag{12}$$

where γ is the surface tension coefficient, n is the interface unit normal, and δ is a Dirac delta function, nonzero only at the fluid interface.

The Lorentz force F in the vertical direction is responsible for the buoyancy force F_z acting on the droplet. The kinetic equation governing the vertical oscillatory motion of the droplet is given as follows:

$$d(u,t) = (F_z - \frac{4}{3}\pi r^3 \rho \cdot g_{const}) / (\frac{4}{3}\pi r^3 \rho) \tag{13}$$

2.2.3. Initial and Boundary Conditions

The titanium–aluminum alloy specimen was initially in a sphere, levitated in a high-frequency harmonic magnetic field between two sets of coils. The weight of the Ti–Al sample was 1.2 g, and its radius was calculated to be about 4.17 mm based on its density, which was placed in a vacuum environment. The initial levitation position was 0 mm away from the center of the coils. The initial and boundary conditions for different physical fields are provided below.

(1) The boundary condition for fluid flow at the droplet surface is given by the following:

$$(-pI + \mu(\nabla u + \nabla u^T)) \cdot n = \gamma \xi \cdot n \tag{14}$$

where ξ is the local average curvature of the surface.

(2) Radiation of heat from the surface to the environment is as follows:

$$-\boldsymbol{n} \cdot \boldsymbol{q}_s = \varepsilon_r \sigma_r (T_{amb}^4 - T^4) \tag{15}$$

where T is the sample temperature, T_{amb} is the ambient temperature, and ε_r and σ_r are the emissivity and Boltzmann constant, respectively.

(3) Initial conditions for the thermal and flow fields include the following: $T_{int} = T_{amb} = 298$ K, $u = 0$, $p_0 = 101$ kPa.

2.3. Model Parameters

The main physical parameters of the numerical simulation are shown in Table 1. The coil material is copper with a conductivity of 5.99×10^7 S/m; the power supply type is alternating current with a current range of 462 A to 546 A and a frequency of 327 kHz. Since the EML process is to control the power of the high-frequency power supply to realize the stable levitation of the titanium–aluminum alloy specimen, we set five groups of levitation power in the experiment and read the corresponding currents accordingly, which were 462 A, 476 A, 504 A, 532 A, and 546 A. Therefore, these five values of currents were selected for the numerical simulation.

Table 1. EML parameters of numerical simulation model [22,23].

Physical Properties of Titanium–Aluminum Alloy	
Melting point [K]	1773
Conductivity [S/m]	1.3×10^7
Relative permeability	5
Density [Kg/m3]	$(-0.0005 \times T + 4.2254) \times 1000$
Dynamic viscosity [Pa·s]	$(0.06071 - (6.89 \times 10-5) \times T + (2.695 \times 10-8) \times T2 - (3.583 \times 10-12) \times T3$

3. Results and Discussion

3.1. Simulation of Electromagnetic Levitation Multiphysics Field

3.1.1. Simulation Validation

In order to assess the reliability of the model used in our study, the same parameters reported in the literature [24] were used in the numerical model, with the same dimensions of the coil structure and identical parameters of the molten droplets, and the same were all considered for the coupling of the electromagnetic field, the flow field, and the temperature field, which was finally verified by the magnitude and distribution of the temperature field. Figure 2 shows the states of the droplets after stabilization of levitation. Figure 2a shows the results of the temperature and fluid fields of the silicon molten droplet in the comparative literature, and Figure 2b shows the results obtained by the numerical model established in this study. It can be seen that the maximum temperature in the molten droplet calculated in this model was 1894.34 K and the maximum flow rate was 0.2912 m/s; the maximum temperature in the comparative literature was 1889 K and the maximum flow rate was 0.2692 m/s. The deviation in temperature was 0.26%, while the deviation in flow rate was 8.2%, which is less than the 10% error margin. The distribution of the thermal field and the fluid field derived from the simulation are essentially the same, indicating that the numerical model in this study is reliable.

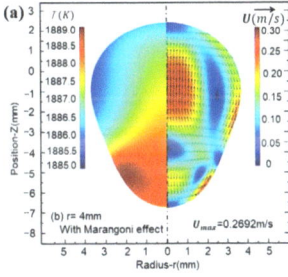 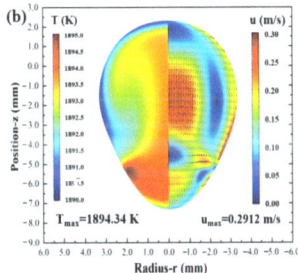

Figure 2. Numerical simulation verification results of temperature and fluid field of silicon droplet: (**a**) relevant literature [24]; (**b**) this study.

3.1.2. Simulation of Electromagnetic Field

In this study, the frequency domain transient is selected, and the simulation calculation time is taken as 5 s, considering the stabilization time of the molten droplet. The distribution of magnetic flux density inside the molten droplet at different currents, calculated from the Equations (1)–(6), is shown in Figure 3. The magnitude of the high-frequency current determines the strength of the induced magnetic field inside the droplet. Consistent with the skin effect, the high-frequency current in the levitation coil also determines the distribution of the induced magnetic field inside the molten drop in the form of a thin wall near the surface, as shown in Figure 3a–e. Since the magnetic flux density is also proportional to the turns of coils, in the longitudinal direction, the distribution in the upper half of the droplet is weaker than that of the lower half, and the strongest regions are located in the lower left and lower right, and the maxima increase almost proportionally with the increase in the current, 0. 122 T, 0.124 T, 0.129 T, 0.133 T, and 0.135 T. The weakest induced magnetic field is found in the central region of the droplet, which is small and negligible. In addition, the distribution of the induced magnetic field generated on the surface of the droplet also determines the force acting on the droplet, which directly affects the stability and deformation of the levitated droplet.

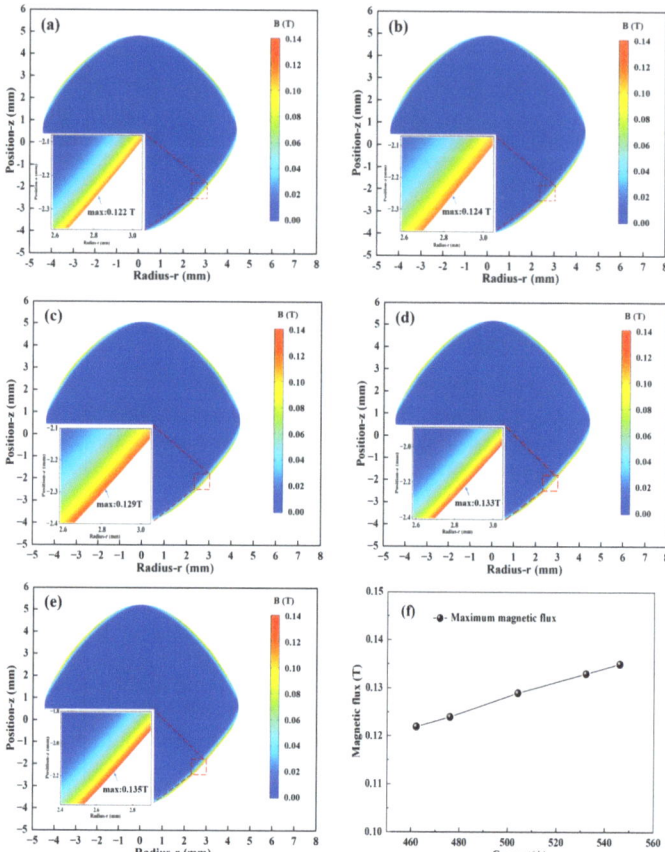

Figure 3. Simulation results of the magnetic field at different currents: (**a**) 462 A; (**b**) 476 A; (**c**) 504 A; (**d**) 532 A; (**e**) 546 A; (**f**) Maximum magnetic flux density.

3.1.3. Simulation of Fluid Field

The effects of different currents on the fluid field inside the molten droplet of Ti-48Al-2Cr alloy are shown in Figure 4. The arrows in Figure 4 indicate the velocity field vectors, so based on the distribution of the velocity field vectors, it can be observed that the regions of higher flux are mainly located on the left, right, and center sides, as well as at the location of the vortices. Based on Equation (13), under the combined effect of Lorentz force and gravity, the fluid in the center region of the droplet flows upward, reaches the top, and then flows downward, two larger vortices are formed in the middle near the two sides, and two smaller vortices are formed in the bottom region. When the current is 462 A, four vortices can be observed, as shown in Figure 4a; and when the current increases from 476 A to 546 A, the left and right sides are increased by two small vortices, and the number of vortices is increased to six, as shown in Figure 4b–e. As shown in Figure 4f, the maximum and minimum flow velocities increase linearly with the current from 0.39 m/s to 0.48 m/s. In addition to the effect of the Lorentz force action, since the viscosity is inversely proportional to the temperature, the shear stress decreases with the temperature increase, which also promotes the increase in the flow velocity.

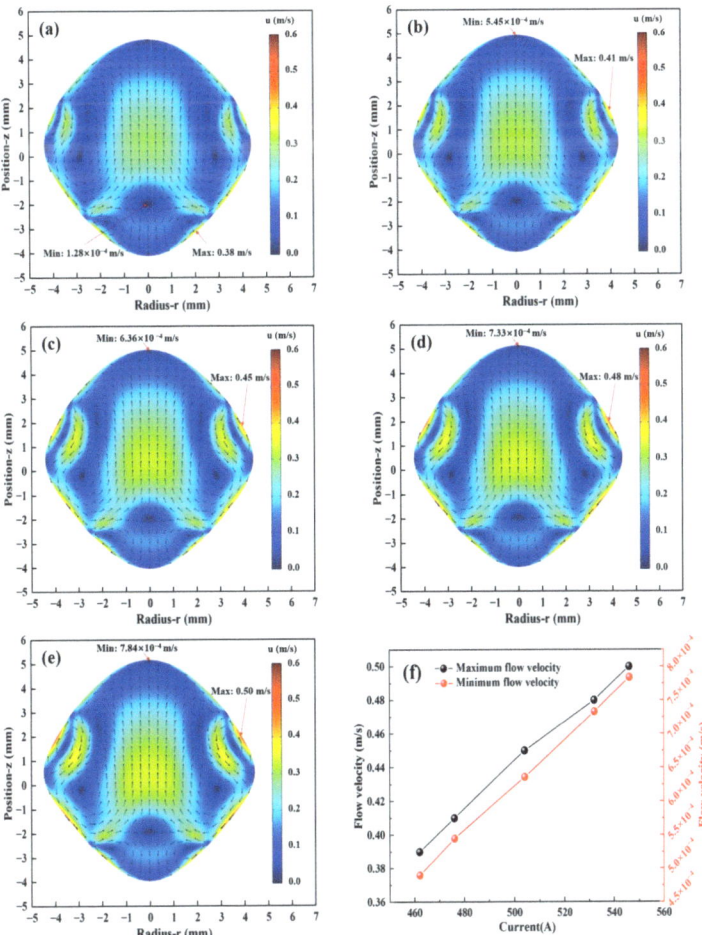

Figure 4. Simulation results of the fluid field at different currents: (**a**) 462 A; (**b**) 476 A; (**c**) 504 A; (**d**) 532 A; (**e**) 546 A; (**f**) Maximum and minimum flow velocity.

3.1.4. Simulation of Thermal Field

Figure 5a–e show the numerical simulation results of the temperature distribution of the molten droplet. Based on the calculations in Equations (8) and (11), it can be seen that the bottom of the droplet near the heating coil obtains the maximum joule heat and the highest temperature, while the area near the two sides in the middle has the lowest temperature. The heat transfer inside the droplet is mainly convection heat transfer. It can be seen from Figure 5f that with the increase in exciting current, the temperature inside the droplet keeps increasing, from 1807 K to 1830 K. This is because the eddy current induced inside the droplet continues to increase, which makes the heating effect of electromagnetic induction more obvious. In addition, the effect of electromagnetic stirring makes the internal temperature distribution of the molten droplet more uniform, and the temperature difference is only about 2 K to 2.4 K. Comparing the surface temperatures of the molten droplets measured during the experiment with the maximum temperatures calculated by the simulation, we can see that the values are very close and the differences between them are less than 2 K, which proves the accuracy of the simulation.

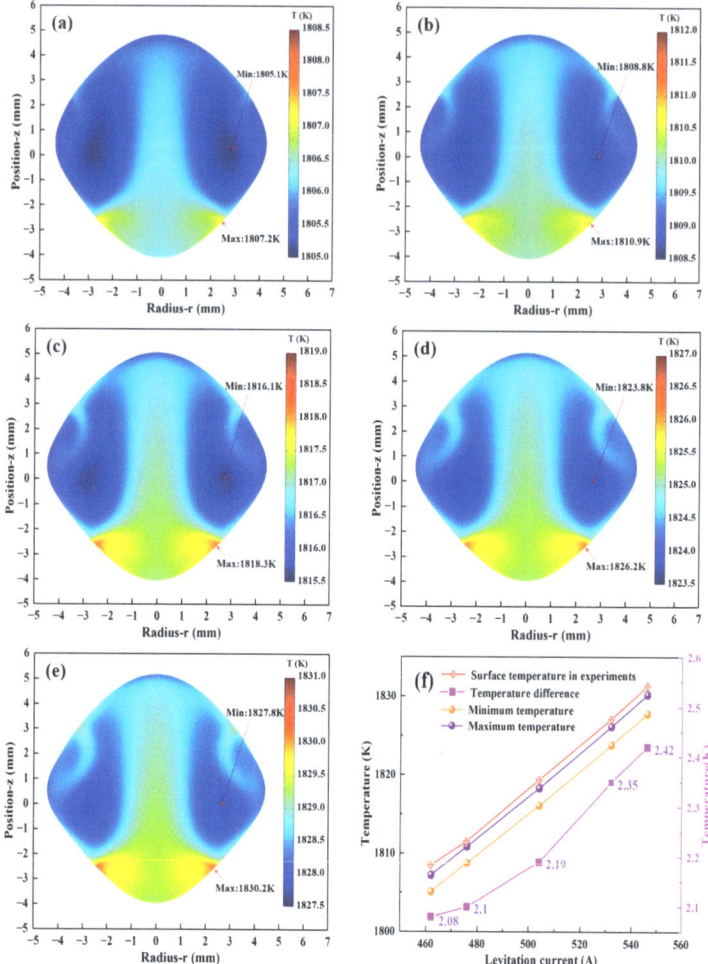

Figure 5. Simulation results of the temperature field at different currents: (**a**) 462 A; (**b**) 476 A; (**c**) 504 A; (**d**) 532 A; (**e**) 546 A; (**f**) temperatures.

3.2. Simulation of the Oscillation Process of Molten Droplets

In the initial stage of levitation melting, there is a transient oscillation process that is affected by a variety of factors, such as the initial position, coil current, number of turns of the coil, and material properties [25,26]. The main influence on the oscillation process is the magnitude of the current when the initial position, coil structure, and current frequency are known. The droplet is mainly subjected to the Lorentz force and gravity, and the direction of the combined force is parallel to the z-axis. During the oscillation process, the Lorentz force on the droplet also changes due to the change in the levitation position. The oscillatory behavior of the electromagnetically levitated droplets can be approximated by the spring model theory. The relation between the displacement of the molten droplet near the equilibrium position and the combined force is consistent with the relation between the spring force applied to the suspension load F_s and the spring deformation x_s described in Hooke's law, that is $F_s = -kx_s$, where x_s is the spring displacement and k is the elastic coefficient. Figure 6 shows the curve of the electromagnetic force on the molten droplet at different currents, and it can be seen that the electromagnetic force in the z-axis direction gradually decreases with time until the stable trend, which is known as a damped oscillation. When the current is 462 A, the Lorentz force on the molten droplet stabilizes after 4.34 s, which is the longest time, and when the current is further increased to 546 A, the corresponding stabilization time decreases to 2.86 s. It can be concluded that the molten droplets perform the damped oscillatory motion in the electromagnetic levitation system, and with the increase in the current, the electromagnetic force on the molten droplets increases, the oscillation damping of the molten droplets gradually increases, and its stabilization time tends to decrease.

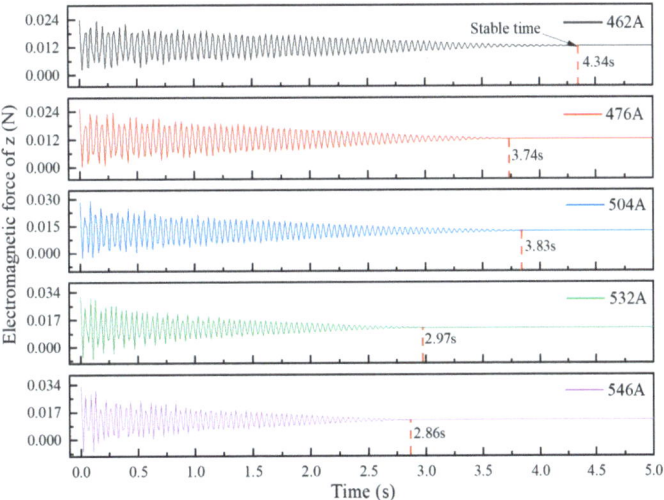

Figure 6. Variation curves of electromagnetic force on the molten droplets at different currents.

3.3. Simulation of the Deformation Process of Molten Droplets

The distributions of the Lorentz force at different currents obtained by numerical simulation are shown in Figure 7a–e. As the current increases, the maximum Lorentz force increases from 1.27×10^7 N/m^3 to 1.58×10^7 N/m^3, and the Lorentz force on the droplet is mainly distributed on its surface. On the contrary, the Lorentz force inside the droplet tends to be close to 0 N/m^3 due to the skinning effect of the high-frequency current induced. The Lorentz force on the upper and lower surfaces of the droplet is directed toward the inside of the droplet, and the arrangement of the coil structure makes the Lorentz force on the lower surface of the droplet significantly higher than that on the upper surface, and the maximum value occurs on the two sides near the lower part of the coil. Figure 7f shows the maximum Lorentz force corresponding to different levitation currents and the variation in the diameter in both the transverse and longitudinal directions. The transverse and longitudinal diameters of the droplets increase, but the change in the transverse diameter tends to flatten, and the shape eventually remains "near rhombus".

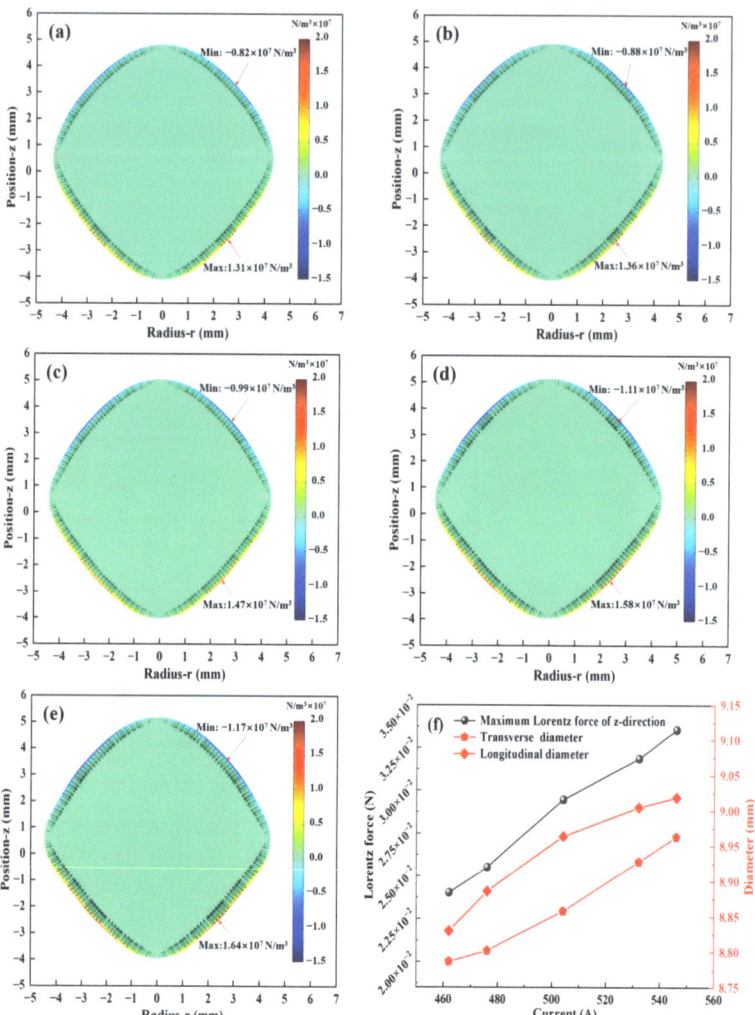

Figure 7. Lorentz force within the molten droplets at different currents: (**a**) 462 A; (**b**) 476 A; (**c**) 504 A; (**d**) 532 A; (**e**) 546 A; (**f**) maximum Lorentz force and the diameter of the molten droplets in different directions.

To deeply analyze the deformation behavior during the oscillation of the molten droplet, the deformation process of the molten droplet was simulated from 0 s to 4 s at a current of 532 A, and the results are shown in Figure 8. From the previous analysis, it can be seen that the Lorentz force pushes the surface of the molten droplet inward and upward, and the component of the Lorentz force in the z-axis direction is opposite to gravity at the equilibrium position. At the initial stage of oscillation and deformation, the molten droplet starts from the initial position, the up and down oscillation amplitude is large, and the deformation is severe, as shown in Figure 8a–d. With the increase in time, the degree of oscillation and deformation is gradually weakened, and the molten droplet tends to be stable in shape at 2.97 s, showing a "near rhombus", as shown in Figure 8e. Then, the shape at 4 s is unchanged compared with that at 2.97 s, as shown in Figure 8f.

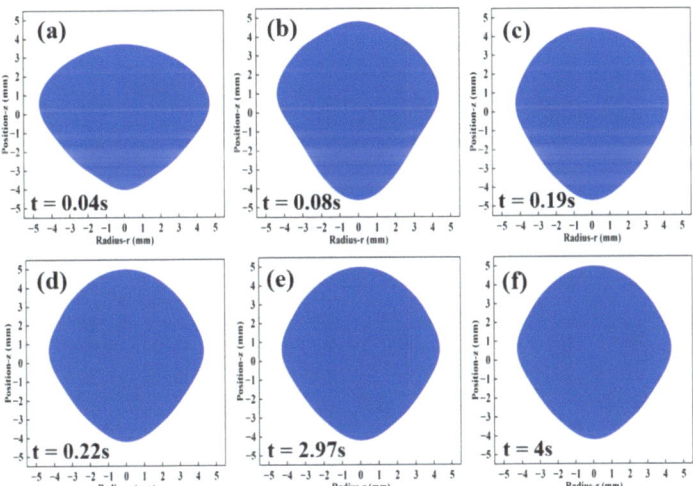

Figure 8. Deformation of the molten droplet in the interval from 0 s to 4 s at a current of 532 A: (**a**) 0.04 s; (**b**) 0.08 s; (**c**) 0.19 s; (**d**) 0.22 s; (**e**) 2.97 s; (**f**) 4 s.

As shown in Figure 9, the color photographs of the molten droplet of Ti-48Al-2Cr alloy in the process of EML (at the current of 532 A) were taken using a video camera. Figure 9a–c show the sample melting in the oscillating deformation stage. By comparison, it can be seen that the deformation is almost consistent with the simulation process in Figure 8, as shown by the red contour curves in each image. After about 3 s to 4 s, the oscillation and deformation tend to stabilize, and the shape of the molten droplet remains "near rhombus", as shown in Figure 9d.

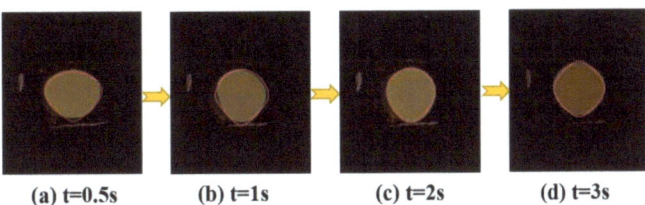

Figure 9. Deformation of the Ti-48Al-2Cr alloy molten droplet at different stages in the EML process (at the current of 532 A): (**a**) 0.5 s; (**b**) 1 s; (**c**) 2 s; (**d**) 3s.

4. Conclusions

In this paper, numerical simulations of multiphysics field coupling and the oscillatory deformation of Ti-48Al-2Cr alloy under different current conditions during EML melting were investigated by the FEM and ALE methods. Finally, the comparative validation was carried out by EML experiments, and the following conclusions were obtained:

(1) The maximum velocity zones of the electromagnetically levitated droplets are located on the left and right sides near the droplet surface. Then, the number of vortices also rises from four to six with the current increase. The maximum temperature is observed at the bottom of the molten droplet, while the temperature near the center is relatively lower. The application of electromagnetic stirring results in a more uniform temperature distribution within the molten droplet, with a temperature difference of approximately 2 K.

(2) The titanium–aluminum alloy molten droplet is damped and oscillates from its initial position due to the combined effects of gravity and the Lorentz force. The amplitude

of the oscillation increases with the current, but when the current increases from 462 A to 546 A, the levitation time to reach a steady state decreases gradually, from 4.34 s to 2.86 s.

(3) The integrated force field exerts a linear influence on the longitudinal diameter of the titanium–aluminum alloy molten droplet, which increases with the current. In contrast, the transverse diameter undergoes a gradual flattening. At the onset of oscillation, the molten droplet exhibits greater deformation, before reaching a stabilization period where its shape remains unchanged and assumes a "near-rhombus" configuration.

Author Contributions: Conceptualization, G.Z. and X.W.; methodology, X.W.; software, X.W.; validation, P.Y. and X.W.; formal analysis, P.Y.; investigation, G.Z. and P.Y.; resources, P.Y. and X.W.; data curation, X.P. and X.W.; writing—original draft preparation, P.Y. and X.W.; writing—review and editing, X.W. and G.Z.; visualization, X.W. and Z.X.; supervision, G.Z. and P.Y.; project administration, G.Z.; funding acquisition, G.Z. All authors have read and agreed to the published version of the manuscript.

Funding: This research was funded by the Yunnan Provincial Department of Education Scientific Research Fund Project (grant number 2023J0130); National Natural Science Foundation of China (grant number 52074140) and Analysis and Testing Foundation of Kunming University of Science and Technology (grant number 2022T20090040).

Institutional Review Board Statement: Not applicable.

Informed Consent Statement: Not applicable.

Data Availability Statement: Data are contained within the article.

Conflicts of Interest: We declare that the research was conducted in the absence of any commercial or financial relationships that could be construed as a potential conflict of interest.

References

1. Genc, O.; Unal, R. Development of gamma titanium aluminide (γ-TiAl) alloys: A review. *J. Alloys Compd.* **2022**, *929*, 167262. [CrossRef]
2. Lim, H.P.; Liew, W.Y.H.; Melvin, G.J.H.; Jiang, Z.-T. A short review on the phase structures, oxidation kinetics, and mechanical properties of complex Ti-Al alloys. *Materials* **2021**, *14*, 1677. [CrossRef] [PubMed]
3. Kastenhuber, M.; Rashkova, B.; Clemens, H.; Mayer, S. Enhancement of creep properties and microstructural stability of intermetallic β-solidifying γ-TiAl based alloys. *Intermetallics* **2015**, *63*, 19–26. [CrossRef]
4. Gupta, R.K.; Pant, B.; Sinha, P.P. Theory and practice of γ+ α2 Ti aluminide: A review. *Trans. Indian Inst. Met.* **2014**, *67*, 143–165. [CrossRef]
5. Ostrovskaya, O.; Badini, C.; Baudana, G.; Padovano, E.; Biamino, S. Thermogravimetric investigation on oxidation kinetics of complex Ti-Al alloys. *Intermetallics* **2018**, *93*, 244–250. [CrossRef]
6. Reiplinger, B.; Plevachuk, Y.; Brillo, J. Surface tension of liquid Ti, V and their binary alloys measured by electromagnetic levitation. *J. Mater. Sci.* **2022**, *57*, 21828–21840. [CrossRef]
7. Egry, I.; Diefenbach, A.; Dreier, W.; Piller, J. Containerless processing in space—Thermophysical property measurements using electromagnetic levitation. *Int. J. Thermophys.* **2001**, *22*, 569–578. [CrossRef]
8. Wu, P.; Yang, Y.; Barati, M.; McLean, A. Electromagnetic levitation of silicon and silicon-iron alloy droplets. *High Temp. Mater. Process.* **2014**, *33*, 477–483. [CrossRef]
9. Brillo, J.; Lohofer, G.; Hohagen, F.S.; Schneider, S.; Egry, I. Thermophysical property measurements of liquid metals by electromagnetic levitation. *Int. J. Mater. Prod. Technol.* **2006**, *26*, 247–273. [CrossRef]
10. Bracker, G.; Schneider, S.; Matson, D.; Hyers, R. Dynamic nucleation in sub-critically undercooled melts during electromagnetic levitation. *Materialia* **2022**, *26*, 101623. [CrossRef]
11. Reinartz, M.; Kolbe, M.; Herlach, D.M.; Rettenmayr, M.; Toropova, L.V.; Alexandrov, D.V.; Galenko, P.K. Study on anomalous rapid solidification of Al-35 at% Ni in microgravity. *JOM* **2022**, *74*, 2420–2427. [CrossRef]
12. Watanabe, M.; Takahashi, Y.; Imaizumi, S.; Zhao, Y.; Adachi, M.; Ohtsuka, M.; Chiba, A.; Koizumi, Y.; Fukuyama, H. Thermophysical properties of liquid Co–Cr–Mo alloys measured by electromagnetic levitation in a static magnetic field. *Thermochim. Acta* **2022**, *708*, 179119. [CrossRef]
13. Liu, Y.; Zhang, G.F.; Qi, X.; Shi, Z.; Yan, P.; Jiang, Q. Research Progress on Application and Numerical Simulation of Electromagnetic Levitation Melting Metal Alloy. *Mater. Rep.* **2022**, *36*, 20080265-8.

14. Matson, D.M.; Battezzati, L.; Galenko, P.K.; Gandin, C.-A.; Gangopadhyay, A.K.; Henein, H.; Kelton, K.F.; Kolbe, M.; Valloton, J.; Vogel, S.C.; et al. Electromagnetic levitation containerless processing of metallic materials in microgravity: Rapid solidification. *NPJ Microgravity* **2023**, *9*, 65. [CrossRef]
15. Kermanpur, A.; Jafari, M.; Vaghayenegar, M. Electromagnetic-thermal coupled simulation of levitation melting of metals. *J. Mater. Process. Technol.* **2011**, *211*, 222–229. [CrossRef]
16. Asakuma, Y.; Sakai, Y.; Hahn, S.H.; Tsukada, T.; Hozawa, M.; Matsumoto, T.; Fujii, H.; Nogi, K.; Imaishi, N. Equilibrium shape of a molten silicon drop in an electromagnetic levitator in microgravity environment. *Metall. Mater. Trans. B* **2000**, *31*, 327–329. [CrossRef]
17. Liang, C.; Wang, H.P.; Zhang, P.C.; Wei, B. Liquid dripping dynamics and levitation stability control of molten Ti–Al–Nb alloy within electromagnetic fields. *Phys. Fluids* **2022**, *34*, 055113. [CrossRef]
18. Braess, H.; Wriggers, P. Arbitrary Lagrangian Eulerian finite element analysis of free surface flow. *Comput. Methods Appl. Mech. Eng.* **2000**, *190*, 95–109. [CrossRef]
19. Saadat, M.H.; Karlin, I.V. Arbitrary Lagrangian–Eulerian formulation of lattice Boltzmann model for compressible flows on unstructured moving meshes. *Phys. Fluids* **2020**, *32*, 046105. [CrossRef]
20. Feng, L.; Shi, W.Y. The influence of eddy effect of coils on flow and temperature fields of molten droplet in electromagnetic levitation device. *Metall. Mater. Trans. B* **2015**, *46*, 1895–1901. [CrossRef]
21. Yan, P.; Zhang, G.; Yang, Y.; Mclean, A. Numerical investigation of the position and asymmetric deformation of a molten droplet in the electromagnetic levitation system. *Metall. Mater. Trans. B* **2020**, *51*, 247–257. [CrossRef]
22. Wunderlich, R.K.; Hecht, U.; Hediger, F.; Fecht, H. Surface Tension, Viscosity, and Selected Thermophysical Properties of Ti48Al48Nb2Cr2, Ti46Al46Nb8, and Ti46Al46Ta8 from Microgravity Experiments. *Adv. Eng. Mater.* **2018**, *20*, 1800346. [CrossRef]
23. Brillo, J.; Wessing, J.J. Density, Molar Volume, and Surface Tension of Liquid Al-Ti. *Metall. Mater. Trans. A* **2017**, *48*, 868–882.
24. Yan, P.; Zhang, G.; Yang, Y.; Mclean, A. Numerical simulation of Marangoni effect on the deformation behavior and convection of electromagnetic levitated silicon droplets under static magnetic fields. *Int. J. Heat Mass Transf.* **2020**, *163*, 120489. [CrossRef]
25. Adachi, M.; Aoyagi, T.; Mizuno, A.; Watanabe, M.; Kobatake, H.; Fukuyama, H. Precise density measurements for electromagnetically levitated liquid combined with surface oscillation analysis. *Int. J. Thermophys.* **2008**, *29*, 2006–2014. [CrossRef]
26. Wang, S.; Li, H.; Yuan, D.; Wang, S.; Wang, S.; Zhu, T.; Zhu, J. Oscillations and size control of titanium droplet for electromagnetic levitation melting. *IEEE Trans. Magn.* **2017**, *54*, 1–4. [CrossRef]

Disclaimer/Publisher's Note: The statements, opinions and data contained in all publications are solely those of the individual author(s) and contributor(s) and not of MDPI and/or the editor(s). MDPI and/or the editor(s) disclaim responsibility for any injury to people or property resulting from any ideas, methods, instructions or products referred to in the content.

Article

Hot Deformation Constitutive Analysis and Processing Maps of Ultrasonic Melt Treated A5052 Alloy

Sun-Ki Kim [1], Seung-Hyun Koo [1], Hoon Cho [2] and Seong-Ho Ha [2,*]

[1] NICE LMS Co., Ltd., Yesan 32446, Republic of Korea
[2] Korea Institute of Industrial Technology, Incheon 21999, Republic of Korea
* Correspondence: shha@kitech.re.kr

Abstract: Hot deformation constitutive analysis and processing maps of ultrasonic melt treated (UST) A5052 alloy were carried out based on a hot torsion test in this study. The addition of the Al–Ti master alloy as a grain refiner with no UST produced a finer grain size than the UST and pure Ti sonotrode. The Al3Ti phase particles in the case of the Al–10Ti master alloy acted as a nucleus for grain refinement, while the Ti atoms dissolved in the melt from the sonotrode were considered to have less of a grain refinement effect, even under UST conditions, than the Al3Ti phase particles in the Al–Ti master alloy. The constitutive equations for each experimental condition by torsion test were derived. In the processing maps examined in this study, the flow instability region was not present under UST in the as-cast condition, but it existed under the no UST condition. The effects of UST examined in this study are considered as (i) the uniform distribution of Ti solutes from the sonotrode and (ii) the reduction of pores by the degassing effect. After the homogenization heat treatment, most instability regions disappeared because the microstructures became uniform following the decomposition of intermetallic compounds and distribution of solute elements.

Keywords: A5052; ultrasonic treatment; grain refinement; hot deformation; processing map

Citation: Kim, S.-K.; Koo, S.-H.; Cho, H.; Ha, S.-H. Hot Deformation Constitutive Analysis and Processing Maps of Ultrasonic Melt Treated A5052 Alloy. *Materials* **2024**, *17*, 3182. https://doi.org/10.3390/ma17133182

Academic Editor: Frank Czerwinski

Received: 25 April 2024
Revised: 9 June 2024
Accepted: 12 June 2024
Published: 28 June 2024

Copyright: © 2024 by the authors. Licensee MDPI, Basel, Switzerland. This article is an open access article distributed under the terms and conditions of the Creative Commons Attribution (CC BY) license (https://creativecommons.org/licenses/by/4.0/).

1. Introduction

Al alloys are widely utilized in various industries, such as aerospace and transportation, because they have high strength, low density, corrosion resistance, good electrical conductivity, and thermal conductivity, as well as machinability. Al–Mg alloys belonging to the 5xxx series of Al alloys show an improved strength and corrosion resistance due to the effect of Mg as a crucial strengthening element. A5052 is an Al–Mg alloy supplemented with a small Cr addition, and has moderate strength, good workability, and very good corrosion resistance [1–4]. The mechanical properties of Al alloys can be controlled by various microstructural characteristics, such as refined grains. In the grain refinement process of Al alloys, the intermetallic compounds such as Al_3Ti contained in chemical grain refiners, like Al–10Ti and Al–5Ti–1B master alloys, act as nuclei to refine the grains [5]. In a previous study [6], ultrasonic melt treatment (UST) using a commercial purity Ti sonotrode for A5052 alloy melts was examined. As a result, the Ti content dissolved in the A5052 melt was confirmed in the range between 0.06 and 0.07 mass%, which is considered a proper amount to lead to a grain refinement effect. The UST, even only for just a few minutes, refined the columnar zone-prevailing coarse grains into the fine equiaxed grains in the A5052 billet. The refinement by the UST also affected microstructures such as grains, Al–Fe-based particles, dendrites, and pores, leading to a remarkable improvement of mechanical properties in the as-cast condition. Such improvement in billet quality possibly results in an increase in plastic workability.

Hot deformation and dynamic softening affect not only flow stress but also microstructure. In general, the deformation process at high temperatures promotes material flow as well as high dynamic softening. In this process, if the deformation speed is high, the thermal energy generated during the deformation will increase [7–9], and many studies in the past have examined the effect of softening by thermal energy generated during deformation. And also, many researchers have proceeded with the premise that the effects must be taken into consideration by analyzing the hot working characteristics [7–12]. In general, studies on hot deformation of metallic materials have been carried out through the build-up of processing maps based on the dynamic material model (DMM) under hot working conditions [7,13–19]. Power dissipation efficiency is known to be related to the microstructure of the material after hot deformation. The stable region in the processing map has been considered to optimize deformation conditions for hot working. The processing maps on the various ferrous and nonferrous materials such as the Al–Si alloy [20], molybdenum [21], NIMONIC 80A [22], Ti–6Al–4V alloy [23], stainless steel [24], and Mg alloys [25–27] have been examined. In the case of A5052 alloys, He et al. [4] studied the hot deformation behavior of cast alloy between 300 and 500 °C. In their study, the optimum working parameters in the ranges of strain and temperature for cast 5052 alloy were determined. Son et al. [19] investigated hot deformation characteristics of A5052 modified with CaO added Mg master alloy as an Mg additive. The alloy modification using CaO added Mg was to improve the oxidation resistance of Mg in the A5052 alloy. However, no research on the effect of UST with a Ti sonotrode on the workability of A5052 based on processing maps has been reported. In this study, hot deformation constitutive analysis and processing maps of the ultrasonic melt treated A5052 alloy were carried out based on a hot torsion test.

2. Materials and Methods

2.1. Fabrication of Materials

The composition of the A5052 alloy examined in this study is shown in Table 1. A5052 melts of approximately 10 kg at 720 °C were prepared using an induction furnace. To conduct the UST for the A5052 alloy melts, a 5 kW ultrasonic power supply (Generator, USGC-5-22 MS) and a water-cooled magnetostrictive transducer (MST-5-18) with 20 kHz were used. A commercially pure Ti was employed for the sonotrode material, which was immersed for 3 min to be preheated to the same temperature as that of the A5052 melts. When the temperature of the melts was at 720 °C, the UST was performed at an output of 4.4 kHz for 1 min to 1 min and 30 s. After the UST, the alloy melts were poured into a mold preheated to 150 °C. The addition of Al–10mass%Ti master alloy, which is a common chemical grain refiner, was examined for comparison. The target Ti content was set in the range of 200 to 300 ppm, considering that of the Ti sonotrode in the UST and the minimum content for the grain refinement effect by Ti. The Al–10Ti master alloy ingots were added and held at 720 °C for approximately 10 min. After that, the melts were poured into the same mold as in the UST process and preheated to 150 °C. And then, for comparison between as-cast and as-homogenized conditions, the homogenization heat treatment was performed at 450 °C for 8 h and followed by furnace cooling. The measured compositions of alloy billets with or without the UST process are shown in Table 1. The contents of Mg and Cr, which are the main constituent elements of the A5052 alloy, did not change significantly. Furthermore, the Ti content was found to be 0.025 mass% and 0.02 mass% in the billets with and without the UST, respectively. Therefore, it is thought that the manufactured billets are suitable for use in this study. The experimental process conducted in this study is shown in Figure 1.

The macrostructures and microstructures for the as-cast alloys were compared. The samples for the macrostructures were processed into a disk shape with a thickness of 10 mm and etched using Poulton's reagent. The microstructures were electrochemically etched by Barker's regent and observed through a polarizing microscope. The grain size was measured using the linear intercept method specified in ASTM E-112 [28].

Table 1. Analyzed compositions of A5052 alloy ingot and billets examined in this study.

Sample	Analyzed Compositions (Mass%)								
	Si	Fe	Cu	Mn	Mg	Cr	Zn	Ti	Al
A5052 (Ingot)	0.0524	0.0958	0.0026	0.0048	2.71	0.238	<0.001	0.0028	Bal.
A5052 (UST)	0.0682	0.101	0.0039	0.0057	2.64	0.248	<0.001	0.0255	Bal.
A5052 (no UST)	0.0481	0.111	0.0025	0.0058	2.74	0.248	<0.001	0.0274	Bal.

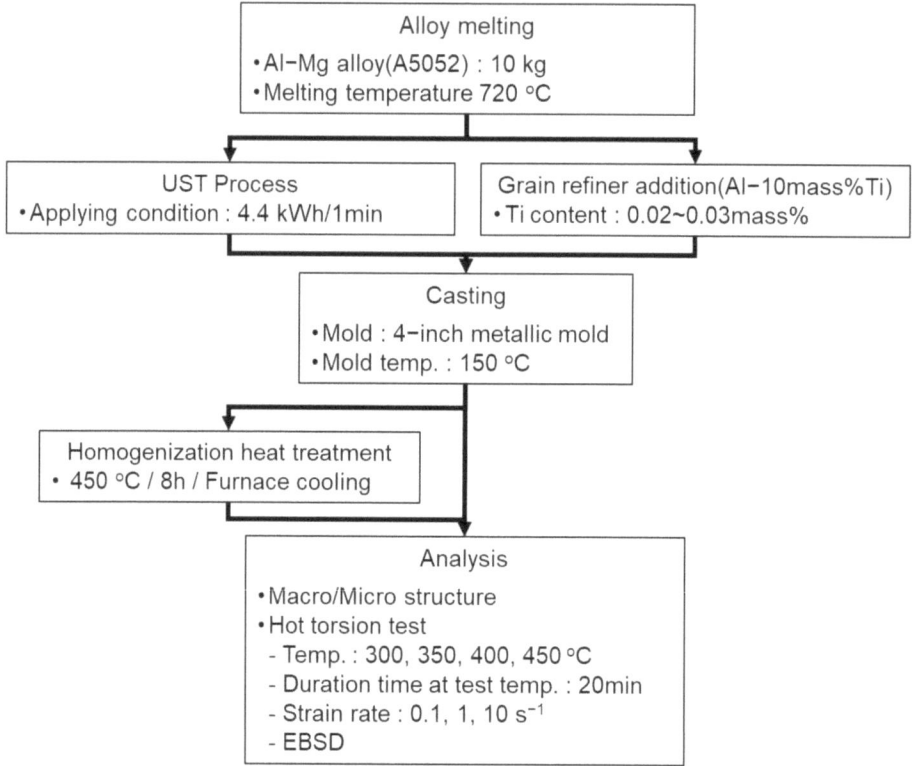

Figure 1. Experimental process conducted in this study.

2.2. Hot Torsion Test

The torsion test specimen was processed into a shape with a length of 20 mm and diameter of 10 mm in the test section from both the as-cast and as-homogenized materials. In the torsion test, the temperature was increased up to the target temperature at 1 °C/s. using an infrared heater. The test began after holding at the target temperature for 10 min to ensure that the temperature of the entire specimen was kept uniform. The test temperatures ranged between 300 and 450 °C, and the strain rate was set from 0.1 to 10 s^{-1}. The torsion test continued until the effective strain reached 15 or the specimen fractured. After the test was completed, the specimens were immediately water-cooled to obtain plastic deformation structures. The effective stress (σ) and strain (ε) of each test specimen were calculated from the torque (M) and angular displacement ($\dot{\theta}$) obtained during the torsion test by using the von Mises criterion and the method proposed by Field and Backofen [16]:

$$\theta = \frac{\sqrt{3}(3+p+q)}{2\pi r} \quad (1)$$

$$\varepsilon = \frac{r\theta}{\sqrt{3}L} \quad (2)$$

$$P = \frac{\partial \ln M}{\partial \ln \dot{\theta}} \mid \theta, T \quad q = \frac{\partial \ln M}{\partial \ln \theta} \mid \dot{\theta}, T \quad (3)$$

where p is strain rate sensitivity, q is the strain hardening coefficient, r is the radius of the gauge section, L is the gauge length of the sample, and $\dot{\theta}$ is the twist rate.

Microstructures deformed during the hot torsion test were observed and analyzed by electro backscattering diffraction (EBSD) on sections parallel to the torsion axis at the center of the specimens. The samples for EBSD were prepared by electropolishing in a solution of 20% perchloric acid in methanol at $-30\ °C$.

3. Results

3.1. Macro/Microstructures of Billets

The macrostructures of the disks taken from each billet are shown in Figure 2. Equiaxed grains were observed, and cast structures such as chill zones and columnar tablets did not appear in both disks. Therefore, it can be said that the Ti addition of about 200 ppm even without the UST can lead to the significant grain refinement of the as-cast microstructures. Figure 3 shows the microstructures depending on locations in each disk. They exhibited non-uniform structures with a mixture of fine or relatively coarse grains. However, the deviation of grain size was larger in the disk with the UST rather than in that with no UST. As a result of grain size measurement, the average grain sizes of center and R/2 under the UST conditions were 200 μm and 230 μm, and was 180 μm under no UST, respectively. Therefore, it was confirmed that the addition of the Al–Ti master alloy as a grain refiner with no UST produced a finer grain size than the UST and pure Ti sonotrode. In the case of the addition of the Al–10Ti master alloy, the Al_3Ti phase particles act as a nucleus for grain refinement. However, the Ti atoms dissolved in the melt from the sonotrode are considered to have less of a grain refinement effect, even under the UST conditions, than the Al_3Ti phase particles in the Al–Ti master alloy.

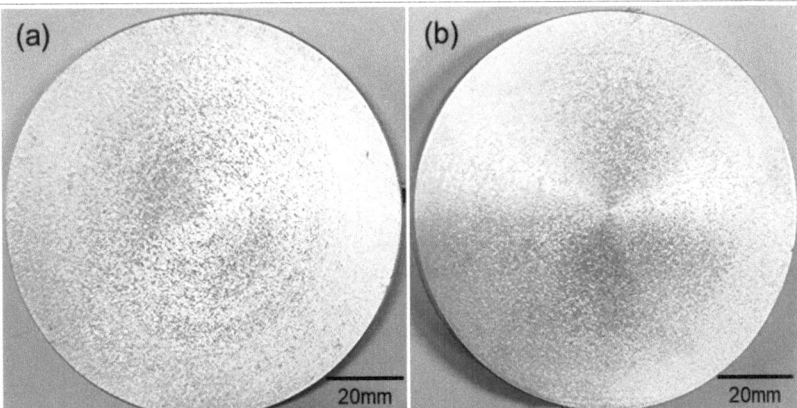

Figure 2. Macrostructures of cross-sectioned disks: (**a**) UST and (**b**) no UST.

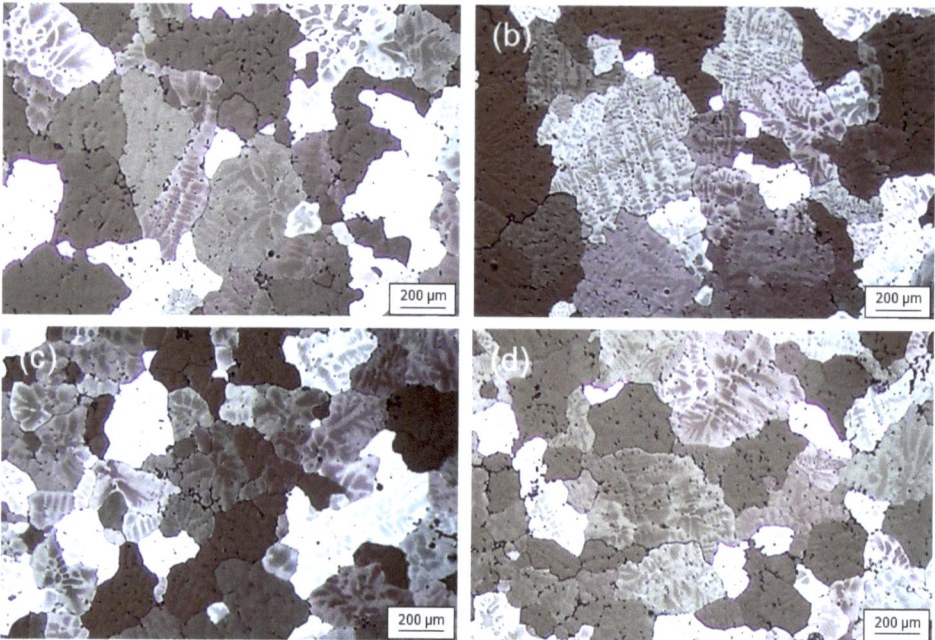

Figure 3. Microstructures of cross-sectioned disks: (**a**) center and (**b**) R/2 with UST, and (**c**) center and (**d**) R/2 with no UST. The average grain sizes are (**a**) 200 μm, (**b**) 230 μm, (**c**) 180 μm, and (**d**) 180 μm, respectively.

3.2. Constitutive Analysis

The constitutive equation was calculated for each condition as a method for analyzing the relationship among the parameters based on the results of the torsion test. It is generally known that, as the deformation temperature rises, the yield stress and flow stress decrease. Similarly, a decrease in work hardening mainly causes a decrease in stress. When work hardening occurs, the flow stress increases as the strain increases. Therefore, the relationship between the flow stress and the test temperature can be described in order to express the given strain (ε) and strain rate ($\dot{\varepsilon}$), which appear in the following Arrhenius-type equation.

$$\sigma_{\varepsilon, \dot{\varepsilon}} = C_1 \exp(Q/RT)$$

where $\sigma_{\varepsilon, \dot{\varepsilon}}$ is the true stress at the given stain and strain rate, Q is the activation energy during deformation, R is a gas constant, T is the test temperature, and C_1 is a constant.

When this relationship is established, a straight line with a slope of Q/R can be created in a graph with respect to 1/T. However, this equation is only suitable when the mechanism to determine the flow stress works in the same temperatures. The determination of flow stress is limited by one equation at a wide temperature range in which other mechanisms operate. Sellars and McTegart [29] proposed three Arrhenius-type equations that can describe the relationship among the parameters in hot deformation as follows:

$$\dot{\varepsilon} \exp(Q/RT) = A_1 \sigma_p^n = Z \tag{4}$$

$$\dot{\varepsilon} \exp(Q/RT) = A_2 \exp(\beta \sigma_p) = Z \tag{5}$$

$$\dot{\varepsilon} \exp(Q/RT) = A(\sinh \alpha \sigma_p)^n = Z \tag{6}$$

where A_1, A_2, A, n', n, $β$, and $α$ ($=β/n'$) are material constants (Table 1); and $\dot{ε}$, $σ_p$, Q, T, R, and Z are strain rate, peak stress, the activation energy for deformation, temperature, gas constant and the Zener–Hollomon parameter, respectively. Equations (4) and (5) break down at high stress and high temperature, respectively. The hyperbolic sine law, Equation (6), is used as a general form suitable for a wide range of applications [30,31].

The material constants were calculated from the effective stress–strain data to build the constitutive equations. Taking natural logarithms on both sides of Equations (4) and (5), the following equations can be obtained:

$$\ln \dot{ε} = \ln A_1 + n' \ln σ_p - Q/RT \tag{7}$$

$$\ln \dot{ε} = \ln A_2 + β σ_p - Q/RT \tag{8}$$

And then constants n' and $β$ were determined from the slope of the plot for $\ln\dot{ε}$–$\ln σ_p$ and $\ln\dot{ε}$–$σ_p$, respectively. The average value of the constants can be determined by the linear fitting method at a given temperature, as shown in Figure 4a,b. The value of $α$ ($=β/n'$) was calculated by the determined n' and $β$. Then, Equation (6) could be rewritten as:

$$\ln \dot{ε} = \ln A_1 + n(\sinh α σ_p) - Q/RT \tag{9}$$

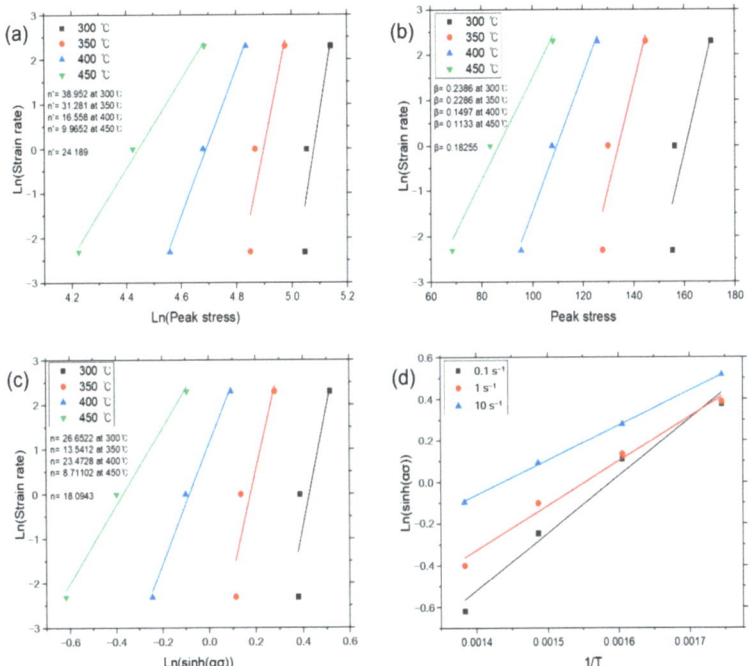

Figure 4. Constitutive analysis of as-cast A5052 alloys with UST according to Equations (4)–(7). (**a**) $\dot{ε}$–$σ_p$ dependence, (**b**) $\dot{ε}$–exp($σ_p$) dependence, (**c**) $\dot{ε}$–sinh($ασ_p$) dependence, and (**d**) Arrhenius dependence.

Taking partial differential equations of Equation (8), the activation energy for deformation can be expressed as:

$$Q = R \left[\frac{\partial \ln \dot{ε}}{\partial \ln \sinh(α σ_p)} \right] \left[\frac{\partial \ln \sinh(α σ_p)}{\partial (1/T)} \right] = RnS \tag{10}$$

where s is a slope of the relationship of $\ln(\sinh\alpha\sigma_p) - 1/T$ at a given strain rate.

The linear relationships of $\ln\dot\varepsilon - \ln(\sinh\alpha\sigma_p)$ and $\ln(\sinh\alpha\sigma_p) - 1/T$ at a given strain rate were fitted from Figure 4c,d. Constants n and s were determined by the average value of slope in the plots. Each constant for deriving the constitutive equation was calculated using the above method and is given in Table 2, and the graphs for each experimental condition are shown in Figures 4–7.

Table 2. Values of each constant depending on temperatures examined in Figures 4–7.

	Temp. (°C)	n′	n′ (avg.)	β	β (avg.)	n	n (avg.)
Figure 3	300	38.952	24.189	0.2386	0.18255	26.6522	18.0943
	350	31.281		0.2286		13.5412	
	400	16.558		0.1497		23.4728	
	450	9.9652		0.1133		8.71102	
Figure 4	300	33.27	18.92	0.2224	0.158815	22.504	14.1
	350	14.49		0.1155		10.761	
	400	16.01		0.1458		12.807	
	450	11.89		0.1516		10.315	
Figure 5	300	42.088	19.7949	0.2966	0.1668	29.2963	13.14416
	350	16.054		0.1339		12.1642	
	400	11.256		0.1143		9.21828	
	450	9.7819		0.1668		8.51829	
Figure 6	300	20.475	13.568	0.1517	0.1252	13.919	9.78
	350	13.644		0.1200		10.153	
	400	11.518		0.1209		9.2397	
	450	8.6338		0.1081		7.3370	

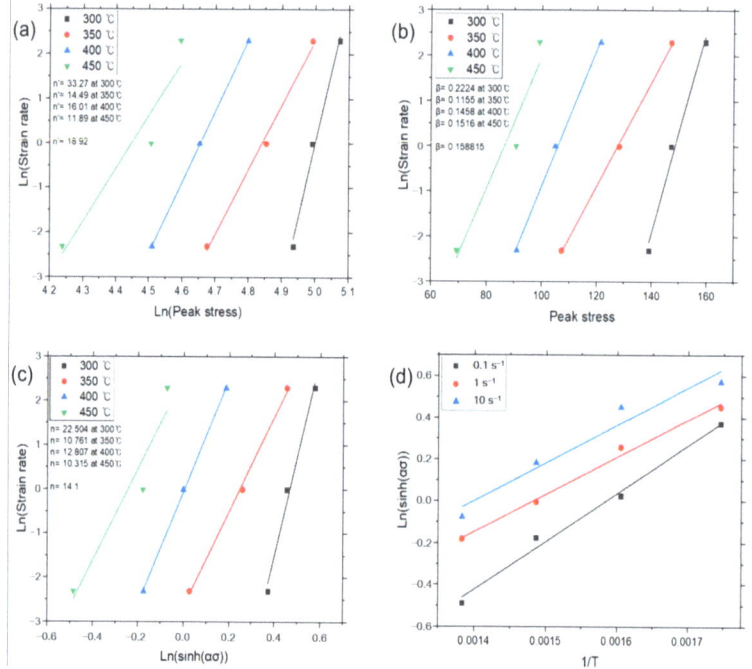

Figure 5. Constitutive analysis of as-cast A5052 alloys with no UST according to Equations (4)–(7). (a) $\dot\varepsilon$–σ_p dependence, (b) $\dot\varepsilon$–$\exp(\sigma_p)$ dependence, (c) $\dot\varepsilon$–$\sinh(\alpha\sigma_p)$ dependence, and (d) Arrhenius dependence.

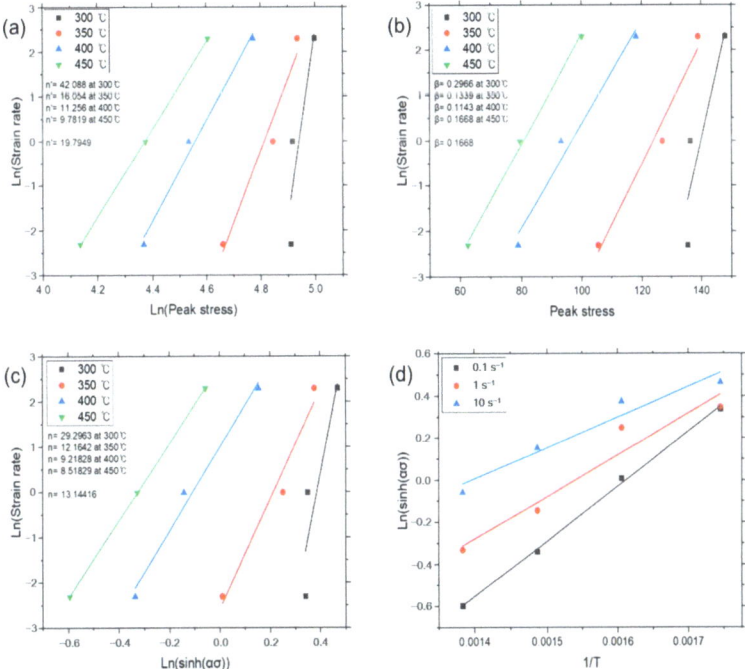

Figure 6. Constitutive analysis of as-homogenized A5052 alloys with UST according to Equations (4)–(7). (a) $\dot{\varepsilon}$–σ_p dependence, (b) $\dot{\varepsilon}$–exp(σ_p) dependence, (c) $\dot{\varepsilon}$–sinh($\alpha\sigma_p$) dependence, and (d) Arrhenius dependence.

The constitutive equations of the experimental condition using the derived constants and graphs are shown in Table 3. As a result of examining the activation energy for the constitutive equation, in the case of the as-cast billets, the activation energies under UST and no UST were 284 kJ/mol and 230 kJ/mol, respectively. After the homogenization heat treatment, the activation energies under UST and no UST were 248 kJ/mol and 166 kJ/mol, respectively. The activation energies throughout the entire conditions appear to be significantly high compared with that of pure Al. In general, the activation energy of pure Al is approximately 142 kJ/mol [32,33] to 150 kJ/mol [34], and, when some alloying elements are added, it increases to 200 kJ/mol [35–43]. According to the literature [4], the activation energy of the A5052 alloy without Ti was approximately 207 kJ/mol, which is lower than the energies in all conditions except in the homogenization heat treatment under no UST. This possibly occurred because the dislocation was formed during the deformation process in the work-hardening alloy. The deformation resistance is generated and increases depending on the energy storage in the microstructure as the number of dislocations increase [20]. Furthermore, the lower activation energy under no UST means that the energy barrier to dislocation movement is relatively low, making the alloy more favorable for high-temperature plastic working. And also, this possibly occurred because the microstructure under the no UST condition had a relatively small deviation of grain size and a higher fraction of equiaxed grains compared with the UST condition [44,45].

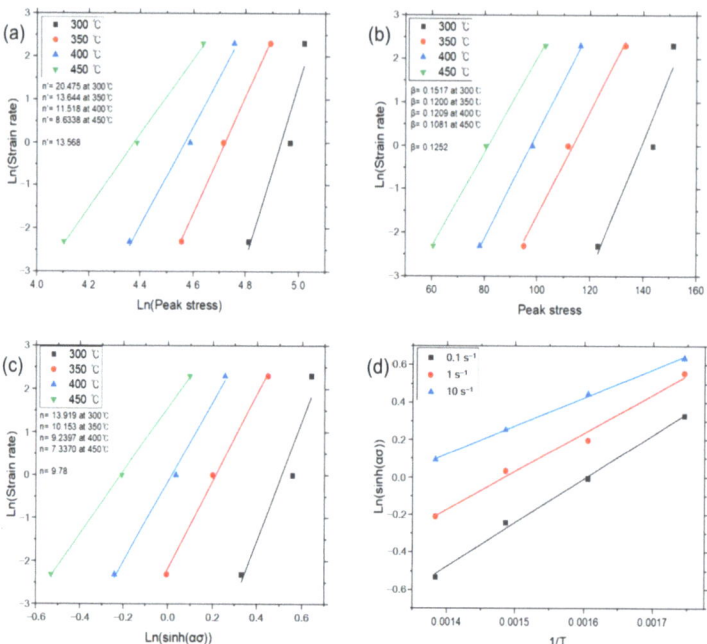

Figure 7. Constitutive analysis of as-homogenized A5052 alloys with no UST according to Equations (4)–(7). (**a**) $\dot{\varepsilon}$–σ_p dependence, (**b**) $\dot{\varepsilon}$–exp(σ_p) dependence, (**c**) $\dot{\varepsilon}$–sinh($\alpha\sigma_p$) dependence, and (**d**) Arrhenius dependence.

Table 3. Constitutive equation for each condition.

UST	As-cast	$\dot{\varepsilon} = 1.64 \times 10^{23} \left(\sinh 0.0075\sigma_p\right)^{16.9} \exp\left(-\frac{284,308}{RT}\right)$
	As-homogenized	$\dot{\varepsilon} = 1.26 \times 10^{18} \left(\sinh 0.0084\sigma_p\right)^{14.1} \exp\left(-\frac{230,154}{RT}\right)$
No UST	As-cast	$\dot{\varepsilon} = 1.9 \times 10^{20} \left(\sinh 0.0084\sigma_p\right)^{13.25} \exp\left(-\frac{248,654}{RT}\right)$
	As-homogenized	$\dot{\varepsilon} = 9.614 \times 10^{12} \left(\sinh 0.0092\sigma_p\right)^{9.78} \exp\left(-\frac{166,141}{RT}\right)$

The activation energy distribution depending on strain temperature and rate based on the activation energies determined above was calculated and is shown in Figure 8. The low activation energy in the as-cast and UST conditions corresponds to ≤200 kJ/mol at high strain throughout the entire temperature range. On the other hand, the as-homogenized alloy did not depend significantly on strain and had low activation energies at temperatures around 300 °C and 450 °C. The as-cast alloy with no UST had low activation energies at high strain and around 300 °C, and at low strain and temperatures higher than 350 °C. And also, low activation energy was found at temperatures higher than 350 °C, regardless of the strain rate in the homogenization heat-treated condition.

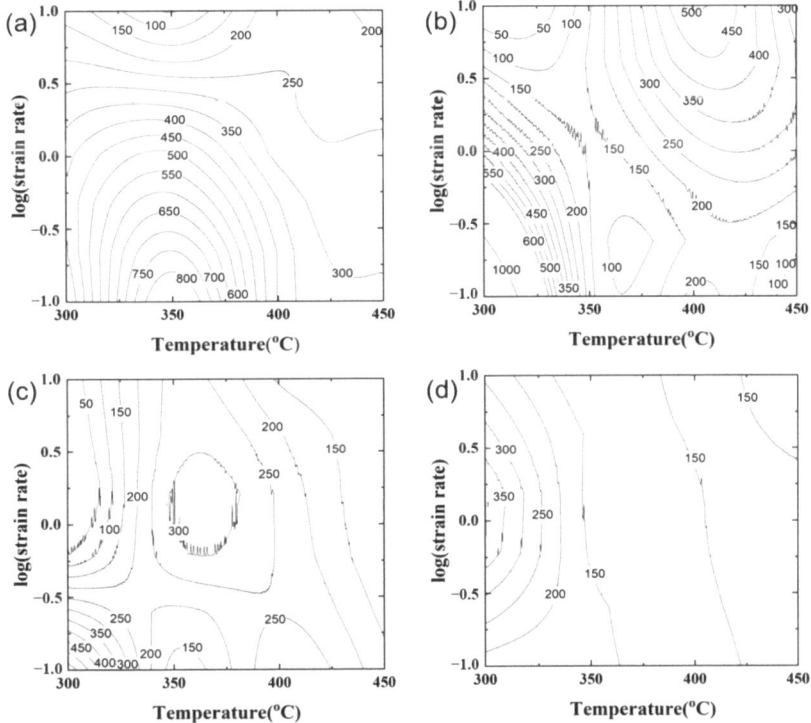

Figure 8. Activation energy maps depending on experimental conditions. (**a**) UST and (**b**) no UST in as-cast condition, and (**c**) UST and (**d**) no UST in as-homogenized condition.

3.3. Processing Map

Power dissipation maps were constructed using experimental data and the principles of the dynamic material model (DMM) [8]. The procedures adopted for the construction of the map are as follows. The relationship between $\ln\dot{\varepsilon}$ and $\ln\sigma$ at constant temperature and strain was fitted using cubic spline interpolation, and the interpolated curves were fitted by a 3rd order-polynomial. And then, the value of strain rate sensitivity, m, was calculated as a function of the strain rate. This procedure was repeated at different deformation temperatures. From the calculated value of m at a given temperature and strain rate, power dissipation efficiency was calculated using the following equation:

$$\eta = J/J_{max} = 2m/(m+1) \tag{11}$$

where η is the power dissipation efficiency through microstructural changes.

The calculated power dissipation efficiency for the given strain and temperature for the test conditions in this study is shown in Figure 9. The η in the power dissipation efficiency map is represented in a percentage. As shown in Figure 9, the power dissipation efficiency has a lower value under the UST condition than that under no UST in all regions. In addition, the same tendency was observed even after the homogenization heat treatment. Moreover, the flow instability region did not appear at both strain values in the as-cast condition with the UST. However, the wide flow instability regions in the as-cast condition with no UST were observed as the strain and temperature increased. Regardless of UST, the flow instability regions appeared at the low temperatures and high strain rates after the homogenization heat treatment. In particular, the power dissipation efficiency under no UST is high in both as-cast and homogenization conditions at high temperatures and low strain rates. However, in the case of the UST condition, it did not appear to be sensitive to the strain rate as the temperature increased.

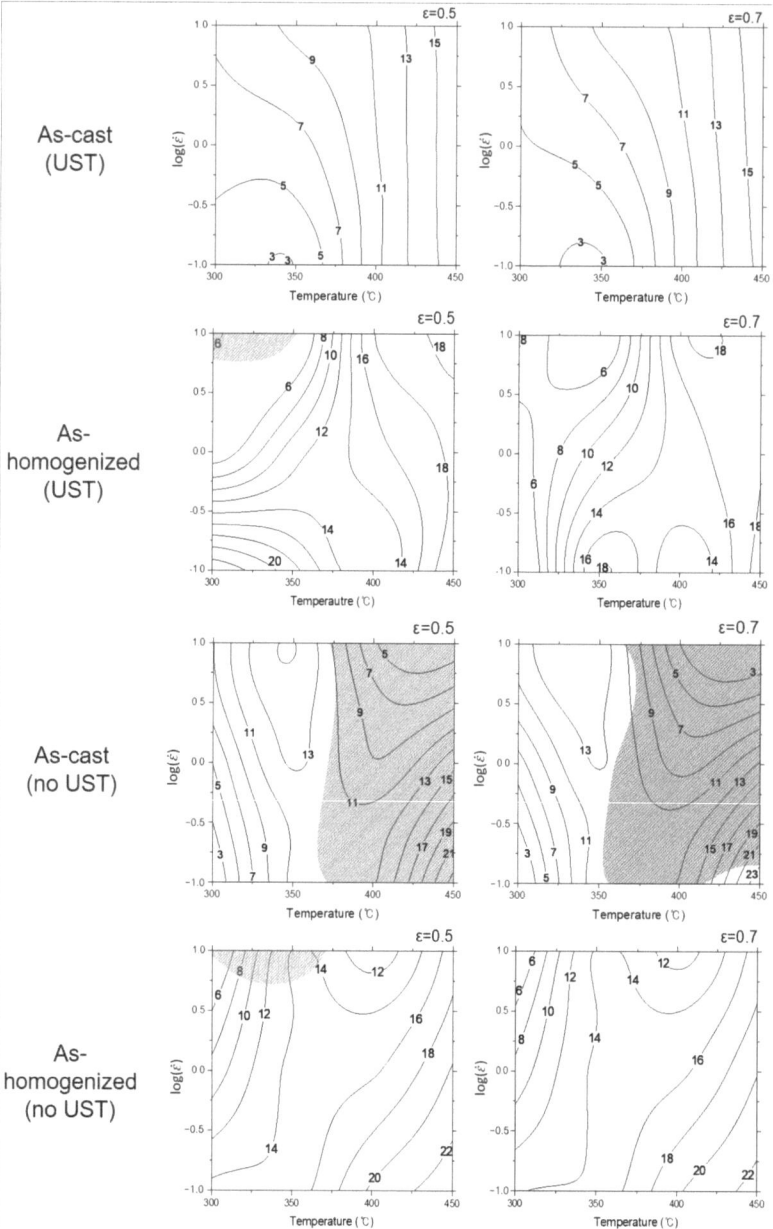

Figure 9. Power dissipation efficiency maps of experimental alloys at a strain of 0.5 and 0.7. The numbers on contour lines represent the power dissipation efficiency calculated from Equation (11).

3.4. EBSD Analysis

The results from the EBSD analysis are given in Figures 10 and 11. They indicate that the presence of a low angle boundary (LAB, $2° < \theta < 15°$) and high angle boundary

(HAB, $15° < \theta$) increases with increasing strain rates at the same temperature under all experimental conditions. At the same strain rate, the amount of increase in misorientation decreased as the temperature increased. This is possibly associated to stress relief and recrystallization caused by an increase in temperature, although misorientation occurs to form and relieve stress due to plastic deformation. In addition, because Al has low stored energy [46], the energy generated during the deformation should be rapidly eliminated. The thermal energy from such eliminated stored energy possibly induced the recrystallization.

Figure 10. Inverse pole figure map of experimental alloys at 350 °C and 450 °C with 0.1 s^{-1} and 1 s^{-1}.

In the as-cast condition, the presence and distribution of LAB and HAB under the UST condition were higher than those under no UST. This is considered to be the different grain size deviation between the UST and no UST conditions, although the average grain size appeared similar in the initial as-cast microstructures. In the case of the UST condition, the grain size deviation was larger than that in the no UST condition, so it may affect the deformation behavior by grain boundary sliding. Therefore, it is thought that the difficulty in deformation led to the higher amounts of LAB and HAB in the torsion test. However, after the homogenization heat treatment, the amounts of LAB and HAB were similar in both the UST and no UST conditions. It is considered that they were not significantly affected by differences in grain size deviation, as the microstructures were stabilized and uniform through the homogenization heat treatment.

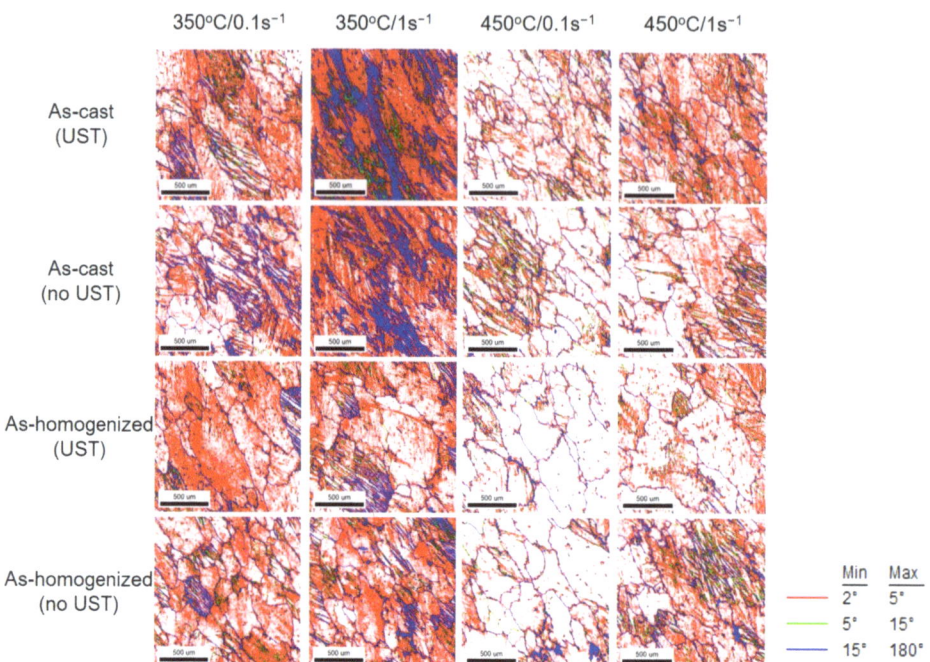

Figure 11. Misorientation maps of the experimental alloys at 350 °C and 450 °C with 0.1 s^{-1} and 1 s^{-1}.

The relationships between the power dissipation efficiency (%) and activation energy (Q) under each condition, and the kernal average misorientation (KAM) and equivalent stress, are shown in Figures 12 and 13. In the as-cast condition under the UST, the activation energy increased as the power dissipation efficiency decreased (see Figure 12a). At the same time, the KAM and equivalent stress increased (see Figure 12c). This implies that higher strain energy is required at the same strain, and the activation energy increases with the reduction in power dissipation efficiency, which contributes to changing the microstructure. In addition, a flow instability zone was observed in the region of approximately 400 to 450 °C of the processing map in the case of the no UST condition. In the aforementioned instability region, the misorientation was non-uniformly distributed, which could be associated with the presence of the Al$_3$Ti phase by the addition of the Al–10Ti master alloy. In the case of the no UST condition, the Al$_3$Ti phase by the addition of the Al–10Ti master alloy as a grain refiner is distributed in melts and remains even after solidification, and begins to decompose at temperatures above 400 °C. However, since the decomposition of the Al$_3$Ti phase does not occur within a short period of time, its presence and distribution are possibly not uniform throughout the specimens, thus causing non-uniform deformation.

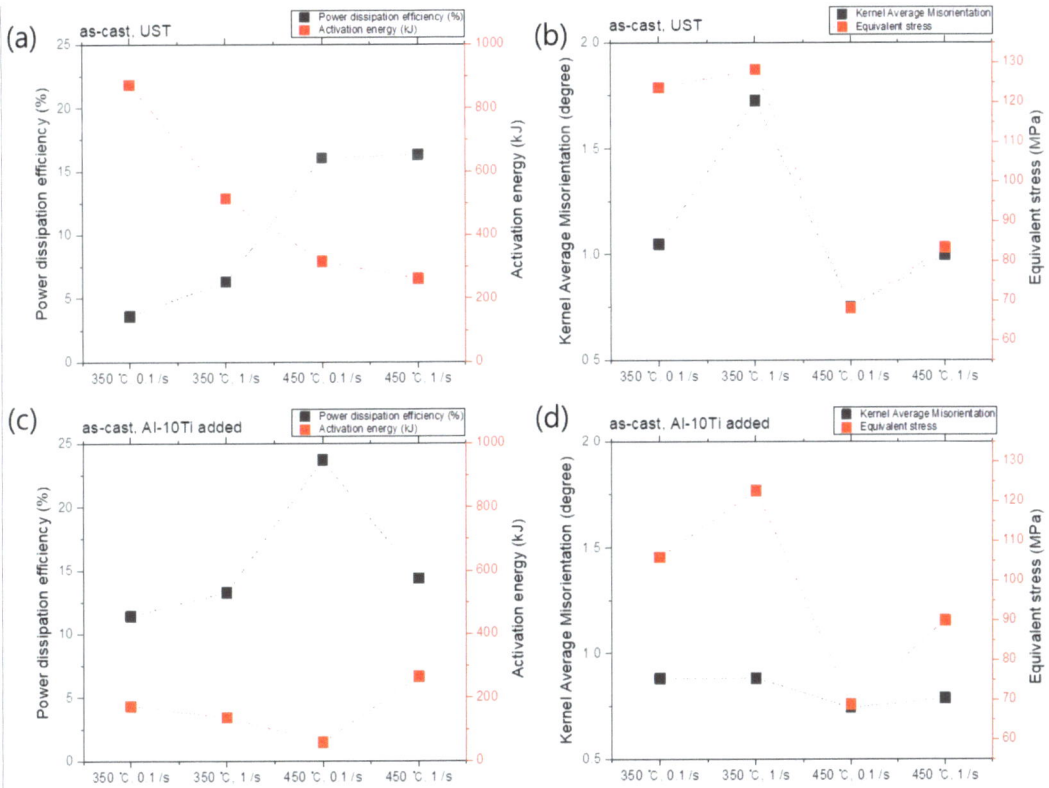

Figure 12. (**a**,**b**) Q-value (activation energy) versus power dissipation efficiency and (**c**,**d**) kernel average misorientation (KAM) versus equivalent stress graphs for (**a**,**c**) UST and (**c**,**d**) no UST in the as-cast condition.

In the as-cast condition under the UST condition, the power dissipation efficiency was relatively low. However, as the Ti atoms in the Al melt were uniformly distributed and the micropores decreased with the degassing effect during the UST, there should be fewer factors that induced non-uniform deformation. These findings imply that (i) an instability region was not found in all the regions of the processing map, (ii) the deformation at high temperature is not affected by speed, and (iii) the high-speed deformation at high temperature is advantageous. After the homogenization heat treatment, the deformation behavior of the UST and no UST conditions appeared to be similar. Based on these results, it can be said that a high power dissipation efficiency and uniform deformation can be expected in the hot deformation process by maximizing the effect of grain refinement and inducing uniform distribution of Ti solutes when the UST is applied to the Al melt.

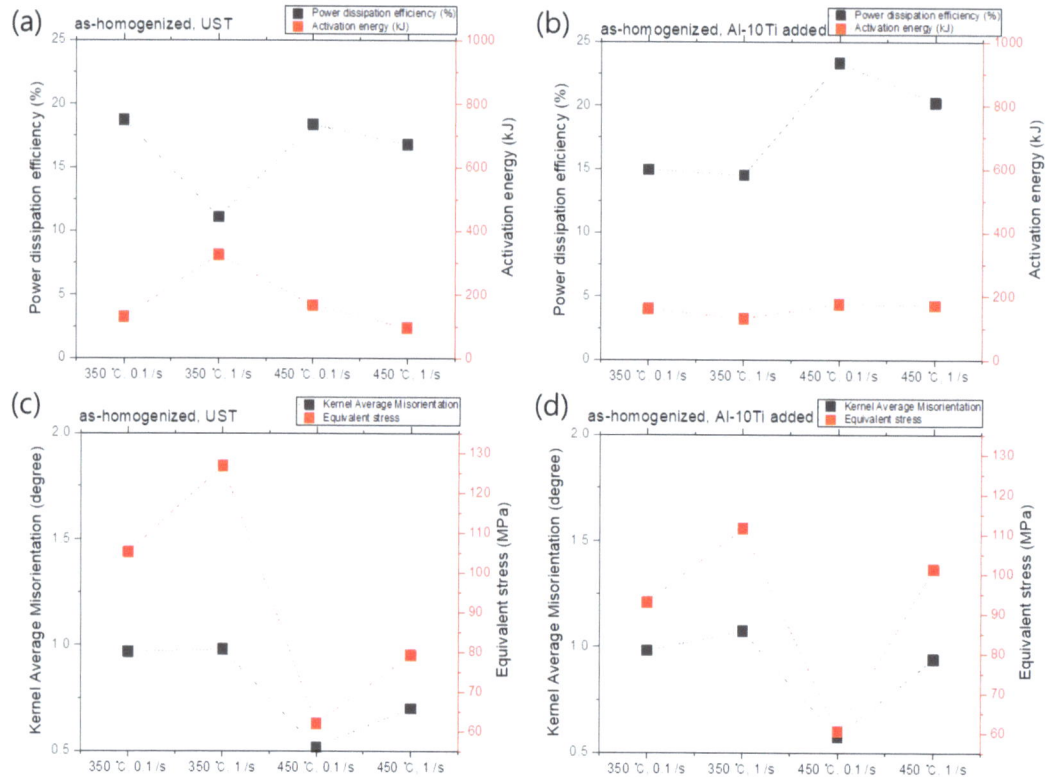

Figure 13. (**a**,**b**) Q-value (activation energy) versus power dissipation efficiency and (**c**,**d**) kernel average misorientation (KAM) versus equivalent stress graphs for (**a**,**c**) UST and (**c**,**d**) no UST in the as-homogenized condition.

4. Conclusions

The constitutive equations for each experimental condition by torsion test were derived in this study. From the processing maps, it can be seen that the flow instability region did not exist under the UST in the as-cast condition, but it appeared under the no UST condition. As a result of UST, the uniform distribution of Ti solutes from the sonotrode and the reduction of pores by the degassing effect possibly facilitated the plastic deformation. However, in the case of the no UST condition, the partially undecomposed Al_3Ti phase and gas pores still remained in the matrix, leading to the creation of instability regions. After the homogenization heat treatment, most instability regions disappeared because the microstructures became uniform following the decomposition of intermetallic compounds and distribution of solute elements. Based on the experimental results in this study, when grain refiners are added to alloys and followed by UST, uniform deformation and high process efficiency can be expected in hot working by maximizing the effect of grain refinement and making the distribution of solutes uniform, and controlling gas pores in the melts.

Author Contributions: Conceptualization: S.-K.K. and H.C.; methodology: S.-K.K.; data curation: S.-K.K. and S.-H.H.; writing—review and editing: S.-K.K. and S.-H.H.; supervision: H.C. and S.-H.K. All authors have read and agreed to the published version of the manuscript.

Funding: This research received no external funding.

Institutional Review Board Statement: Not applicable.

Informed Consent Statement: Not applicable.

Data Availability Statement: The raw data supporting the conclusions of this article will be made available by the authors on request.

Acknowledgments: This study has been conducted with the support of the Ministry of Trade, Industry, and Energy as "Materials & Parts Technology Development Program (20011420)".

Conflicts of Interest: Authors Sun-Ki Kim and Seung-Hyun Koo were employed by NICE LMS Co., Ltd. Authors Hoon Cho and Seong-Ho Ha were employed by Korea Institute of Industrial Technology. The authors declare that the research was conducted in the absence of any commercial or financial relationships that could be construed as potential conflicts of interest.

References

1. Yukawa, H.; Murata, Y.; Morinaga, M.; Takahashi, Y.; Yoshida, H. Heterogeneous distributions of Magnesium atoms near the precipitate in Al-Mg based alloys. *Acta Metall.* **1995**, *43*, 681–688. [CrossRef]
2. Gubicza, J.; Chinh, N.; Horita, Z.; Langdon, T. Effect of Mg addition on microstr-ucture and mechanical properties of aluminum. *Mater. Sci. Eng A* **2004**, *387*, 55–59. [CrossRef]
3. Horvath, G.; Chinh, N.; Gubicza, J.; Lendvai, J. Plastic instabilities and dislocation densities during plastic deformation in Al-Mg alloys. *Mater. Sci. Eng. A* **2007**, *445*, 186–192. [CrossRef]
4. He, J.; Wen, J.; Zhou, X.; Liu, Y. Hot deformation behavior and processing map of cast 5052 aluminum alloy. *Procedia Manuf.* **2019**, *37*, 2–7. [CrossRef]
5. Gruzleski, J.E.; Closset, B.M. *The Treatment of Liquid Aluminum-Silicon Alloys*, 1st ed.; American Foundrymen's Society: Des Plaines, IL, USA, 1999; pp. 132–137.
6. Kim, S.; Shin, J.; Cho, H.; Kim, Y.; Yi, S. Microstructural Refinement of As-Cast Al Mg Alloy by Ultrasonic Melt Treatment Using a Titanium Sonotrode under Fully Liquid Condition. *Mater. Trans.* **2022**, *63*, 1469–1476. [CrossRef]
7. Prasad, Y.V.R.K.; Sasidhara, S. *Hot Working Guide: A Compendium on Processing Maps*, 2nd ed.; ASM International: Materials Park, OH, USA, 1997.
8. Prasad, Y.V.R.K.; Gegel, H.L.; Doraivelu, S.M.; Malas, J.C.; Morgan, J.T.; Lark, K.A.; Barker, D.R. Modeling of dynamic material behavior in hot deformation: Forging of Ti-6242. *Metall. Mater. Trans. A* **1984**, *15*, 1883–1892. [CrossRef]
9. Prasad, Y.V.R.K. Processing maps: A status report. *J. Mater. Eng. Perform.* **2003**, *12*, 638–645. [CrossRef]
10. Liao, H.; Wu, Y.; Zhou, K.; Yang, J. Hot deformation behavior and processing map of Al–Si–Mg alloys containing different amount of silicon based on Gleeble-3500 hot compression simulation. *Mater. Des.* **2015**, *65*, 1091–1099. [CrossRef]
11. Sarebanzadeh, M.; Mahmudi, R.; Roumina, R. Constitutive analysis and processing map of an extruded Mg–3Gd–1Zn alloy under hot shear deformation. *Mater. Sci. Eng. A* **2015**, *637*, 155–161. [CrossRef]
12. Li, B.; Pan, Q.; Yin, Z. Characterization of hot deformation behavior of as-homogenized Al-Cu-Li-Sc-Zr alloy using processing maps. *Mater. Sci. Eng. A* **2014**, *614*, 199–206. [CrossRef]
13. Zhang, M.; Li, F.; Wang, S.; Liu, C. Characterization of hot deformation behavior of a P/M nickel-base superalloy using processing map and activation energy. *Mater. Sci. Eng. A* **2010**, *527*, 6771–6779.
14. Wang, S.; Hou, L.G.; Luo, J.R.; Zhang, J.S.; Zhuang, L.Z. Characterization of hot workability in AA 7050 aluminum alloy using activation energy and 3-D processing map. *J. Mater. Process. Technol.* **2015**, *225*, 110–121. [CrossRef]
15. Fields, D.S.; Backofen, W.A. Determination of Strain Hardening Characteristics by Torsion Testing. *ASTM Proc. 6th Annu. Meet.* **1957**, *57*, 1259–1272.
16. Cepeda-Jiménez, C.M.; Hidalgo, P.; Carsí, M.; Ruano, O.A.; Carreño, F. Microstructural characterization by electron backscatter diffraction of a hot worked Al-Cu-Mg alloy. *Mater. Sci. Eng. A* **2011**, *528*, 3161–3168. [CrossRef]
17. Kim, S.I.; Yoo, Y.C. Continuous dynamic recrystallization of AISI 430 ferritic stainless steel. *Met. Mater. Int.* **2002**, *8*, 7–13. [CrossRef]
18. Matsumoto, H.; Kitamura, M.; Li, Y.; Koizumi, Y.; Chiba, A. Hot forging characteristic of Ti-Al-5V-5Mo-3Cr alloy with single metastable β microstrucre. *Mater. Sci. Eng. A* **2014**, *611*, 337–344. [CrossRef]
19. Son, K.T.; Kim, M.H.; Kim, S.W.; Lee, J.W.; Hyun, S.K. Evaluation of hot deformation characteristic in modified AA5052 using processing map and activation energy map under deformation heating. *J. Alloys Compd.* **2018**, *740*, 96–108. [CrossRef]
20. Jeon, K.S.; Rho, H.R.; Kim, M.S.; Kim, J.H.; Park, J.P. Microstructure and high temperature plastic deformation behavior of Al-12Si based alloy fabricated by an electromagnetic casting and stirring process. *Korean J. Met. Mater.* **2017**, *55*, 386–395.
21. Kim, Y.M.; Lee, S.H.; Lee, S.; Noh, J.W. Microstructural evolution during hot deformation of molybdenum using processing map approach. *J. Korean Powder Metall. Inst.* **2008**, *15*, 458–465. [CrossRef]
22. Ha, M.C.; Hwang, S.W.; Kim, C.S.; Kim, C.Y.; Park, K.T. High temperature deformation behavior of a NIMONIC80A Ni based superalloy. *Trans. Mater. Process.* **2013**, *22*, 258–263. [CrossRef]
23. Yeom, J.T.; Kim, D.H.; Na, Y.S.; Park, N.K. Characterization of hot deformation behavior of Ti-6Al-4V alloy. *Trans. Mater. Process.* **2001**, *10*, 347–354.

24. Kil, T.D.; Han, S.W.; Moon, Y.H. Studies on Quantitative estimation of hot forgeability by using deformation processing map. *Korean J. Met. Mater.* **2013**, *52*, 731–737.
25. Sivakesavam, O.; Prasad, Y.V.R.K. Hot deformation behaviour of as cast Mg-2Zn-1Mn alloy in compression: A study with processing map. *Mater. Sci. Eng. A* **2003**, *362*, 118–124. [CrossRef]
26. Srinivasan, N.; Prasad, Y.V.R.K.; Rao, P.R. Hot deformation behaviour of Mg-3Al alloy-A study using processing map. *Mater. Sci. Eng. A* **2008**, *476*, 146–156. [CrossRef]
27. Son, H.W.; Jung, T.K.; Lee, J.W.; Hyun, S.K. Hot deformation characteristic of CaO-dded AZ31 based on kinetic model and processing map. *Mater. Sci. Eng. A* **2017**, *695*, 379–385. [CrossRef]
28. *ASTM E-112*; Standard Test Methods for Determining Average Grain Size. ASTM International: West Conshohocken, PA, USA, 2003.
29. Sellars, C.M.; McTegart, W.J. On the mechanism of hot deformation. *Acta Metall.* **1966**, *14*, 1136–1138. [CrossRef]
30. Sellars, C.M.; McTegart, W.J. Hot Workability. *Int. Metall. Rev.* **1972**, *17*, 1–24. [CrossRef]
31. McQueen, H.J.; Ryan, N.D. Constitutive analysis in hot working. *Mater. Sci. Eng. A* **2002**, *322*, 43–63. [CrossRef]
32. McQueen, H.J.; Jonas, J.J. Recovery and recrystallization during hot deformation. *Treatise Mater. Sci. Technol.* **1975**, *6*, 393–493.
33. Lundy, T.; Murdock, J. Diffusion of Al^{26} and Mn^{54} in Aluminum. *J. Appl. Phys.* **1962**, *33*, 1671–1673. [CrossRef]
34. Sherby, O.D.; Klundt, R.H.; Miller, A.K. Flow stress, subgrain size, and subgrain stability at elevated temperature. *Metall. Mater. Trans. A* **1977**, *8*, 843–850. [CrossRef]
35. McQueen, H.J.; Evangelista, E.; Forcellese, A.; Smith, I.C.; Russo, E.D.; Lowe, T.; Rollett, T. Modeling the Deformation of Crystalline Solids. In Proceedings of the TMS-AIME, Warrendale, PA, USA, 17–21 February 1991.
36. McQueen, H.J. Substructural influence in the hot rolling of Al alloys. *JOM* **1998**, *50*, 28–33. [CrossRef]
37. Gouret, S.; Chovet, C.; McQueen, H.J. Microstructure of a Hot Worked 6060 Aluminum Alloy. *Alum. Trans.* **2001**, *3*, 59–68.
38. Verlinden, B.; Suhadi, A.; Delaey, L. A generalized constitutive equation for an AA6060 aluminium alloy. *Scr. Metall. Mater.* **1993**, *28*, 1441–1446. [CrossRef]
39. Sakaris, P.; McQueen, H.J. Aluminum Alloys: Their Physical and Mechanical Properties. In Proceedings of the ICAA3, Trondheim, Norway, 22–27 June 1992; Arnberg, L., Nes, E., Lohne, O., Ryum, N., Eds.; Norwegian University of Science and Technology: Trondheim, Norway, 1992.
40. Verlinden, B.; Wouters, P.; McQueen, H.J.; Aernoudt, E.; Delaey, L.; Cauwenberg, S. Effect of different homogenization treatments on the hot workability of aluminium alloy AA2024. *Mater. Sci. Eng. A* **1990**, *123*, 229–237. [CrossRef]
41. Smith, I.C.; Avramovic-Cingara, G.; McQueen, H.J. *Aluminum-Lithium Alloys*; Sanders, T.H., Starke, E.A., Jr., Eds.; MCE Publication Ltd: Birmingham, UK, 1989.
42. McQueen, H.J.; Hopkins, A.; Jain, V.; Avramovic-Cingara, G.; Sakaris, P. *Hot Workability of Steels and Light Alloys-Composites, Montreal, Canada, 24–28 August 1996*; McQueen, H.J., Konopleva, E.V., Ryan, N.D., Eds.; CIM: Montreal, QC, Canada, 1996.
43. Avramovic-Cingara, G.; Perovic, D.D.; McQueen, H.J. Hot deformation mechanisms of a solution-treated Al-Li-Cu-Mg-Zr alloy. *Metall. Mater. Tran. A* **1996**, *27*, 3478–3490. [CrossRef]
44. Ma, A.; Takagi, M.; Saito, N.; Iwata, H.; Nishida, Y.; Suzuki, K.; Higematsu, I. Tensile properties of an Al-11 mass%Si alloy at elevated temperatures processed by rotary-die equal-channel angular pressing. *Mater. Sci. Eng. A* **2005**, *408*, 147–153. [CrossRef]
45. Jia, Y.; Cao, F.; Guo, S.; Ma, P.; Liu, J.; Sun, J. Hot deformation behavior of spray-deposited Al-Zn-Mg-Cu alloy. *Mater. Des.* **2014**, *53*, 79–85. [CrossRef]
46. Titchener, A.L.; Bever, M.B. The stored energy of cold work. *Prog. Met. Phys.* **1958**, *7*, 247–338. [CrossRef]

Disclaimer/Publisher's Note: The statements, opinions and data contained in all publications are solely those of the individual author(s) and contributor(s) and not of MDPI and/or the editor(s). MDPI and/or the editor(s) disclaim responsibility for any injury to people or property resulting from any ideas, methods, instructions or products referred to in the content.

Article

Enhanced Mechanical Properties of Ti/Mg Laminated Composites Using a Differential Temperature Rolling Process under a Protective Atmosphere

Zichen Qi [1,2], Zhengchi Jia [1], Xiaoqing Wen [3], Hong Xiao [2,*], Xiao Liu [4], Dawei Gu [1], Bo Chen [1] and Xujian Jiang [1]

1. College of Mechanical Engineering, Zhejiang University of Technology, Hangzhou 310023, China; qizichen@zjut.edu.cn (Z.Q.); 2112102229@zjut.edu.cn (Z.J.); goodavid@zjut.edu.cn (D.G.); jiangxujian1519@163.com (X.J.)
2. National Engineering Research Center for Equipment and Technology of Cold Strip Rolling, Yanshan University, Qinhuangdao 066004, China
3. Zhejiang YaTong Advanced Materials Co., Ltd., Hangzhou 310030, China; wenxiaoqing1992@163.com
4. College of Mechanical and Vehicle Engineering, Taiyuan University of Technology, Taiyuan 030024, China; liuxiao@tyut.edu.cn
* Correspondence: xhh@ysu.edu.cn

Citation: Qi, Z.; Jia, Z.; Wen, X.; Xiao, H.; Liu, X.; Gu, D.; Chen, B.; Jiang, X. Enhanced Mechanical Properties of Ti/Mg Laminated Composites Using a Differential Temperature Rolling Process under a Protective Atmosphere. *Materials* **2024**, *17*, 2753. https://doi.org/10.3390/ma17112753

Academic Editors: Seong-Ho Ha, Young-Ok Yoon, Young-Chul Shin and Dong-Earn Kim

Received: 24 April 2024
Revised: 31 May 2024
Accepted: 3 June 2024
Published: 5 June 2024

Copyright: © 2024 by the authors. Licensee MDPI, Basel, Switzerland. This article is an open access article distributed under the terms and conditions of the Creative Commons Attribution (CC BY) license (https://creativecommons.org/licenses/by/4.0/).

Abstract: Addressing the issue of low bonding strength in Ti/Mg laminated composites due to interfacial oxidation, this study employs a differential temperature rolling method using longitudinal induction heating to fabricate Ti/Mg composite plates. The entire process is conducted under an argon gas protective atmosphere, which prevents interfacial oxidation while achieving uniform deformation. The effects of reduction on the mechanical properties and microstructure of the composite plates are thoroughly investigated. Results indicate that as the reduction increases, the bonding strength gradually increases, mainly attributed to the increased mechanical interlocking area and a broader element diffusion layer. This corresponds to a transition from a brittle to a ductile fracture at the microscopic tensile–shear fracture surface. When the reduction reaches 47.5%, the Ti/Mg interfacial strength reaches 63 MPa, which is approximately a 20% improvement compared to the bonded strength with previous oxidation at the interface. Notably, at a low reduction of 17.5%, the bonding strength is significantly enhanced by about one time. Additionally, it was found that a strong bonded interface at a high reduction is beneficial in hindering the propagation of interfacial cracks during tensile testing, enhancing the ability of the Ti/Mg composite plates to resist interfacial delamination.

Keywords: Ti/Mg composite plates; induction heating; protective atmosphere; differential temperature rolling; bonding strength

1. Introduction

Magnesium alloys are the lightest structural metals with a density two-thirds that of aluminum alloys, offering high specific strength and specific stiffness, among other superior properties that make them highly promising materials for automotive and aerospace applications [1–4]. However, the poor corrosion resistance of magnesium alloys has limited their development [5–7]. Therefore, it is crucial to form bimetal plates with other corrosion-resistant metals. Titanium and titanium alloys, as new structural materials, are widely used in the aerospace industry for their low density, high strength, corrosion resistance, wear resistance, and high-temperature impact resistance [8–10]. Utilizing the lightweight nature of magnesium alloys and the excellent corrosion resistance of titanium and titanium alloys, the development of new lightweight Ti/Mg laminated composites has broad application prospects [11–14].

Preparation methods for Ti/Mg composite plates mainly include explosive welding, diffusion welding, explosive + rolling, and hot rolling composite methods. Wu et al. [15] successfully prepared Ti/Mg composite plates through explosive welding, achieving a maximum bonding strength of 64 MPa, and studied the mechanical properties and interfacial bonding mechanisms of the composite plates. However, the explosive welding process generates seismic waves, noise, and toxic gases, which are not conducive to the large-scale production of composite plates. Tan et al. [16] used an AZ91 magnesium-based brazing wire as an intermediate material to achieve a continuously uniform welding joint between magnesium alloy and titanium alloy through laser brazing. Xiong Jiangtao et al. [17] used aluminum foil as an intermediate layer for transient liquid phase diffusion welding of magnesium alloy and titanium alloy, studying their connection mechanism and mechanical properties. However, liquid phase diffusion welding and laser welding processes are only suitable for local connections of plates and not for the overall composite connection of composite plates. Therefore, compared to explosive welding and diffusion welding methods, the rolling composite method has a stable product quality and simple equipment, and can easily achieve automation and large-scale production. In recent years, the rolling method has been widely studied due to its efficiency and economy and has been applied to the preparation of other homogeneous or heterogeneous metal laminated composite plates [18–22].

However, due to the significant difference in mechanical properties between titanium and magnesium, deformation coordination issues arise during hot rolling, and when the deformation is large, necking and fracture of the titanium layer occur, which has become a bottleneck in the preparation of Ti/Mg composite plates by rolling [23]. Therefore, a method of differential temperature rolling of Ti/Mg composite plates by separately heating the titanium plate with a resistance furnace was proposed [24], solving the problem of deformation coordination between titanium and magnesium, and achieving high bonding strength at a high reduction. This method was also successfully applied to the preparation of Ti/Al composite plates [25]. The downside is that the study found that the surface oxidation of the titanium plate was severe during the heating process of the differential temperature rolling, and the brittle oxide layer hindered the combination of fresh metal at a low reduction. Moreover, the presence of the brittle oxide layer has a negative impact on the tensile and bending properties of the composite plates.

To avoid the formation of an oxide layer on the surface of the plates after prolonged heating, which would greatly reduce the bonding performance of the composite plates, this work innovatively proposes a method of differential temperature rolling with longitudinal electromagnetic induction heating under a protective atmosphere. The principle mainly utilizes the thermal effect of eddy current heating, which can rapidly increase the temperature of ferromagnetic materials in a short period of time. A unique billet arrangement is designed, with pure iron plates used as the intermediate layer, taking advantage of the rapid temperature rise of the iron plate, and then transferring the heat sequentially to the titanium plates and the outermost magnesium plates, thus creating a significant plate temperature difference. This method can achieve the purpose of rapid heating of the billet while avoiding oxidation of the plates during the heating process and accurately controlling the temperature difference of the plates. Then, the effects of reduction on the bonding interface, shear strength, fracture morphology, and tensile delamination of the composite plates are studied in detail, the relationship between micromorphology and macroscopic mechanical properties is discussed, and the bonding mechanism of Ti/Mg composite plates is proposed.

2. Materials and Experimental Procedure

2.1. Materials Preparation

The experimental materials are 2 mm thick industrial pure titanium TA1, AZ31B magnesium alloy, and pure iron plates used as the intermediate layer. Rectangular plates with initial rolling direction dimensions of 100 mm × 60 mm are taken, and the chemical element content of the used TA1 and AZ31B magnesium alloy plates is listed in Table 1. The mechanical properties of TA1 and AZ31B plates after annealing are shown in Table 2. The surfaces of the titanium and magnesium plates to be combined are treated to remove surface oil, impurities, and oxides, which is conducive to the combination of fresh metal during rolling. In this experiment, a flat sanding machine equipped with 180-grit sandpaper (manufactured by Zhejiang Minli Power Tools Co., Ltd., Jinhua, China) is first used to remove impurities and oxides from the metal surfaces to be combined. Then, the surface is repeatedly wiped with acetone and alcohol and immediately dried with a hair dryer.

Table 1. Chemical composition of commercial pure Ti-TA1 sheet, Mg alloy-AZ31B sheet (wt%).

Materials	C	N	H	O	Si	Fe	Al	Ca	Zn	Mn	Cu	Ti	Mg
TA1	0.05	0.03	0.015	0.15	0.1	0.15	-	-	-	-	-	Bal.	-
AZ31B	-	-	-	-	0.08	0.03	3.1	0.04	0.9	0.5	0.01	0.15	Bal.

Table 2. Mechanical properties of the used materials in the experiment.

Materials	Ultimate Tensile Strength (MPa)	Yield Strength (MPa)	Shear Strength (MPa)	Fracture Elongation (%)
TA1	434	328	285	25.3
AZ31B	226	160	128	12.5

2.2. Differential Temperature Rolling Process with Induction Heating

2.2.1. Billet Arrangement

The experimental design of the multilayer symmetrical billet arrangement is shown in Figure 1. The plates are stacked in a symmetrical multilayer structure in the order of Mg-Ti-pure iron-Ti-Mg. The advantage of the symmetrical structure is that it can not only make full use of the heat of the iron plate for energy saving, but also effectively reduce the warping deformation of the Ti/Mg composite plate after rolling. Since the thermal effect of eddy current heating can rapidly increase the temperature of ferromagnetic materials in a short time, a pure iron plate is chosen as the intermediate layer. Utilizing the rapid temperature rise of the iron plate, the heat is then sequentially transferred to the titanium plates and the outer magnesium plates, creating a significant temperature difference between the plates. To enable the separation of the iron plate and the Ti/Mg composite plate after rolling, talcum powder is selected as a lubricant between the iron and titanium plates. Additionally, spacers are placed on both sides of the titanium and magnesium plates to control the heat transfer rate between titanium and magnesium by leaving a gap of 0~1 mm. If the gap is too large, it will not be conducive to the plates biting into the rolling mill. Finally, the ends of the billet are drilled and fixed with aluminum rivets.

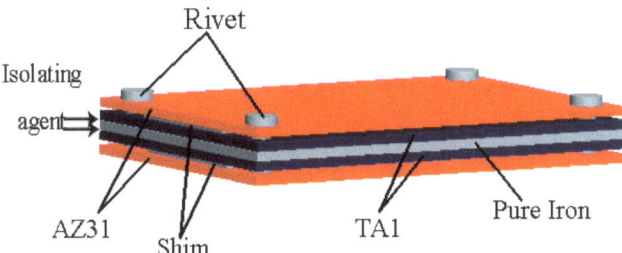

Figure 1. Schematic diagram for the symmetrical structure of multilayer plates.

2.2.2. Differential Temperature Rolling Process

Figure 2 show the schematic diagram for the rolling process of Ti/Mg laminated composites by differential temperature rolling with induction heating. The induction heating equipment mainly consists of a spiral coil, a cooling system, and an electrical control system. The assembled multilayer billet is placed in the center of the induction heating furnace, and an appropriate induction current and heating time are applied. After the heating is completed, the billet is immediately pushed into the rolling mill for differential temperature rolling. The entire process from induction heating to rolling is sealed and protected by argon gas to prevent oxidation of the billet surface during heating. The mid-frequency induction heating furnace used in the experiment has a controllable induction current range of 0~2400 A.

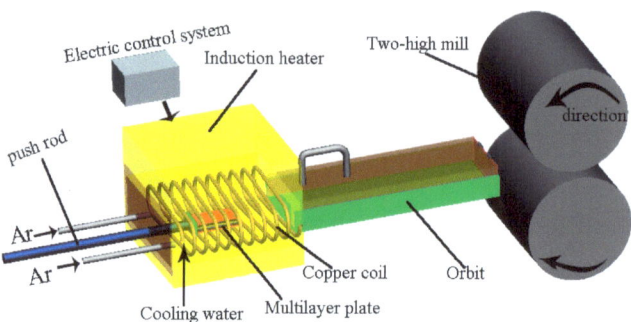

Figure 2. Schematic diagram of differential temperature rolling process with induction heating.

2.2.3. Determination of Rolling Process Parameters

In this experiment, a thermocouple thermometer (Flank F-8855) (Suzhou TASI Electronics Co., Ltd., Suzhou, China) is used to measure the temperature changes of the pure iron plate, titanium plate, and magnesium plate during the heating process. The actual temperature measurement device for the billet is shown in Figure 3. First, a hole with a diameter of 1 mm and a depth of 30 mm is drilled at the midpoint on the edge of the plates. Then, one end of the thermocouple wire (K-type, range $-200\ °C$ to $1372\ °C$, error $\pm 1\ °C$) is inserted into the hole, and the other end is connected to the thermometer. Figure 4 shows the temperature changes of each layer of the plate under a 2100 A induction current and a 1 mm gap between TA1 and AZ31B. After heating for 20 s, the temperature of the titanium plate is 532 °C, and the magnesium plate is 318 °C, with a maximum temperature difference of 214 °C. Therefore, in this experiment, a gap of 1 mm is chosen between the titanium and magnesium through aluminum spacers, a current of 2100 A is applied, and the plates are heated for 20 s before being immediately pushed into the rolling mill for rolling. The rolling reduction amounts are 17.5%, 32.5%, and 47.5%, all of which are single-pass reductions, and the reduction amounts are measured by the thickness of the titanium/aluminum composite plate after rolling. The parameters of the two-roll mill used in the experiment are: the roll size is $\varphi 200\ mm \times 200\ mm$, the rolling speed is 50 mm/s, and there is no lubrication.

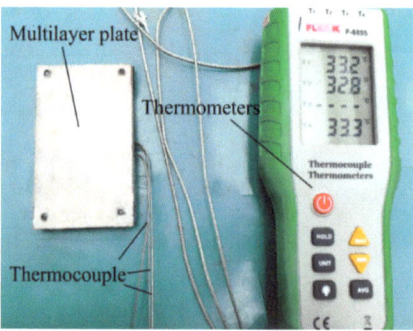

Figure 3. Actual temperature measurement procedure for the laminated plates.

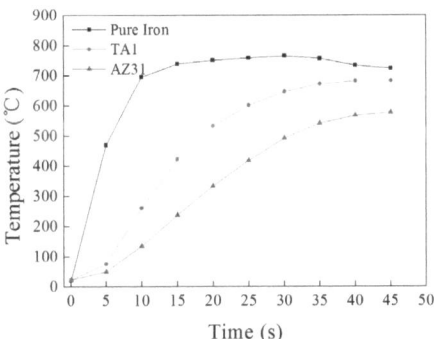

Figure 4. Temperature variation in individual laminated composite under 2100 A current and 1 mm clearance between TA1 and AZ31B.

2.3. Mechanical Properties Test and Microstructure Observation

The bonding area ratio and the shear strength of the bimetal plates are two key parameters for the measurements of the properties of the laminated composites. The specimens for the tensile–shear test were made according to the GB/T 6396-2008 [26] (clad steel plates–mechanical and technological test) and GB/T 8547-2006 [27] (titanium clad steel plate) standards. Three specimens were selected from each plate parallel to the rolling direction and were tested to obtain the average shear strength, which was calculated according to the formula: bonding strength = peak loading/(bond width × bond length). Figure 5a–c show the geometry of the tensile–shear test specimen and fractured surface observed in the RD, TD and ND, respectively. The shear test was conducted at room temperature on an INSPEKT Table 100 kN electronic (Esum Technology Limited., Beijing, China) universal testing machine, with a shear rate of 1 mm/min, as shown in Figure 6. The tensile performance of the composite plate in the rolling direction is tested to evaluate the ductility, overall tensile strength, and the relationship between bonding performance and tensile performance.

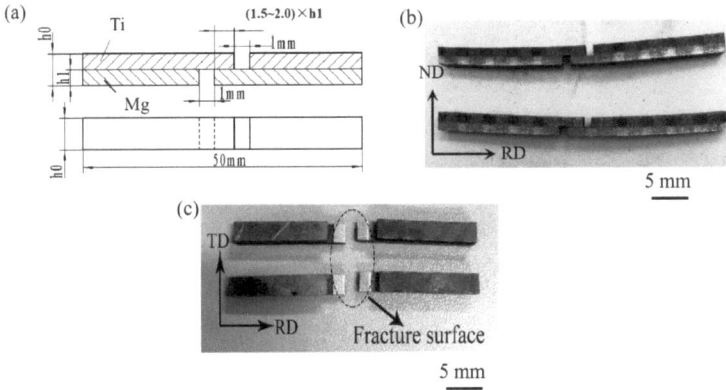

Figure 5. Tensile–shear test specimen and interface observation of the laminated composites: (**a**) schematic of the specimen; (**b**) real image of tensile–shear specimen; (**c**) fractured specimen.

Metallographic specimens are taken parallel to the rolling direction, and they are ground with sandpaper from 400# to 5000#, followed by rough polishing with a 2.5 um grit diamond polishing paste. Finally, a 1.5 um grit diamond polishing paste is used for fine polishing. The bonding interface and the morphology of the shear fracture are observed using an FEI Scios scanning electron microscope (SEM) (Thermo Fisher Scientic, Tokyo, Japan), and the elemental distribution near the bonding interface and the fracture is analyzed using an energy-dispersive spectrometer (EDS) (Thermo Fisher Scientic, Tokyo, Japan).

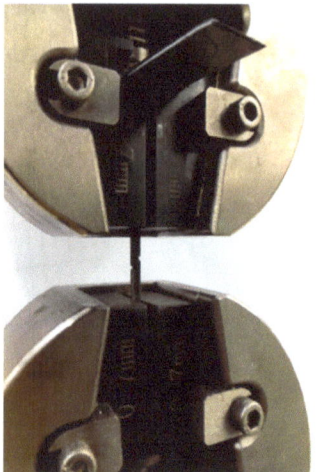

Figure 6. Actual tensile–shear testing process of specimen in the machine.

3. Results and Discussion

3.1. Macroscopic Bonding Property

Figure 7 shows the effect of reduction on the shear strength of the Ti/Mg laminated composites. Due to the symmetrical billet arrangement, the overall thickness of the plates is relatively large, and the rolling capacity of the mill is limited. According to the actual measured values, curve fitting is performed, and the fitted equation is $y = 0.018x^2 - 0.56x + 49.3$, where x represents the value of the reduction, ranging from 17.5 to 47.5, and Y represents the shear strength. R^2 is a statistical measure that assesses the discrepancy between the fitted curve and the actual data, with values ranging from 0 to 1. The closer the value is to 1, the closer the fitted curve is to the actual data; conversely, the further away it is. The R^2 value of the curve fitted in this work is almost close to 1, indicating a good fit between the curve fitting and the measured values. The maximum single-pass reduction achieved for the Ti/Mg composite plate is 47.5%. Therefore, this experiment only compares the bonding strength of the composite plates at a similar reduction via a differential temperature rolling with the resistance furnace heating. From Figure 5, it can be seen that as the reduction increases, the shear strength of the composite plate gradually increases. At a reduction of 47.5%, the shear strength reaches a maximum of 63 MPa, which is nearly a 19% increase compared to the 53 MPa shear strength of the Ti/Mg composite plate at a 46% reduction using differential temperature rolling with resistance furnace heating [24]. More significantly, at a reduction of 17.5%, the shear strength reaches 45 MPa, which is nearly double the 25 MPa shear strength at a 25% reduction using differential temperature rolling with resistance furnace heating [24]. Therefore, the rapid induction heating combined with argon gas protection plays a key role in maintaining the cleanliness of the plate surface, avoiding the formation of oxides, thus allowing the fresh titanium and magnesium metals to come into contact at a low reduction, thereby significantly improving the interface bonding performance. Subsequently, the micro-interfaces and fracture morphologies of the Ti/Mg composite plates are characterized to further illustrate the microstructural features of the composite plates under this differential temperature rolling process.

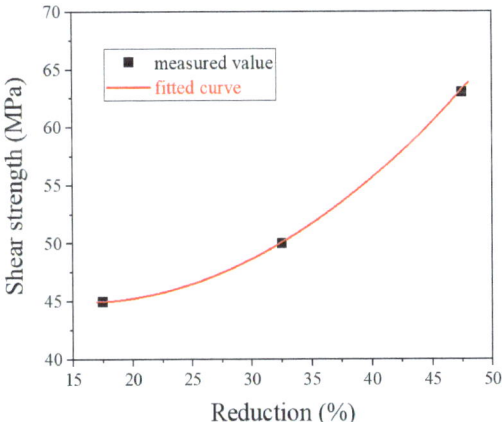

Figure 7. Effect of reduction on the shear strength of the Ti/Mg laminated composites.

3.2. Bonding Interface and Fracture Morphology

Figure 8 shows the SEM images of the bonding interface of the composite plates under different reductions. No obvious unbonded areas such as holes were observed at the bonding interface in the images, indicating that the titanium has undergone relatively large plastic deformation and has good fluidity, resulting in close contact at the interface and achieving a good composite effect. When the reductions are 17.5% and 32.5%, the Ti/Mg bonding interface appears relatively straight in both low and high magnifications, as shown in Figure 8a–d. As the reduction increases to 47.5%, the overall bonding interface in Figure 8e shows a distinct wavy shape, and the enlarged Figure 8f also reveals microscopic wavy interfaces. The change in the interface shape is consistent with the interface change in the differential temperature rolling with resistance furnace heating [24], mainly due to the large deformation rate of titanium at high temperatures in both processes, where the intense plastic deformation of the two metals on either side of the interface leads to the formation of waves. The wavy interface can withstand greater shear force during shearing, which is beneficial for improving the bonding strength of the composite plate. No obvious oxide layer was observed at the interface in Figure 8a–f, and the interface has always been in a clean state, which is different from the resistance furnace heating differential temperature rolling. The clean interface ensures the contact of fresh metals and enhances the inter-diffusion ability of elements, thereby overall improving the bonding performance of the composite plate.

Figure 9 presents the overall morphology of the tensile–shear fracture of the composite plates under different reductions. As can be seen from Figure 8c,d, at a 32.5% reduction, the bonding interface is straight, and after the shear test, the fracture surfaces on the titanium and magnesium sides are shown in Figures 9a and 9b, respectively. Macroscopically, both fracture surfaces appear relatively flat. At a higher reduction of 47.5%, the fracture surfaces on both the titanium and magnesium sides, as shown in Figure 9c,d, exhibit distinct wavy characteristics, which correspond to the morphology of the bonding interface. To further observe the finer microstructures at the fracture, the fracture surfaces were magnified and subjected to EDS elemental scanning tests.

Figure 10 shows the microscopic morphology of the tensile–shear fracture of the composite plates, and Figure 11 presents the EDS elemental scanning results of the titanium side fracture under different reduction rates. Figure 10a,c show that at reduction rates of 17.5% and 32.5%, vertical cracks are produced on the titanium side, perpendicular to the rolling direction, and the fracture surface is characteristic of typical brittle fracture. Additionally, from the elemental scanning results in Figure 11a–h, it can be seen that the cracks are filled with Mg elements. Moreover, apart from the crack areas, the titanium side also contains some Mg elements, indicating that the titanium side has adhered to some magnesium metal, with evidence of magnesium metal squeezing into the titanium cracks. Comparing the distribution of O elements with that of Ti and Mg elements at the fracture, it is found that the distribution of O is random and does not follow any discernible pattern related to the distribution of Ti and Mg, suggesting that the presence of oxygen is not due to oxidation during the heating and rolling processes of the plates, but rather due to oxidation that occurs when the fracture is exposed to air after shearing. Figure 10b,d show that at reductions of 17.5% and 32.5%, the magnesium side also exhibits characteristics of brittle fracture, suggesting that at low reductions, the fracture of the composite plate primarily occurs at the interface between titanium and magnesium. As the reduction increases to 47.5%, the fracture on the titanium side, as shown in Figure 10e,f, is undulating with the formation of some dimples. The EDS scanning test results in Figure 11i–l show that all the dimples on the titanium side are filled with Mg elements, while the remaining non-dimple areas are filled with Ti elements, indicating that the formation of some dimples is due to the fracture of the magnesium matrix. Correspondingly, dimples are also formed on the magnesium side, as shown in Figure 10g,h. Therefore, it is inferred that at a higher reduction, the fracture of the composite plate occurs at a mixed location between the magnesium matrix and the Ti/Mg bonding surface. The proportion of dimples on the entire fracture surface suggests that there is still significant room for improvement in the Ti/Mg composite plate.

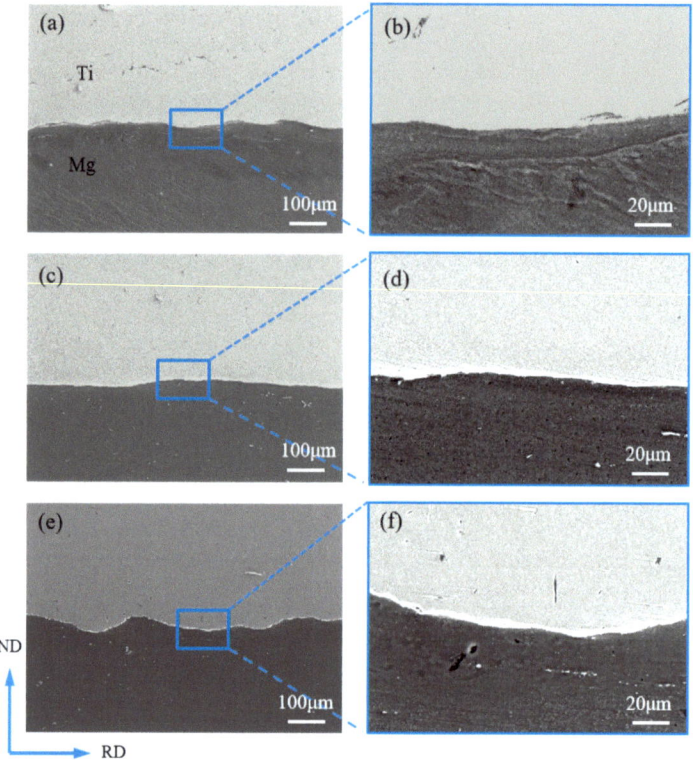

Figure 8. Bonding interface of laminated composites under different reductions: (**a**) 17.5% reduction; (**b**) enlarged area in image (**a**); (**c**) 32.5% reduction; (**d**) enlarged area in image (**c**); (**e**) 47.5% reduction; (**f**) enlarged area in image (**e**).

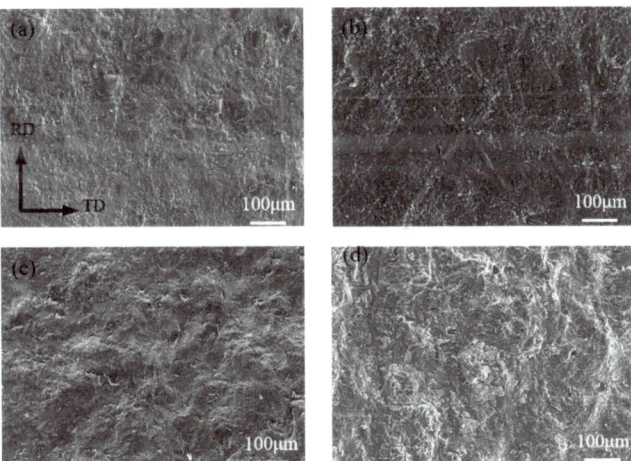

Figure 9. Overall morphology of the tensile–shear fracture of composite plates: (**a**) 32.5%/Ti side; (**b**) 32.5%/Mg side, (**c**) 47.5%/Ti side, (**d**) 47.5%/Mg side.

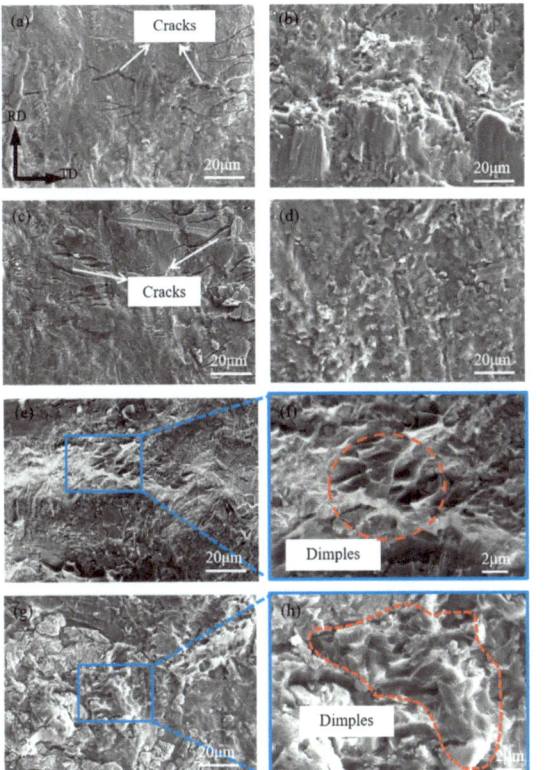

Figure 10. Microscopic morphology of the tensile–shear fracture of composite plates: (**a**) 17.5%/Ti side; (**b**) 17.5%/Mg side; (**c**) 32.5%/Ti side; (**d**) 32.5%/Mg side; (**e**) 47.5%/Ti side; (**f**) enlarged area; (**g**) 47.5%/Mg side; (**h**) enlarged area.

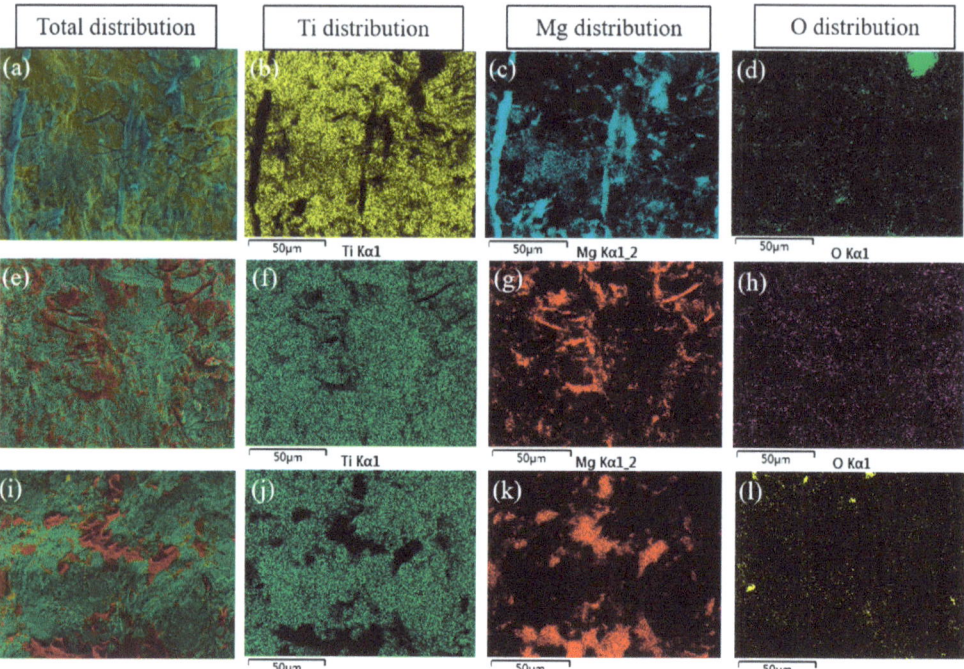

Figure 11. Element mapping scanning in Ti side of the tensile–shear fracture under different processes: (**a**) 17.5%; (**b**–**d**) Ti, Mg and O distributions in image (**a**); (**e**) 32.5%; (**f**–**h**) Ti, Mg and O distributions in image (**e**); (**i**) 47.5%; (**j**–**l**) Ti, Mg and O distributions in image (**i**).

3.3. Tensile Property and Interfacial Delamination of Laminated Composites

The schematic diagram of the tensile specimen is shown in Figure 12. Figure 13 presents the overall tensile performance test of the Ti/Mg composite plates, and the engineering stress–strain curves are shown in the figure. At a reduction of 17.5%, the overall tensile strength of the composite plate is about 300 MPa. As the rolling reduction increases, the tensile strength gradually increases, reaching a maximum of about 350 MPa at a 47.5% reduction. This is because, with the increase in the reduction rate, both the titanium and magnesium layers in the composite plate undergo work hardening, resulting in an increase in the strength of each layer, which in turn increases the overall tensile strength of the composite plate. Since the elongation rate of the magnesium plate is much lower than that of the titanium plate, during the tensile process, it is observed that the magnesium plate breaks first, followed by the continued plastic deformation and subsequent fracture of the titanium plate, as shown in Figure 14. This leads to the noticeable step-like drops in the three tensile curves in Figure 12. The first vertical drop is due to the fracture of the magnesium plate, and the second vertical drop is due to the eventual fracture of the titanium plate. Due to the effect of work hardening, while the tensile strength is improved, the total elongation rate of the composite plate shows a downward trend. However, the tensile curves show that the overall elongation rate of the composite plates at reductions of 32.5% and 47.5% is essentially the same. It is analyzed that this may be due to the increased bonding strength of the composite plate, which promotes the coordinated deformation of the titanium and magnesium plates during the overall tensile process, resulting in the composite plate at a 47.5% reduction having the same elongation rate as the composite plate at a 32.5% reduction.

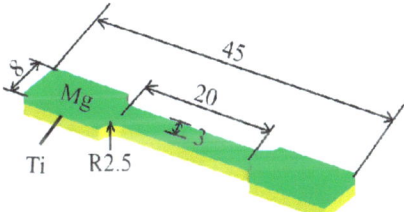

Figure 12. Schematic diagram of the tensile specimen (distances are in mm).

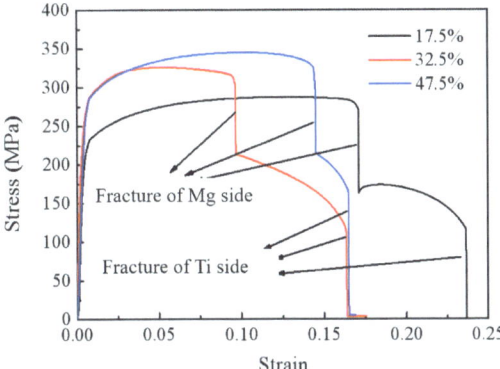

Figure 13. Engineering stress–strain curves of Ti/Mg laminated composites.

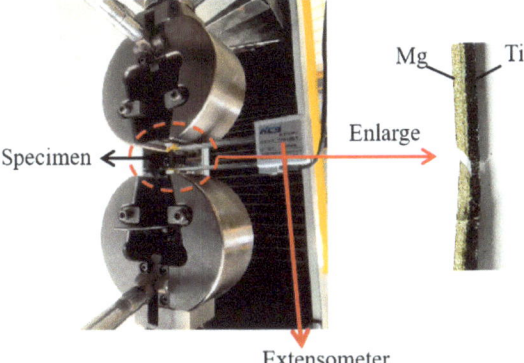

Figure 14. The tensile fracture process of Ti/Mg laminated composite.

After the tensile fracture of the composite plate, the fracture surfaces were spliced together, and the macroscopic appearance of the titanium and magnesium sides after splicing is shown in Figure 15. It can be seen from Figure 15a that under different reductions, the composite plate ultimately fails with the fracture of the titanium plate, and a significant necking phenomenon occurs on the titanium plate. Figure 15b shows that the elongation rate decreases significantly from a 17.5% to a 32.5% reduction rate, and the elongation rate is relatively consistent from a 32.5% to a 47.5% reduction. However, the fracture length of the magnesium side of the composite plate is significantly larger at a 32.5% reduction, indicating that a weaker bonding interface does not play a restraining role on the fracture of the two plates during the tensile process, while a stronger bonding interface allows the titanium plate to pull the magnesium plate, preventing the premature fracture of the magnesium plate and achieving more consistent coordinated deformation.

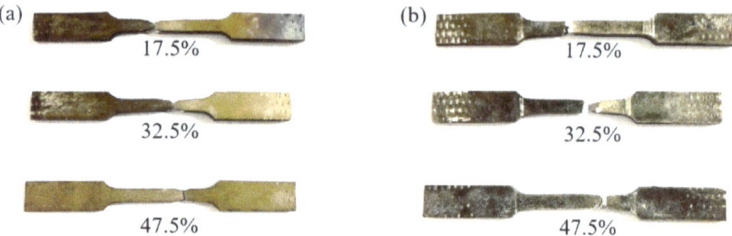

Figure 15. Tensile fracture of Ti/Mg laminated composites: (**a**) Ti side; (**b**) Mg side.

Figure 16 shows the interfacial delamination situation of the tensile fracture. Figure 16a shows that at a 17.5% reduction, the composite plate has the longest interface crack propagation and obvious interfacial delamination. As the reduction increases and the bonding strength increases, the interfacial delamination is gradually improved, as shown in Figure 16b. At a 32.5% reduction, the length of the interface crack in the composite plate becomes shorter, and at a 47.5% reduction, no obvious macroscopic cracks are observed in Figure 16c, indicating that the composite plate has obtained better resistance to interfacial delamination. The reason for the improved resistance to interfacial delamination is analyzed as follows: during the tensile process, the plates undergo necking, and the different necking tendencies of titanium and magnesium will generate shear force at the interface. A weak bonding interface cannot resist this shear force, leading to interface cracking and subsequent crack propagation. However, a strong bonding interface can resist this shear force, allowing the two metals to have the same necking tendency, thus avoiding interfacial delamination of the composite plate.

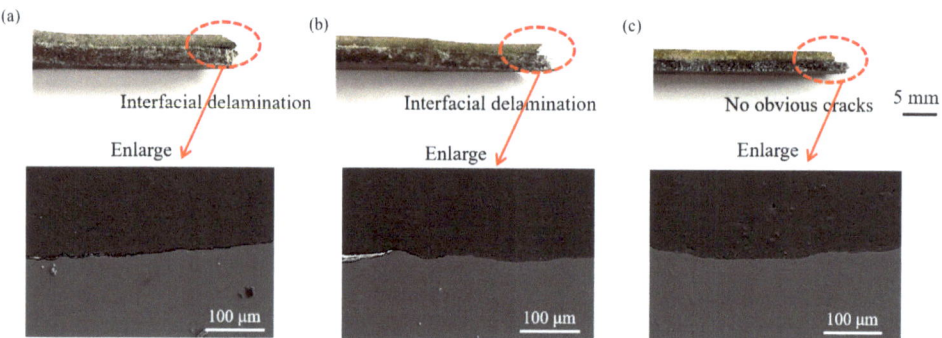

Figure 16. Interfacial delamination of the tensile fracture: (**a**) 17.5%; (**b**) 32.5%: (**c**) 47.5%.

3.4. Bonding Mechanism

Figure 17 presents a schematic diagram of the Ti/Mg bonding mechanism. Based on the above analysis of the tensile–shear fracture surface and the micro-morphology of the composite interface, the bonding mechanism of the differential temperature rolled Ti/Mg composite plate with induction heating is summarized, as shown in the diagram. In the initial stage of low reduction rolling, the titanium matrix, due to its larger plastic deformation, first produces micro-cracks. As the reduction increases, the bonding interface gradually exhibits a typical wavy appearance, accompanied by crack characteristics on the titanium side. When the reduction is further increased, the magnesium side metal squeezes into the cracks of the wavy interface, forming mechanical interlocking. Finally, under the action of pressure and temperature, mutual diffusion of titanium and magnesium elements occurs, achieving a better bonding performance. The difference in the bonding mechanism of the differential temperature rolled Ti/Mg composite plate with induction heating compared to the resistance furnace heating is that there is no oxide formation at the titanium matrix interface, which promotes full contact of fresh metals and mutual diffusion of elements, ultimately achieving the effect of improving the bonding strength of the composite plate.

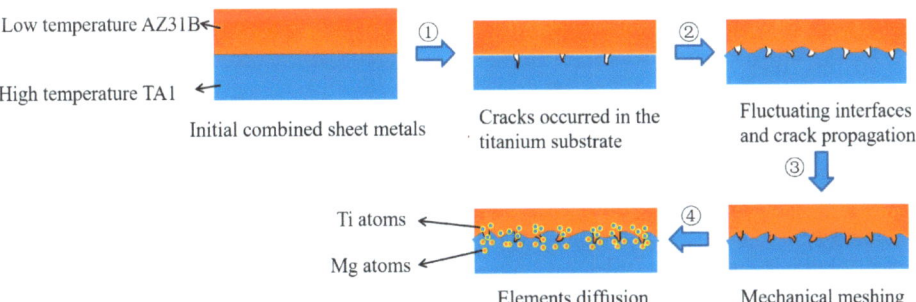

Figure 17. Schematic diagram of Ti/Mg bonding mechanism. (1) cracks occurred in the titanium substrate, (2) fluctuating interfaces and crack propagation, (3) mechanical meshing, (4) elements diffusion.

4. Conclusions

This work has innovated a method for preparing Ti/Mg metal laminated composites by differential temperature rolling with longitudinal electromagnetic induction heating under a protective atmosphere. This method successfully avoids oxidation of the plates during the rapid heating of the billet and has studied the effects of reduction and the presence or absence of an oxide layer on the bonding interface, shear strength, fracture morphology, and tensile interface delamination of the two types of composite plates. It has also explored the relationship between micro-morphology and macroscopic mechanical properties. The detailed conclusions are as follows:

(1) A clean and oxide-free bonding interface promotes contact between fresh metals and mutual diffusion of elements, which improves the bonding strength of the double-layer composite plates compared to those made by resistance furnace heating. Especially, it significantly enhances the bonding strength of the composite plates at a low reduction.

(2) With the increase in bonding strength, the fracture surfaces on both sides of the shear fracture gradually change from brittle fracture to ductile fracture. This is attributed to the fact that at low bonding strengths, the shear fracture occurs at the interface between the two plates, while at high bonding strengths, the shear fracture occurs within the metal matrix.

(3) A strong bonding interface is beneficial in preventing the propagation of interfacial cracks during the tensile testing of Ti/Mg composite plates, thereby enhancing the composite plates' ability to resist interfacial delamination.

(4) The bonding interface of the Ti/Mg composite plates prepared by differential temperature rolling with induction heating is primarily achieved through a combination of mechanical interlocking and elemental diffusion.

Author Contributions: Methodology, Z.Q. and X.W.; Software, B.C.; Validation, D.G.; Formal analysis, Z.Q. and X.L.; Resources, X.J.; Data curation, X.W. and Z.J.; Writing—original draft, Z.Q.; Writing—review & editing, X.W. and X.J.; Supervision, H.X. All authors have read and agreed to the published version of the manuscript.

Funding: This work was supported by the National Key Research and Development Project of China (2018YFA0707302), the National Natural Science Foundation of China (52075472), Natural Science Foundation of Zhejiang Province (LQ23E050014, LQ23E050018), an Open Research Fund from the National Key Laboratory of Metal Forming Technology and Heavy Equipment (S2308100.W05,

S2308100.W01), the China Postdoctoral Science Foundation (2023T160580, 2023M743102), and Special Metallurgical Products, Process Research and Application Technology Services (KYY-HX-20211002).

Institutional Review Board Statement: Not applicable.

Informed Consent Statement: Not applicable.

Data Availability Statement: The raw data supporting the conclusions of this article will be made available by the authors on request.

Conflicts of Interest: Xiaoqing Wen was employed by the company Zhejiang YaTong Advanced Materials Co., Ltd. The remaining authors declare that the research was conducted in the absence of any commercial or financial relationships that could be construed as a potential conflict of interest.

References

1. Song, J.F.; She, J.; Chen, D.L.; Pan, F.S. Latest research advances on magnesium and magnesium alloys worldwide. *J. Magnes. Alloy.* **2020**, *8*, 1–41. [CrossRef]
2. Mark, E.; Aiden, B.; Matthew, B.; Chris, D.; Gordon, D.; Yvonne, D.; Stuart, B.; Tim, H.; Peter, B. Magnesium alloy applications in automotive structures. *JOM* **2008**, *60*, 57–62.
3. Vaira, V.R.; Padmanaban, R.; Mohan, D.K.; Govindaraju, M. Research and development in magnesium alloys for industrial and biomedical Applications: A Review. *Met. Mater. Int.* **2020**, *26*, 409–430.
4. Dobrzański, L.A.; Tański, T.; Čížek, L.; Brytan, Z. Structure and properties of magnesium cast alloys. *J. Mater. Process. Technol.* **2007**, *192–193*, 567–574. [CrossRef]
5. Liu, M.Y.; Zhang, Q.Y.; Tang, X.H.; Liu, C.X.; Mei, D.; Wang, L.G.; Zhu, S.J.; Mikhail, Z.; Sviatlana, L.; Guan, S.K. Effect of medium renewal mode on the degradation behavior of Mg alloys for biomedical applications during the long-term in vitro test. *Corros. Sci.* **2024**, *229*, 111851. [CrossRef]
6. Song, G.L.; Atrens, A. Corrosion Mechanisms of magnesium alloys. *Adv. Eng. Mater.* **1999**, *1*, 11–13. [CrossRef]
7. Bender, S.; Goellner, J.; Heyn, A.; Boese, E. Corrosion and corrosion testing of magnesium alloys. *Mater. Corros.* **2007**, *58*, 977–982. [CrossRef]
8. Li, Z.; Mo, H.T.; Tian, J.H.; Li, J.H.; Hu, X.; Xia, S.Q.; Lu, Y.; Jiang, Z.Y. A novel Ti/Al interpenetrating phase composite with enhanced mechanical properties. *Mater. Lett.* **2024**, *357*, 135723. [CrossRef]
9. Huang, H.G.; Chen, P.; Ji, C. Solid-liquid cast-rolling bonding (SLCRB) and annealing of Ti/Al cladding strip. *Mater. Des.* **2017**, *118*, 233–244. [CrossRef]
10. Ma, M.; Huo, P.; Liu, W.C.; Wang, G.J.; Wang, D.M. Microstructure and mechanical properties of Al/Ti/Al laminated composites prepared by roll bonding. *Mat. Sci. Eng. A* **2015**, *636*, 301–310. [CrossRef]
11. Zhao, H.; Zhao, C.C.; Yang, Y.; Wang, Y.Z.; Sheng, L.Y.; Li, Y.X.; Huo, M.; Zhang, K.; Xing, L.W.; Zhang, G. Study on the microstructure and mechanical properties of a Ti/Mg alloy clad plate produced by explosive welding. *Metals* **2022**, *12*, 399. [CrossRef]
12. Blanco, D.; Rubio, E.M.; Manuel, J.; Marta, M. Thicknesses/roughness relationship in Mg-Al-Mg and Mg-Ti-Mg hybrid component plates for drilled aeronautical lightweight parts. *Appl. Sci.* **2020**, *10*, 8208. [CrossRef]
13. Feng, Y.; Tang, Y.Z.; Chen, W.H.; He, W.J.; Jiang, B.; Pan, F.S. Fabrication of Mg/Ti composite with excellent strength and ductility by hot rolling. *Mat. Sci. Eng. A* **2023**, *888*, 145783. [CrossRef]
14. Ouyang, S.H.; Liu, Y.; Huang, Q.L.; Gan, Z.Y.; Tang, H.C. Effect of composition on in vitro degradability of Ti–Mg metal-metal composites. *Mat. Sci. Eng. C* **2020**, *107*, 110327. [CrossRef] [PubMed]
15. Wu, J.Q.; Wang, W.X.; Cao, X.Q.; Zhang, N. Interface bonding mechanism and mechanical behavior of AZ31B/TA2 composite plate cladded by explosive welding. *Rare Met. Mater. Eng.* **2017**, *46*, 640–645.
16. Tan, C.W.; Chen, B.; Meng, S.H.; Zhang, K.P.; Song, X.G.; Zhou, L.; Feng, J.C. Microstructure and mechanical properties of laser welded-brazed AZ31/TA1 joints with AZ91 Mg based filler. *Mater. Des.* **2016**, *99*, 127–134. [CrossRef]
17. Xiong, J.T.; Zhang, F.S.; Li, J.L.; Huang, W.D. Transient liquid phase bonding of magnesium alloy (AZ31B) and titanium alloy (Ti6Al4V) using aluminum interlayer. *Rare Met. Mater. Eng.* **2006**, *35*, 1677–1680. (In Chinese)
18. Yu, C.; Xiao, H.; Yu, H.; Qi, Z.C.; Xu, C. Mechanical properties and interfacial structure of hot-roll bonding TA2/Q235B plate using DT4 interlayer. *Mater. Sci. Eng. A* **2017**, *695*, 120–125. [CrossRef]
19. Wu, Y.; Wang, T.; Ren, Z.K.; Liu, Y.M.; Huang, Q.X. Evolution mechanism of microstructure and bond strength based on interface diffusion and IMCs of Ti/steel clad plates fabricated by double-layered hot rolling. *J. Mater. Process. Technol.* **2022**, *310*, 117780. [CrossRef]
20. Wang, T.; Zhao, W.; Yun, Y.; Li, Z.; Wang, Z.; Huang, Q. A dynamic composite rolling model based on Lemaitre damage theory. *Int. J. Mech. Sci.* **2024**, *269*, 109067. [CrossRef]
21. Guo, S.; Wang, F.R.; Wang, Y.Q.; Xie, G.M. Microstructural evolution and properties of Ti/Al clad plate fabricated by vacuum rolling and heat treatment. *Mat. Sci. Eng. A* **2023**, *882*, 145445. [CrossRef]
22. Hosseini, M.; Manesh, H.D. Bond strength optimization of Ti/Cu/Ti clad composites produced by roll-bonding. *Mater. Des.* **2015**, *81*, 122–132. [CrossRef]

23. Mo, T.Q.; Xiao, H.Q.; Lin, B.; Li, W.; Ma, K. Improving ductility and anisotropy by dynamic recrystallization in Ti/Mg laminated metal composite. *Acta Metall. Sin.-Engl.* **2022**, *35*, 1946–1958. [CrossRef]
24. Qi, Z.C.; Yu, C.; Xiao, H. Microstructure and bonding properties of magnesium alloy AZ31/CP-Ti clad plates fabricated by rolling bonding. *J. Manuf. Process.* **2018**, *32*, 175–186. [CrossRef]
25. Xiao, H.; Qi, Z.C.; Yu, C.; Xu, C. Preparation and properties for Ti/Al clad plates generated by differential temperature rolling. *J. Mater. Process. Technol.* **2017**, *249*, 285–290. [CrossRef]
26. *GB/T 6396-2008*; Clad Steel Plates—Mechanical and Technological Test. Standards Press of China: Beijing, China, 2008.
27. *GB/T 8547-2006*; Titanium Clad Steel Plate. Standards Press of China: Beijing, China, 2006.

Disclaimer/Publisher's Note: The statements, opinions and data contained in all publications are solely those of the individual author(s) and contributor(s) and not of MDPI and/or the editor(s). MDPI and/or the editor(s) disclaim responsibility for any injury to people or property resulting from any ideas, methods, instructions or products referred to in the content.

Article

Research on the Influence of Cold Drawing and Aging Heat Treatment on the Structure and Mechanical Properties of GH3625 Alloy

Ji Li [1], Yujie Wo [1], Zhigang Wang [2], Wenhao Ren [3], Wei Zhang [4], Jie Zhang [1,*] and Yang Zhou [2]

1 School of Mechanical Engineering, Zhejiang University of Technology, Hangzhou 310014, China
2 Hubei Key Laboratory of High-Quality Special Steel, Daye Special Steel Co., Ltd., Huangshi 435001, China
3 Research Institute for Special Steel, Central Iron and Steel Research Institute, Beijing 100081, China
4 CITIC Metal Co., Ltd., Beijing 100004, China
* Correspondence: zhangjie0231@zjut.edu.cn

Citation: Li, J.; Wo, Y.; Wang, Z.; Ren, W.; Zhang, W.; Zhang, J.; Zhou, Y. Research on the Influence of Cold Drawing and Aging Heat Treatment on the Structure and Mechanical Properties of GH3625 Alloy. *Materials* **2024**, *17*, 2754. https://doi.org/10.3390/ma17112754

Academic Editors: Seong-Ho Ha, Young-Ok Yoon, Dong-Earn Kim and Young-Chul Shin

Received: 19 March 2024
Revised: 26 April 2024
Accepted: 27 April 2024
Published: 5 June 2024

Copyright: © 2024 by the authors. Licensee MDPI, Basel, Switzerland. This article is an open access article distributed under the terms and conditions of the Creative Commons Attribution (CC BY) license (https://creativecommons.org/licenses/by/4.0/).

Abstract: With the development of the petroleum industry, the demand for materials for oilfield equipment is becoming increasingly stringent. The strength increase brought about by time strengthening is limited in meeting the needs of equipment development. The GH3625 alloy with different strength levels can be obtained through cold deformation and heat treatment processes. A study should be carried out to further develop the potential mechanical properties of GH3625. In this study, the GH3625 alloy was cold drawn with different reductions in area (0–30%) and heat treated, and its mechanical properties were tested. The microstructure of the alloy during deformation and heat treatment was characterized by methods such as optical microscopy (OM), scanning electron microscopy (SEM), and transmission electron microscopy (TEM) based on the principles of physical metallurgy. The strength increase caused by dislocation strengthening was calculated from the dislocation density, tested by X-ray diffraction (XRD). The calculated value was compared to the measured value, elucidating the strengthening effect of cold deformation and heat treatment. The results showed that the yield strength and yield ratio of the cold-drawn alloy significantly reduced after aging at 650 °C and 760 °C. Heat treatment can make a cold-deformed material recover, ablate dislocations, and greatly reduce the dislocation density in the microstructure of the GH3625 alloy, which was the main factor in the decrease in yield strength. The work-hardening gradient of the cold-drawn material varied greatly with different reductions in area. When the reduction in area was small (10%), the hardness gradient was obvious. When it increased to 30%, the alloy was uniformly strengthened as the deformation was transmitted to the axis. This study can provide more mechanical performance options for GH3625 alloy structural components in the petrochemical industry.

Keywords: GH3625 alloy; cold drawing; mechanical properties; dislocation strengthening

1. Introduction

The GH3625 superalloy is a solid-solution-strengthened nickel-based deformation superalloy, with Mo and Nb as the main strengthening elements. With excellent corrosion resistance and oxidation resistance, it has good tensile and fatigue properties from low temperature to 980 °C, as well as resistance to salt spray corrosion. GH3625 is often used in the chemical processing, power, aerospace, and automobile industries owing to its unique combination of high strength, excellent fabricability and weldability, and outstanding corrosion resistance [1–4].

With the development of the petroleum industry in recent years, the service conditions of materials have become increasingly rigorous. As the depth of oil wells increases, oil extraction equipment is exposed to higher temperatures and pressures, as well as acidic gas. GH3625 superalloy bars produced by the cold drawing process are commonly made into piston rods for oil drilling [5–7]. The original intention of the GH3625 material's design was to have excellent corrosion resistance, while mechanical properties were not its main advantage. However, for structural materials, higher strength and more alternative strength grades can significantly expand the application range. The strength of conventional GH3625 can be appropriately enhanced through aginge strengthening. Because the effect of the second phase is limited by its chemical composition, the mechanical properties of the material can be significantly improved through cold deformation from another dimension. Cold deformation has both advantages and disadvantages. Its advantages are its high processing efficiency and obvious strengthening effect. However, its disadvantages are that the deformed steel bar has a great residual stress that cannot be removed, and is prone to deformation as the stress is released during service. Aging treatment after cold drawing may be a good solution. After a large number of dislocations are generated after cold drawing, the residual stress can be appropriately released by heat treatment. Although some dislocations may ablate at this time, the second phase precipitated during aging can compensate for the lost strength. Therefore, elucidating the balance between the size of deformation and the aging treatment process based on the principles of physical metallurgy is the key to achieving microstructure and performance control. However, researchers have paid more attention to welding and hot working for the GH3625 superalloy and similar alloys, compared to the effects of cold drawing deformation and subsequent heat treatment on the microstructure and properties [8–15].

Cold deformation can not only change the shape and size of the material, but also alter the microstructure, thereby affecting its properties. Ding studied the influence of cold drawing with different reductions in area on the location of δ-phase precipitation, indicating that when the strain was 35% the δ phase first nucleated and precipitated at grain boundaries and deformation twin boundaries, and then, nucleated and grew within the grain. When the strain was $\geq$50%, the δ phase first nucleated and grew at the deformation twin boundaries, grain boundaries, and deformation bands, and then, nucleated and precipitated within the grains. And as the strain increased, the δ phase precipitated more in the deformation bands [16–18]. Zhao et al. studied the effect of different cold deformation on the mechanical properties, it was found that an increase in the cold deformation rate could result in more cold work hardening, leading to a significant improvement in tensile strength and a decrease in the elongation of the alloy at room temperature [19]. Wang Zhigang et al. studied the effects of different reductions in area and deformation passes on the microstructure and properties of alloys. It was found that the grains were elongated during cold deformation, resulting in deformation structures such as dislocation cells and deformation twins, which increased the resistance of dislocation movement and led to work hardening. Cold drawing was a gradual diffuse process from the outside to the inside. When the reduction in area varied between 19% and 32%, as the reduction in area increased, the hardness of the alloy steadily, but not significantly, increased [20,21]. The relationship between the performance at room temperature and the reduction in performance of the cold-drawn GH3625 alloy was also revealed. However, the yield strength ratio of the produced bars was up to 0.99 or above, greatly reducing their safety.

A few researchers have studied the effect of heat treatment after cold deformation on alloy properties. Qin et al. studied the effects of different heat treatment temperatures on the microstructure, mechanical properties, and intergranular corrosion resistance of the GH3625 alloy after cold deformation. It was found that as the heat treatment temperature increased, grains engulfed each other, grain boundaries moved, and grains grew. The amount of the precipitated phases in the matrix gradually decreased with increasing temperature, but some precipitated phases increased [22].

Gao Yubi et al. studied the microstructure evolution of cold-deformed GH3625 alloy pipes during the intermediate annealing process. It was found that with the degree of cold deformation increasing, the uniformity of the alloy microstructure gradually improved, and the hardness value significantly increased, especially at cold deformation degrees from 0 to 50%. It was also found that at an annealing temperature of 1120 °C, complete recrystallization occurred and the recrystallized grain size decreased as the cold deformation rate increased. As above, the research data on the mechanical properties of the GH625 alloy after cold drawing and heat treatment with different reductions in area is not sufficient. Therefore, it is necessary to systematically study the microstructure and mechanical properties of the GH3625 alloy after cold drawing and aging treatment [23,24].

In this paper, the microstructure and mechanical properties of GH3625 alloy bars in different conditions were studied by adjusting the reduction in area and aging temperatures, in order to optimize their microstructure and performance. Meanwhile, the study can also lay the basis for further research on other high-temperature alloys, provide solutions for selecting GH3625 alloys with different mechanical indicators in the subsequent equipment design process, and provide experimental data and a theoretical basis for designing and manufacturing more advanced and safer oil and gas equipment. The process provided by this research can also improve the problem of residual stress causing bending of cold-drawn parts during service.

2. Material and Experimental Procedure

For this study, GH3625 ingots of Φ360 mm were built with vacuum induction melting (VIM) + vacuum arc remelting (VAR), using niobium chromium master alloys (the chemical composition is shown in Table 1, andsupplied by CITIC Metal Co., Ltd., Beijing, China), nickel plates (≥99.96 wt%), metallic molybdenum (≥99.85 wt%), etc. as raw materials. The chemical composition of GH3625 was measured by a fluorescence spectrometer and calibrated with a GH3625 standard sample. The chemical composition of the alloy is shown in Table 2.

Table 1. Main chemical composition of niobium chromium master alloys (wt%).

Nb	C	S	Al	Si	Fe	Cr
23.65	0.05	0.001	0.11	0.10	0.23	Bal

Table 2. Main chemical composition of GH3625 (wt%).

C	Cr	Mo	Al	Ti	Nb	Fe	Ni
0.04	21.36	9.09	0.24	0.25	3.48	0.45	Bal

The GH3625 ingot was held at 1150 °C in a natural gas heating furnace, and formed into square steel billets of 100 mm × 100 mm by a pneumatic hammer. The billets were forged and rolled into bars of Φ20 mm. The bars were annealed at 980 °C for 1 h, followed by air cooling and polishing into bars of Φ18 mm, which were acidized and saponified, and cold-drawn in a 20-ton double-chain cold drawing machine with different reductions in area (10%, 20%, 30%) at a speed of 10.32 m/min. (Deformation at room temperature is called cold deformation or cold drawing in this paper). The 12 cold-drawn bars were divided into 3 groups, with each group consisting of 4 bars. The first group was kept in the cold-drawn state, and the other two groups were aged in resistance furnaces. One group was held at 760 °C in air for 1 h, followed by natural air cooling. The other group was held at 650 °C in air for 24 h, followed by natural air cooling. These three groups of bars were tensile tested once per experiment, with an error range of no more than ± 1%. The hardness of the samples before cold drawing, after cold drawing without heat treatment, and after heat treatment was determined by an EV500-2A semi-automatic Vickers hardness tester with a test loading of 10 kg. Three pieces of 3 mm thick sheets were taken parallel to their cross-section from three different state bars for the experiments, followed by statistical analysis of their mean and variance. Points at intervals of 0.5 mm from the center to the edge on the cross-section of the samples were tested to analyze the hardness distribution. The tensile performance and hardness at room temperature were measured and the grain structure, precipitated phase, and dislocation density of the samples were analyzed.

The tensile performance includes tensile strength, yield strength, elongation, and section shrinkage. The sampling position was at the center of the bar in the axial direction. The samples for tensile testing were uniformly processed according to the specified proportion of samples with an available part diameter of 5 mm and a gauge distance of 25 mm. The tensile test was carried out by a Zwick-Z400-type tensile testing machine at room temperature. The tensile rate within the elastic deformation range was 3 mm/min, which increased to 10 mm/min when the stress exceeded the yield strength.

A 20 mm steel rod was selected and dissected along the central axis, and polished with abrasive paper. And then, it was etched using a mixture of copper chloride and hydrochloric acid for characterization by OM and SEM. The metallographic structure was observed using a Zeiss 40MAT optical metallographic microscope with a magnification of 500 times. The microstructure was characterized using a FEI-Quanta650 FEG thermal field emission scanning electron microscope at a magnification of 4000 times, with a working distance of 16–18 mm and an electron acceleration voltage of 20 kV. A 0.3 mm thick slice was cut from the cross-section of the bar by wire electric discharge machining, and then, polished to 50 μm with #1000 abrasive paper, and cut into a wafer of Φ3 mm. A mixed solution of $HClO_4$ and C_2H_5OH was used for electrolytic double spraying to make a TEM sample with an area of a certain thinckness.

At the same time, TEM (FEl Tecnai G2 F20) operating at a voltage of 200 kV was utilized to observe the microstructure of samples, and a Bruker energy spectrometer was equipped to semi-quantitatively characterize the composition. An X-ray diffraction test was performed using a German Brooke D8 ADVANCE X-ray diffractometer, with the following process parameters: Co target with wave length of 1.7889 Å, tube current of 40 mA, tube voltage of 35 kV, scanning speed of 2°/min, and Lynxeye XE detector.

3. Results and Discussion

3.1. Microstructure

Thermal-Calc (TCNI10: Ni-Alloys v10.0) was used to analyze the microstructure of the GH3625 alloy in a complete equilibrium state. The composition system used in the calculation is shown in Table 1, with a temperature range of 500–1500 °C and a total amount of 1 g. The thermodynamic calculation results are shown in Figure 1.

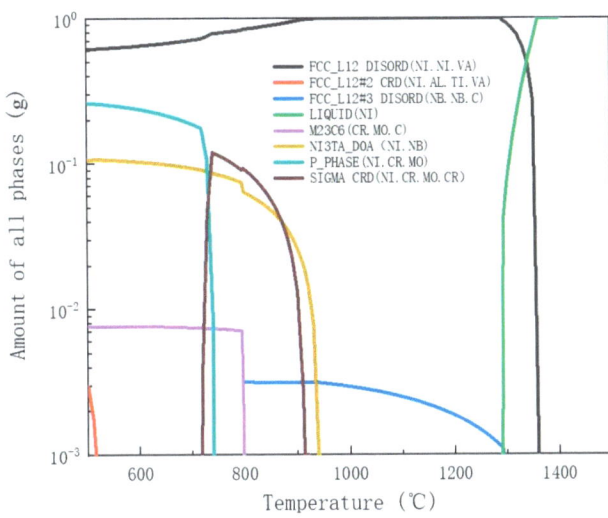

Figure 1. Equilibrium content of phases in GH3625 alloy calculated by Thermal-Calc.

Florren S drew a time–temperature–transformation (TTT) plot of the GH3625 alloy, as shown in Figure 2.

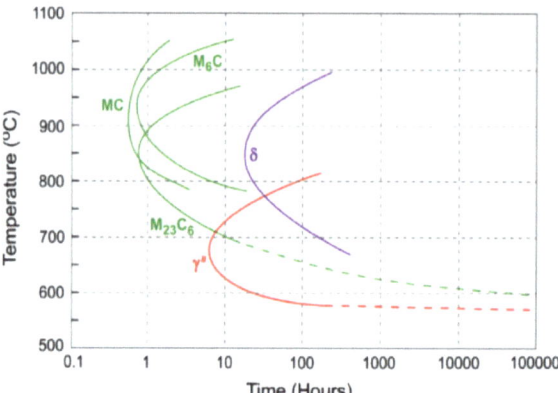

Figure 2. TTT of GH3625 alloy [25–27].

The microstructure of the GH3625 alloy is analyzed based on Figures 1 and 2. Figure 1 shows that the GH3625 matrix is austenite. In a fully equilibrium state, if the temperature of the alloy liquid drops to 1355 °C, the austenite begins to form, and the liquid phase completely solidifies at 1295 °C. When the temperature is above 940 °C, there is a small amount of MX phases which have a face-centered cubic (FCC) structure, mainly composed of niobium carbide (NbC), with a mass fraction not exceeding 0.3% (shown as the blue line in Figure 1, FCC-L12 # 3). The MX phase is commonly referred to as primary carbide. Residual carbide and niobium distributed on the interdendritic zone combine in the early stages of solidification. If there are nitride and oxide inclusions in the liquid phase, it will further promote the nucleation of NbC. However, the NbC phase is unstable at temperatures of 700–950 °C and easily decomposes into the M23C6 or M6C phase, while the NbC phase could also decompose into the M6C phase at temperatures of 800–980 °C. Some researchers believe that the M6C phase is transformed from M23C6, and the final form is M6C [28], while the author believes that the M6C phase and M23C6 may coexist, and their content ratio depends on the aging temperature and the Cr/Mo ratio of the material itself. When the temperature is less than 800 °C, 0.08 wt% M23C6 phase always exists in a thermodynamic equilibrium (shown as the purple line in Figure 1) [29].

Except for carbides, the main precipitates in the microstructure are the δ phase and the γ'' phase. When the temperature drops to 940 °C, the δ phase forms in a thermodynamic equilibrium state, composed of Ni$_3$Nb, with a maximum content of 10 wt% (shown as the brown line in Figure 1, NI3TA). Because the γ'' phase is the metastable phase of the δ phase, the γ'' phase does not exist in the thermodynamic equilibrium state. The δ phase has an orthorhombic structure (D0a), and the γ'' phase has an ordered body-centered tetragonal structure (D022). The γ'' phase is the major strengthening phase of the GH3625 alloy, and usually distributes dispersedly. The γ'' phase with Ni3Nb stoichiometry, has lattice parameters of a = 0.362 nm and c = 0.740 nm. The formation of the δ phase slightly reduces the intergranular corrosion resistance. The δ phase precipitates when the samples age at above 700 °C and below 950 °C for a long time. If an aging temperature of 650 °C is selected, from the perspective of precipitation kinetics this temperature is exactly the optimal precipitation temperature of the γ'' phase. There are still two types of intermetallic phases in thermodynamic equilibrium, the P phase and the σ phase (Figure 1), which have similar compositions. The σ phase transforms into the P phase as the temperature drops below 720 °C. The σ phase, commonly harmful for corrosion-resistant alloys, is the topologically close packed (TCP) phase that theoretically exists in the GH3625 alloy. Its tendency to precipitate is relatively less if there is reasonable heat treatment of the solution. The P phase only exists theoretically in thermodynamic calculations and cannot meet its kinetic precipitation conditions in practice.

Figure 3 shows the microstructure of the GH3625 alloy with different reductions in area. As seen, cold drawing has a relatively small effect on the precipitation of GH3625, slightly refining the grain size, and significantly changing the carbides. As shown in Figure 3a, the grains without cold drawing are mostly equiaxed, and the overall microstructure retains the flow left by the hot deformation process, and the carbide is nodular. As shown in Figure 3b, when the area decreases by 10%, the carbides have a slight tendency to crack, but the overall microstructure changes little. Figure 3c shows that when the reduction in area increases to 20%, great tensile cracking (above 10 μm) occurs in the large-sized carbides, and the number of twins increases. Figure 3d shows that when the reduction in area increases to 30%, slightly smaller carbides (3–10 μm) are also broken, and multiple grains are elongated, resulting in an overall increase in grain size by 1 level (ASTM). As Figure 3e,f show, different aging treatments were performed on the samples with a 30% reduction in area after cold drawing, and was found that the precipitates at the grain boundaries only slightly increased. This is because the aging at 760 °C for 1 h makes the precipitates change slightly: $M_{23}C_6$ does not precipitate, while aging at 650 °C for 24 h separates out some of the γ phase, which is so small that it cannot be observed by optical microscope.

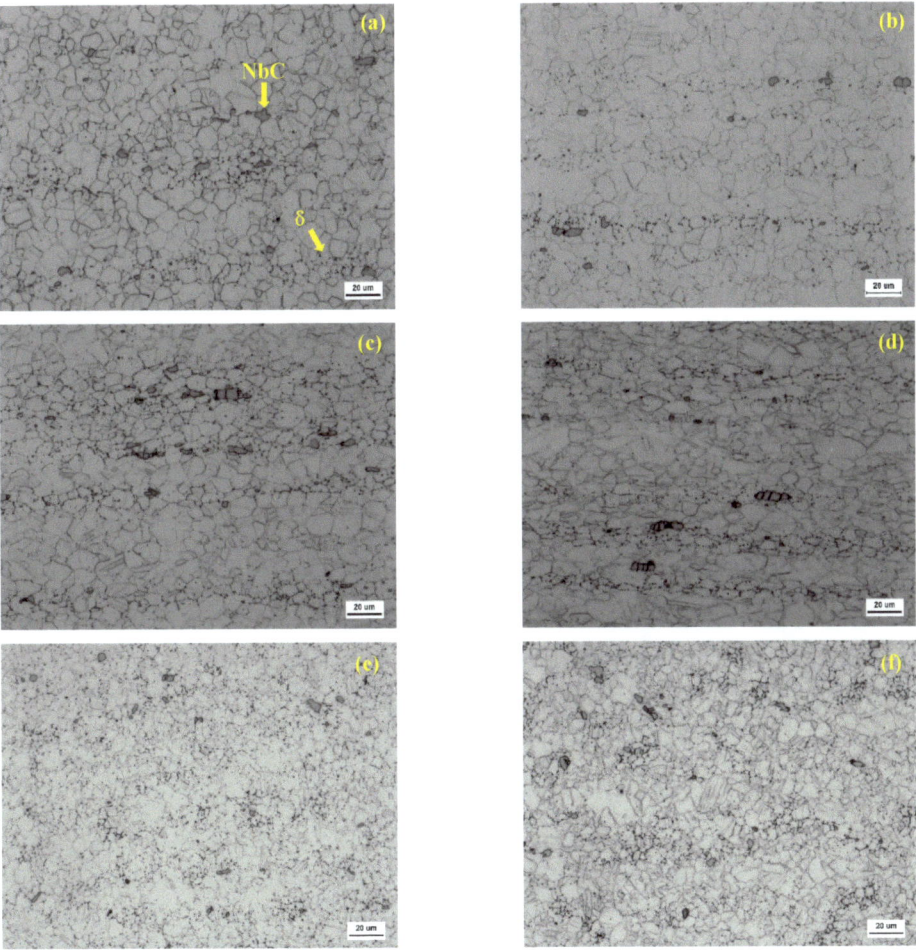

Figure 3. OM microstructure of cold-drawn GH3625 alloy with different reductions in area (500×): (**a**) 0%, (**b**) 10%, (**c**) 20%, (**d**) 30%, (**e**) 30% + aging at 760 °C for 1 h, and (**f**) 30% + aging at 650 °C for 24 h.

Figure 4 shows the change in microstructure caused by cold deformation can be more clearly demonstrated by SEM, as described above. And it also clearly shows that the precipitates of secondary carbide, shown as spots, are almost entirely at grain boundaries, distributed parallel to the direction of hot work along the segregation bands, which is almost unrelated to cold work.

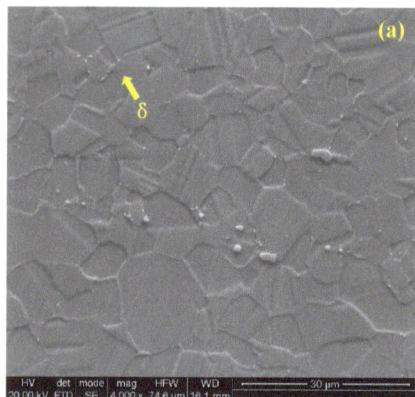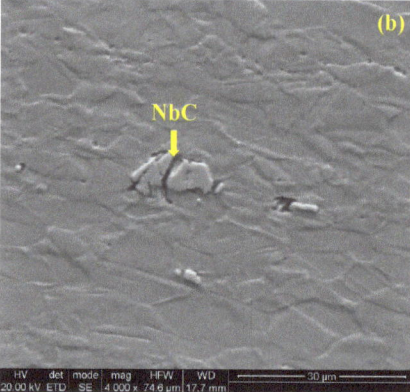

Figure 4. SEM microstructure of cold-drawn GH3625 alloy with different reductions in area (4000×): (**a**) 0%, (**b**) 30%.

The composition of precipitates of the samples after cold drawing with a 30% reduction in area and aging at 650 °C for 24 h was analyzed by energy dispersive X-ray spectroscopy (EDS). As shown in Figure 5 and Table 3, the precipitates are mainly NbC, and the semi-quantitative analysis shows a 61.5% mass fraction of Nb. The only visible precipitates in the structure are NbC, because the holding temperature of 760 °C is low and the holding time is short. In addition, the low carbon content results in no precipitation of the M23C6 phase. Theoretically, the γ'' phase should be precipitated in the microstructure after aging for 24 h, but it is small to nanoscale, making it difficult to characterize.

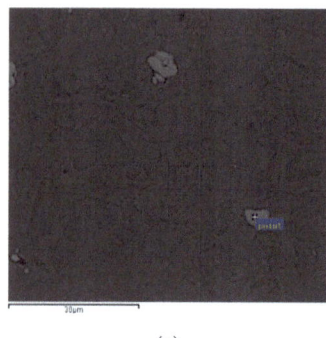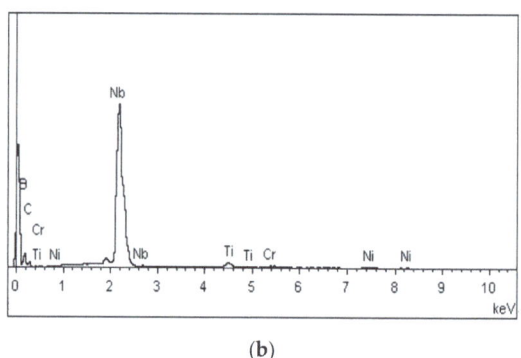

(**a**) (**b**)

Figure 5. Carbide EDS of cold-drawn GH 3625 alloy: (**a**) back-scattered electron (BSE) image of carbides, (**b**) EDS spectrum of selected points.

Table 3. Semi-quantitative chemical composition of carbides.

Element	Wt%	At%
Nb	61.5	17.5
C	25.1	55.4

Table 3. *Cont.*

Element	Wt%	At%
B	10.4	25.6
Ti	1.3	0.7
Cr	0.6	0.3
Ni	1.1	0.5

Figure 6 shows the bright-field images (by TEM) of the microstructure before and after cold drawing. Figure 6a shows that there are some dislocation lines within the grains, and the distribution of the dislocations at the grain boundaries is denser. The grain boundaries may be the origin of dislocations. This is because the experimental bar was initially hot-rolled from alloy ingots, and although heat treatment was carried out after rolling, it was not enough to completely eliminate all dislocations in the structure. Therefore, there were some dislocations before cold drawing. As shown in Figure 6b,c, a large number of dislocation lines occur in the microstructure of the samples. Figure 6b shows a lot of dislocation pile-up and tangling at grain boundaries, while Figure 6c shows the dislocation pile-up near the second phase. As can be seen from the above, the numerous dislocation pile-ups at grain and phase boundaries lead to the deformation structures.

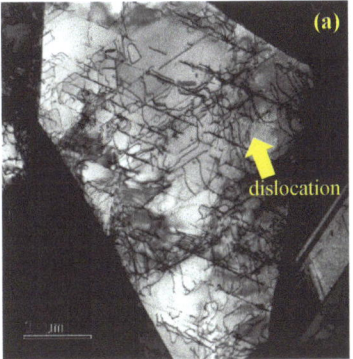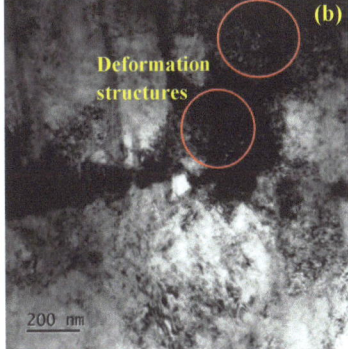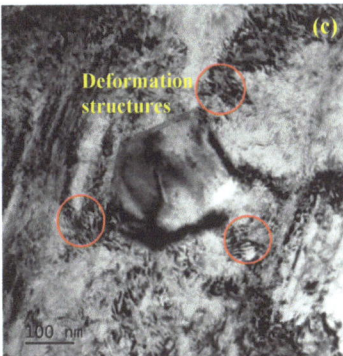

Figure 6. Bright-field images (TEM) of the microstructure before and after cold drawing. (**a**) Dislocation microstructure within grains and at grain boundaries before cold drawing, and the g vector indicates the dislocations within the grains before cold drawing. (**b**) Deformation of dislocation pile-up at grain boundary for the sample with 30% reduction in area. (**c**) Deformation of dislocation pile-up near second phase for the sample with 30% reduction in area.

Figure 7 shows the selected area electron diffraction patterns (SAED) of the precipitate phases. Combined with EDS analysis (Table 4), it can be found that the main precipitates are in the NbC phase. Comparing Figure 7a,b with Figure 7c,d, it can be seen that there is a small change in the types of precipitates before and after drawing, mainly in MC-type carbides. In the sample without cold drawing, the MC phase (in Figure 7a) and a small amount of the M6C phase (in Figure 7b) are found. In the cold-drawn sample, the MC phase (in Figure 7c) and M6C phase are found, as well as a small amount of the M23C6 phase (in Figure 7d).

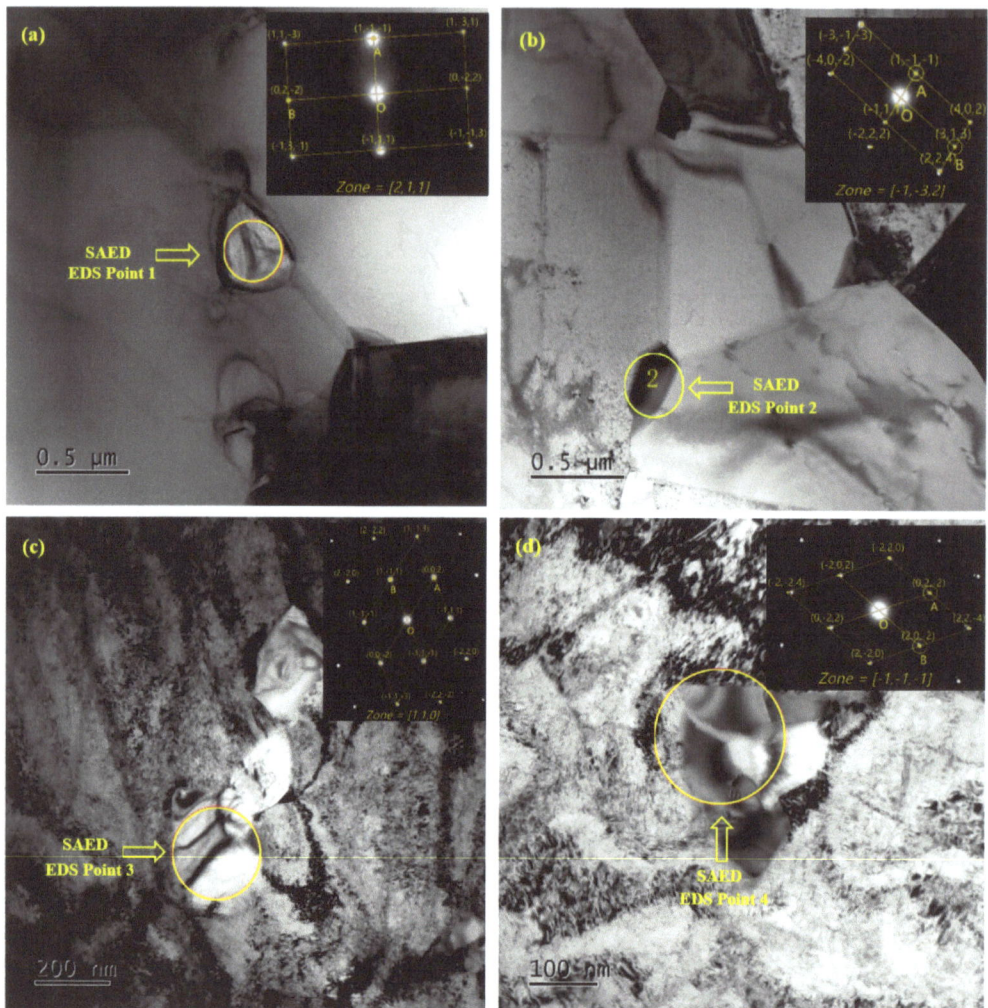

Figure 7. Images (TEM) before and after cold drawing. (**a**) Bright-field image with 0% reduction in area, and corresponding SAED of MC phase; (**b**) bright-field image and corresponding SAED of M6C phase before cold drawing; (**c**) bright-field image with 30% reduction in area, and corresponding SAED of MC phase after cold drawing; (**d**) bright-field image with 30% reduction in area, and corresponding SAED of M23C6 phase after cold drawing.

Table 4. EDS analysis results.

Element	Point 1		Point 2		Point 3		Point 4	
	At%	Wt%	At%	Wt%	At%	Wt%	At%	Wt%
Nb	86.2	91.6	9.8	11.6	66.6	76.2	0.6	0.9
Ni	5.1	3.4	31.9	24.5	18.7	13.5	22.8	22.3
Cr	1.8	1.1	14.5	9.9	6.2	4	60.3	52
Fe	1.7	1	1.9	1.4	2	1.4	0.7	0.6
Ti	5.2	2.9	-	-	4.6	2.7	-	-
Mo	-	-	41.9	52.6	1.8	2.2	15.1	24
Al	-	-	-	-	-	-	0.5	0.2

A small number of dispersed and tiny γ'' phases are observed in the microstructure of the cold-drawn samples after aging at 650 °C for 24 h, which has some consistency with the results displayed by the dynamic curve. The morphology is shown in Figure 8.

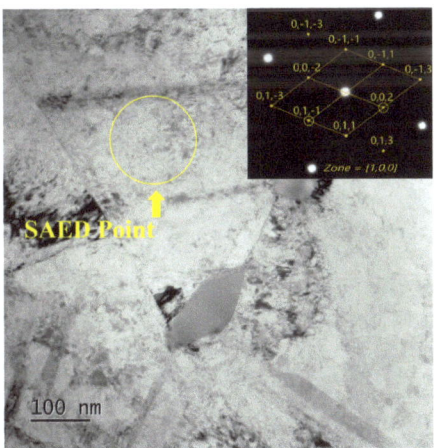

Figure 8. Morphology and SAED image of γ'' phases.

3.2. Tensile Properties

The tensile properties of samples with different deformation and heat treatment states were tested at room temperature. The yield strength (Rp0.2), tensile strength (Rm), elongation (A%), and section shrinkage (Z%) are shown in Table 5 and Figure 9.

Table 5. Tensile properties before and after cold drawing and heat treatment.

State	Reduction in Area	Rp0.2	Rm	A%	Z%
Cold drawn	0	582	994	46.5	58
	10	954	1128	32.5	52
	20	1254	1302	22.5	48
	30	1505	1505	9.5	42
650 °C for 24 h	0	608	911	43	57
	10	923	1152	33	52
	20	1135	1315	25	48
	30	1296	1427	21	42
760 °C for 1 h	0	632	1012	46	56
	10	873	1127	32.5	53
	20	1074	1264	26.5	51
	30	1267	1399	19.5	47.5

Figure 9a shows that cold deformation significantly improves the strength, with a strength increase of 130–200 MPa for every 10% increase in reduction in area. Heat treatment has little effect on the tensile strength of the samples with ≤20% reduction in area. For the sample with a 30% reduction in area, recovery occurs following heat treatment, resulting in a decrease in strength of about 100 MPa. Figure 9b shows that the strength increase caused by cold drawing is more pronounced in the yield strength, which increases by 250–370 MPa for every 10% increase in reduction in area. Meanwhile, aging can slightly enhance the yield strength of the undeformed sample by the second phase, and its strengthening mechanism is based on the formula for the second-phase strength increase [30].

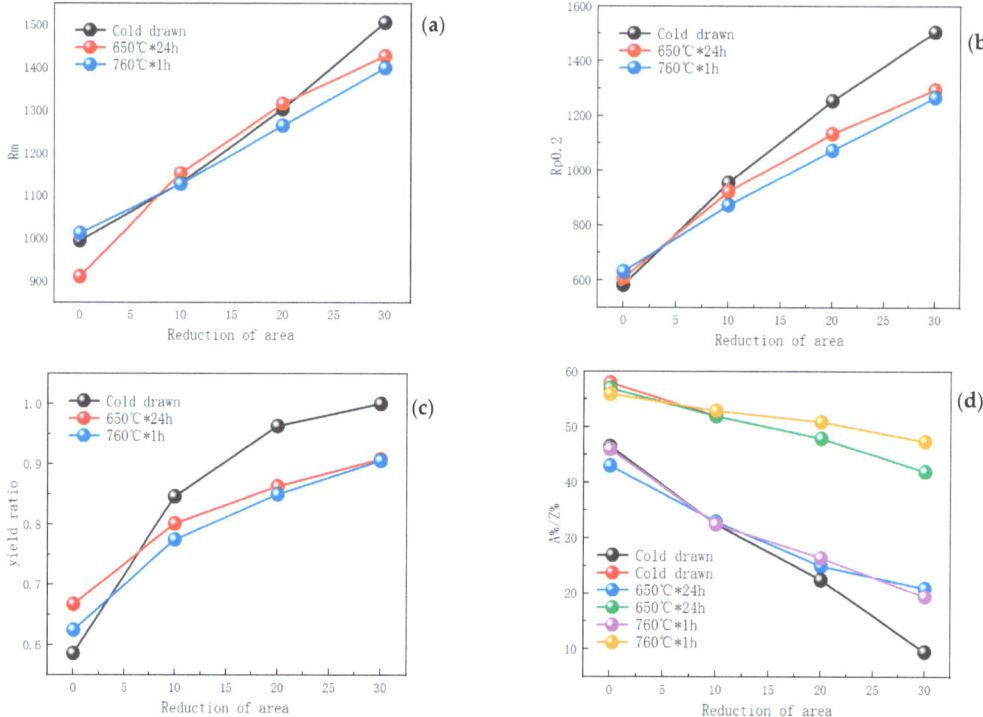

Figure 9. Tensile properties of cold-drawn bars with different reductions in area at room temperature. (**a**) Tensile strength, (**b**) yield strength, (**c**) yield ratio, and (**d**) elongation and reduction in area.

$$\tau = 0.8 * M \frac{Gb}{\lambda}$$

τ is the strength increase, M is the Taylor factor (=3), G is the shear modulus, b is the Burgess vector, and λ is the particle spacing in the second phase.

Heat treatment has a little effect on the mechanical properties of the alloy with 10% strain, while it has a greater impact on the alloys with 20% and 30% deformation. This is because during the cold drawing process, when the strain is small (10%), the strengthening effect from the edge to the center decreases, so the strengthening effect in the center is weak. However, the tensile samples could only be chosen from the center of the bar. Therefore, the heat treatment has little effect on the core of the sample with a 10% reduction in area. Meanwhile, it can be seen that the yield strength of the sample held at 650 °C is higher than that of the sample held at 760 °C for 1 h. This is because the γ'' phase starts nucleation at 650 °C, and although it has not yet grown due to the relatively short time, it will strengthen the second phase from a microscopic perspective. At 760 °C, there are fewer precipitated secondary carbides, and the main carbides in the microstructure are still the large primary carbides during solidification, which have little effect on improving strength. Therefore, the smelting temperature should be properly controlled, which could reduce the number of primary carbides and make the carbides precipitate in the form of small secondary carbides, thus benefiting the performance of the alloy. Figure 9c shows that as the reduction in area increases, the yield strength ratio approaches 1, which is very unfavorable for the safe service of the material. When the yield strength ratio is appropriately reduced and the stress is greater than the yield strength, the material will undergo plastic deformation in advance, which will make the failure signal more obvious, providing reaction time for component replacement and reducing the losses caused by material failure in the entire system. Both heat treatment processes can significantly reduce the yield ratio from 1 to around 0.9. Figure 9d shows that the plasticity significantly decreases as the reduction in area increases. Meanwhile, the two heat treatment processes improve the elongation with 30% strain, which is attributed to the recovery effect. However, the change in dislocation does not damage the grain boundaries. After necking occurs, the alloy can still continue necking, which also makes the influence of heat treatment on the reduction in area small.

The stress–strain curve reflects the deformation behavior of the material at various stages. Figure 10 shows that the curves of the elastic stage almost overlap for samples with different reductions in area, but there is an obvious difference in the plastic strain stage. The two main factors affecting the plasticity index are the plastic uniform extension after elastic strain, and the effect of necking after exceeding the maximum stress. In this experiment, the necking of all samples was similar, so the elongation was significantly affected by the former factor. The main factors determining the plastic deformation ability of the alloy in the uniformly elongation section are its dislocation movement and work-hardening ability. When the sample is subjected to tensile stress load, the parallel segments gradually become thinner, then a large number of dislocations occur. The first part that becomes thinner bears the maximum stress, but due to the large increase in the number of dislocations and their movement, work hardening occurs, which enables the alloy to have a self-healing mechanism. After local strengthening, the fracture ultimate strength at this point increases, and continues to extend uniformly until fracture. The premise of this experiment is cold drawing, which has led to the formation of a large number of dislocations in advance. The pile-up of a large number of dislocations obviously hinders the further increase and movement of dislocations, slowing down the production of new dislocations and their movement to the weakest position in the cold-deformed sample, reducing its ability to uniformly extend, and thus, reducing the overall elongation of the material after forging.

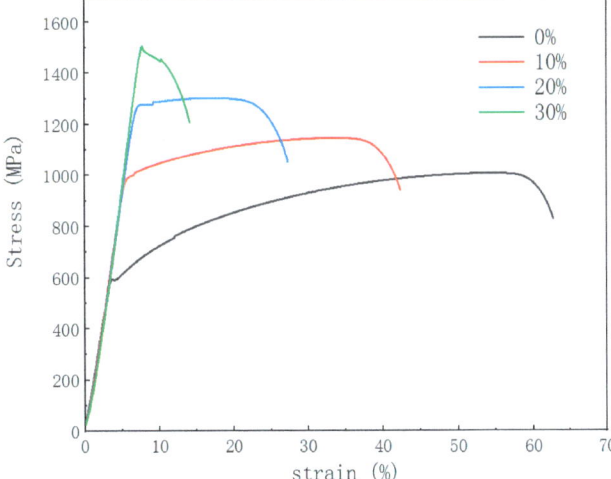

Figure 10. Stress–strain curves with different reductions in area.

Figure 11 shows the influence of heat treatment on the stress–strain curve, which is mainly reflected in samples with large reductions in area. Both heat treatment processes can restore the stress–strain curve of a sample that has been strengthened by a large number of dislocations to the shape of the stress–strain curve before cold drawing. The only difference is that the turning point (yield strength) improves. It can also be seen that the sample aged at 650 °C has a larger elastic strain under the same load.

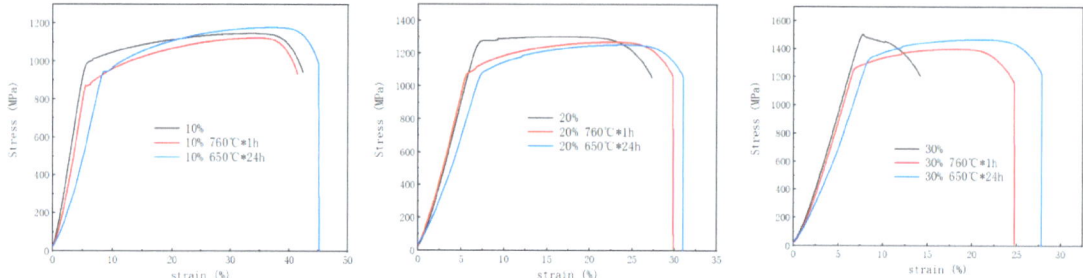

Figure 11. The influence of heat treatment on the mechanical properties with various reductions in area.

The changes in the mechanical properties of GH3625 after cold drawing were studied. Ding et al. studied the relationship between cold deformation and the mechanical properties of GH3625. It was found that cold deformation was the main factor affecting work hardening [31]. Zhao studied the effect of different degrees of cold deformation on the mechanical properties. It was found that when the cold rolling deformation was 20%, the tensile strength could reach 1050 MPa. With an increase in reductions in area, the strength increased. The condition of the material was different from that in this study, so the experimental data obtained are not the same. However, the overall trend and strengthening mechanism of the materials obtained from the research are similar to the direction of this study [19].

3.3. Hardness Gradient

The Vickers hardness of samples before cold drawing, after cold drawing, and after heat treatment were tested from the edge to the center. Figure 12 shows that due to the annealing following the hot rolling, the hardness tested from the center to the edge of the steel bar is uniform before cold drawing. When the reduction in area is 10%, the strengthening effect of the edge of the cold-drawn steel bar is stronger, and the hardness values from the edge to the center decrease monotonically, with a difference of 41 HV between the edge and the center. When the reduction in area is 20%, the hardness of the steel bar after cold drawing increases, but the deformation still cannot be fully transmitted to the center. When the reduction in area is 30%, the strength of the steel bar after cold drawing is uniform, and the hardness values at the center and edges are close. This indicates that when the total cold deformation is small, the strain at the edges of the steel bar is larger and the strain at the center is smaller. As the total deformation increases, the strain at the center increases, approaching that at the edge.

The hardness of samples after heat treatment was tested. For samples aged at 760 °C for 1 h with 10% and 30% reductions in area, it is found that heat treatment cannot completely eliminate the work hardening influence. The hardness value at different positions of the sample after heat treatment decreases by about 30 HV, and the hardness value from the edge to the center of the sample still keep decreasing, which is consistent with that of the sample after cold drawing without heat treatment.

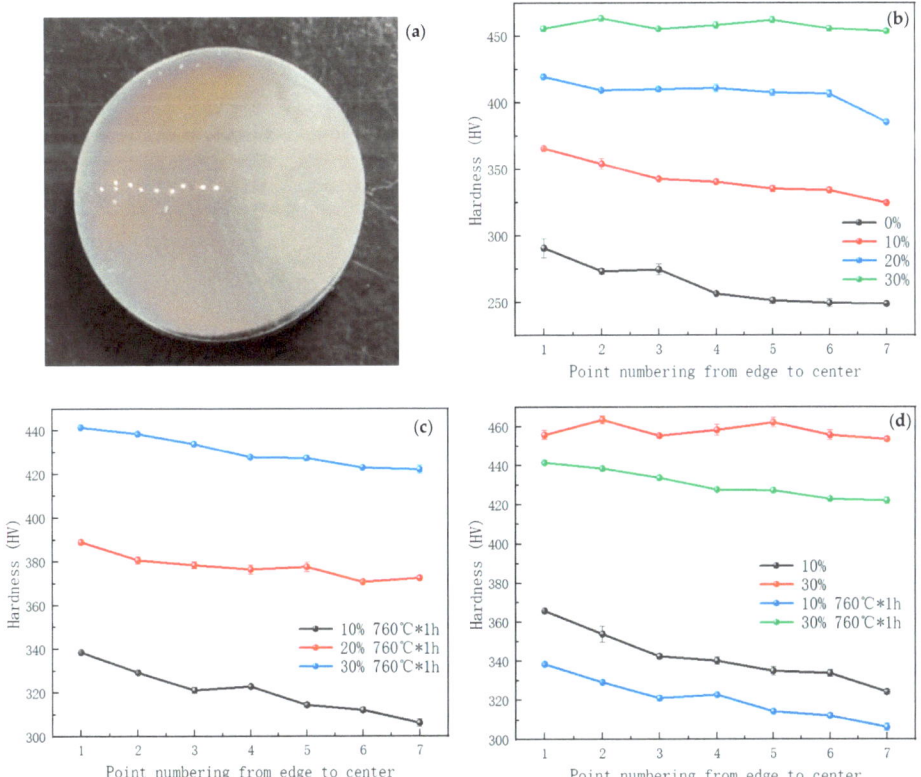

Figure 12. Hardness after cold drawing and heat treatment. (**a**) Macroscopic photo; (**b**) hardness with different reductions in area; (**c**) hardness after aging at 760 °C; and (**d**) hardness before and after aging treatment with 10% and 30% reductions in area.

3.4. Dislocation Density

There are usually two methods for calculating dislocation density. One is to use TEM for microscopic statistics, which uses the ratio of the total length of all dislocation lines per unit field of view to the area to represent the facial density of dislocations. This method is inaccurate and it can only indicate the facial density of dislocations in a certain microscopic area, with units of nm^{-1}. Another method is to use XRD to test and calculate the average dislocation density. The test results are shown in Figure 13.

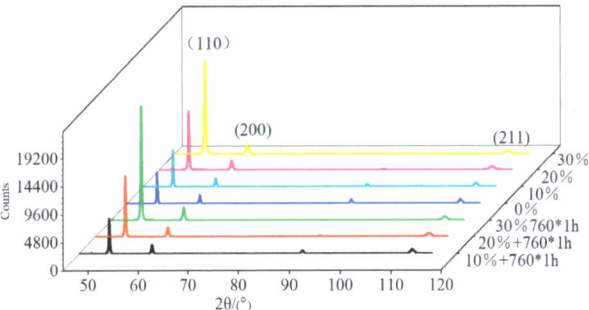

Figure 13. XRD map of samples with different reductions in area.

Firstly, the average thickness of the crystallites perpendicular to the crystal face is calculated, then the D value calculated by different 2 θ angles is averaged, and finally, the volume density of dislocations δ is calculated, with units of cm^{-2}. This is calculated by the Scherer equation [32,33]:

$$D = \frac{k\lambda}{\beta cos\theta}$$

where D is the crystallite size, k is the Scherrer constant (=0.89) [34], β is the measured width at half maximum of the diffraction peak of the sample, θ is the Bragg diffraction angle, λ is the wavelength of the X-ray, Co target, with a wavelength of 1.7889 Å.

$$\delta = \frac{1}{D^2}$$

Then, the equation above can be used to calculate the dislocation density δ [35]. Test results of the dislocation density with different reductions in area and aged at 760 °C for 1 h are shown in Table 6.

Table 6. Dislocation density in different states.

Reduction in Area	Cold Drawn	After Heat Treatment
0	1.1201×10^{10}	
10	8.6161×10^{10}	6.2325×10^{10}
20	2.9209×10^{11}	1.4214×10^{11}
30	9.1703×10^{11}	2.3126×10^{11}

It can be seen in the table that a 10% reduction in area increases the dislocation density by 8 times, and for every 10% increase in reduction in area thereafter, the dislocation density increases by 3 times. After heat treatment, the dislocation density significantly decreases, and the larger the reduction in area, the greater the decrease in dislocation density. The dislocation density of the sample with a 10% reduction in area decreases to 72%. The dislocation density of the sample with a 20% reduction in area decreases to 50%. The dislocation density of the sample with a 30% reduction in area decreases to 25%. This is related to the heat treatment temperature, where 760 °C is much higher than the temperature at which the material is recover. During the heat treatment process, a large number of dislocations ablate, and the lattice distortion energy is fully released due to static recovery at this temperature.

Deformation strengthening is one of the main methods to improve the strength of material. The yield strength values strengthened by four different strengthening methods are the superposition of effects caused by the different methods. When there is fine-grained strengthening and dislocation strengthening, the two are superimposed using the sum of square roots [36,37]. In this experiment, solution strengthening and the change in grain size are not obvious. The strength increase brought about by the second-phase strengthening is estimated to be 50 MPa based on the strength increase in the annealed and undeformed samples before and after aging, mainly calculating the contribution of dislocation strengthening.

The contribution of dislocations to the strength increase before and after cold drawing and heat treatment is calculated by the following formula [38,39]:

$$\sigma_\rho = aGb\left(\rho^{1/2}\right)$$

where σ_ρ is the strength increase; α is a constant, taken as 0.88; G is the shear modulus, taken as 79 GPa; b is the Burgess vector, taken as 0.25 nm; and ρ is the dislocation density. Tables 7–9 show the calculation results of samples with different dislocation densities.

Table 7. Strength increase.

Reduction in Area	Cold Drawn/MPa	Aging at 760 °C/MPa
0	183	
10	508	432
20	936	701
30	1658	918

Table 8. Calculated yield strength.

Reduction in Area	Cold Drawn/MPa	Aging at 760 °C/MPa
0	582	
10	907	831
20	1335	1100
30	2057	1317

Table 9. Measured yield strength.

Reduction in Area	Cold Drawn/MPa	Aging at 760 °C/MPa
0	582	632
10	954	873
20	1254	1074
30	1505	1267

Tables 7–9 show that the yield strength of the sample without cold drawing increases by 50 MPa after aging at 760 °C. It is believed that this strength increase is mainly caused by the strengthening of the second-phase precipitation. The increase in dislocation strength of the sample without cold deformation is 183 MPa, while the measured strength is 582 MPa. Therefore, the basic strength is considered to be 399 MPa in the calculations.

The calculation results show that the strength increase caused by 10–20% reductions in area and the strength reduction after heat treatment have a close relationship with the measured values. Although the specific theoretical values are slightly different from the measured values, the trend is very close. The theoretical dislocation strengthening of the material with a 30% reduction in area is enhanced sharply, with the yield strength reaching 2057 MPa, while the measured ultimate strength is only 1505MPa, and the yield ratio is 1 at this time. This is because materials with a yield ratio greater than or equal to 1 have extremely low plasticity, and the tested yield strength is forced to decrease. The material does not reach its yield strength during the tensile process, and due to the influence of ultimate strength, it fractures prematurely before yielding, resulting in a decrease in material safety. In summary, the good consistency between the calculated values and the measured values confirms that the strength increase is mainly provided by dislocation strengthening, reflecting the role of heat treatment in adjusting the mechanical properties of materials.

4. Conclusions

(1) The cold drawing process with 10–30% reductions in area can significantly improve the tensile strength and yield strength of the GH3625 alloy. After aging treatment at 650 °C and 760 °C, the yield strength decreased and the yield strength ratio greatly decreased. This study can provide more mechanical performance options for GH3625 alloy structural components in the petroleum and petrochemical industry.

(2) Proper heat treatment processes can ablate the residual dislocations after cold deformation and reduce the dislocation density in the microstructure of the GH3625 alloy, which is the main factor in reducing yield strength. It is clarified that the main strengthening mechanism in this study is dislocation strengthening by testing and calculating dislocation density under different states and the resulting strength increase. The precipitation

strengthening effect caused by the second-phase precipitate during heat treatment provides an additional strength increase. Affected by γ'', the yield strength after aging at 650 °C for 24 h is higher than that after aging at 760 °C for 1 h. If further aging research can be carried out to study the trend in dislocation density changes after long-term holding, and to clarify and quantify the kinetics of the γ'' phase, the practical significance of this research work will be further deepened.

(3) The work-hardening gradient of cold-drawn material varies greatly with different reductions in area. When the deformation is small (10%), the gradient is obvious. When the deformation increases to 30%, the material is uniformly strengthened as the deformation is transmitted to the axis. Therefore, the actual strength provided by the workpieces with low reductions in area will be higher than that of the sample.

5. Prospects

This paper conducted a study on the mechanical properties of cold-drawn materials with 10–30% reductions in area and under aging at 650 °C and 760 °C. The research on cold deformation with a smaller reduction in area (3–10%) is not sufficient yet, so the effect of small cold deformation on surface strengthening should be further studied. In addition, the precipitation kinetics and mechanical performance after long-term holding at 650 °C have not been fully studied, and research on multi-step heat treatments at different temperatures is not sufficient. Afterwards, further adjustments can be made to the heat treatment process to provide more mechanical performance options.

Author Contributions: Conceptualization, J.L. and Z.W.; Software, W.Z., J.Z. and Y.Z.; Validation, W.Z., J.Z. and Y.Z.; Investigation, Z.W.; Resources, Z.W. and W.R.; Data curation, J.L., Y.W. and W.R.; Writing—original draft, J.L.; Writing—review & editing, J.L.; Visualization, J.L. and Y.W.; Supervision, W.Z. and J.Z.; Funding acquisition, W.Z. and J.Z. All authors have read and agreed to the published version of the manuscript.

Funding: This research was funded by [Major Scientific and Technological Innovation Project of CITIC Group] grant number [2022ZXKYA06100].

Institutional Review Board Statement: Not applicable.

Informed Consent Statement: Not applicable.

Data Availability Statement: Data are contained within the article.

Acknowledgments: Thank you to CITIC Group for the strong support of the "Development of New Materials Based on Rare Metal Applications" in the major technological innovation project for this research work.

Conflicts of Interest: Wenhao Ren was employed by the company Research Institute for Special Steel, Central Iron and Steel Research Institute; Wei Zhang was employed by the company CITIC Metal Co., Ltd. The remaining authors declare that the research was conducted in the absence of any commercial or financial relationships that could be construed as a potential conflict of interest.

References

1. Huang, J.; Yao, R.; Yao, S. Precipitation transformation and influencing factors of nickel-based 625 alloy. *Dev. Appl. Mater.* **2016**, *31*, 91–97.
2. Yao, M. *Study on Aging Microstructure and Creep Rupture Behavior of Cold Deformed GH3625 Alloy*; Yanshan University: Qinhuangdao, China, 2018.
3. Cui, L.; Lin, X.; Zhu, Q.; Wang, C.; Wang, P. Research progress of heat treatment process of high temperature alloy. *Mater. Rep.* **2016**, *30*, 106–110+132.
4. Guo, J. A low-cost directionally solidified nickel-based superalloy with excellent properties. *Acta Metall. Sin.* **2016**, *38*, 1163–1174.
5. Yu, Z. *Study on Composition Optimization and Heat Treatment Process of IN625 Superalloy*; Shenyang Ligong University: Shenyang, China, 2014.
6. Paraschiv, A.; Matache, G.; Constantin, N.; Vladut, M. Investigation of Scanning Strategies and Laser Remelting Effects on Top Surface Deformation of Additively Manufactured IN 625. *Materials* **2022**, *15*, 3198. [CrossRef] [PubMed]
7. Alena, K.; Vladimir, B. Effect of Fe and C Contents on the Microstructure and High-Temperature Mechanical Properties of GH3625 Alloy Processed by Laser Powder Bed Fusion. *Materials* **2022**, *15*, 6606.

8. Kopytowski, A.; Świercz, R.; Oniszczuk-Świercz, D.; Zawora, J.; Kuczak, J.; Żrodowski, Ł. Effects of a New Type of Grinding Wheel with Multi-Granular Abrasive Grains on Surface Topography Properties after Grinding of Inconel 625. *Materials* **2023**, *16*, 716. [CrossRef] [PubMed]
9. Liu, W.; Li, L.; Mi, G.; Wang, J. Pan Effect of Fe Content on Microstructure and Properties of Laser Cladding Inconel 625 Alloy. *Materials* **2022**, *15*, 8200. [CrossRef] [PubMed]
10. Li, J.; He, X.; Xu, B.; Tang, Z.; Fang, C.; Yang, G. Effect of Silicon on Dynamic/Static Corrosion Resistance of T91 in Lead-Bismuth Eutectic at 550 °C. *Materials* **2022**, *15*, 2862. [CrossRef] [PubMed]
11. Li, H.; Hsu, E.; Szpunar, J.; Utsunomiya, H.; Sakai, T. Deformation mechanism and texture and microstructure evolution during high-speed rolling of AZ31B Mg sheets. *J. Mater. Ence* **2008**, *43*, 7148–7156. [CrossRef]
12. Kumar, S.S.; Raghu, T.; Bhattacharjee, P.P.; Rao, G.A.; Borah, U. Work hardening characteristics and microstructural evolution during hot deformation of a nickel superalloy at moderate strain rates. *J. Alloys Compd.* **2017**, *709*, 394–409. [CrossRef]
13. Gao, H.; Huang, Y.; Nix, W.D.; Hutchinson, J.W. Mechanism-based strain gradient plasticity—I. Theory. *J. Mech. Phys. Solids* **1999**, *47*, 1239–1263. [CrossRef]
14. Kubin, L.P.; Mortensen, A. Geometrically necessary dislocations and strain-gradient plasticity: A few critical issues. *Scr. Mater.* **2003**, *48*, 119–125. [CrossRef]
15. Wu, Z.; Li, D.; Guo, S.; Guo, Q.; Peng, H.; Hu, J. Study on dynamic recrystallization model of GH625 nickel-based superalloy. *Rare Met. Mater. Eng.* **2012**, *41*, 235–240.
16. Ding, Y.; Gao, Y.; Dou, Z.; Gao, X.; Jia, Z. Microstructure evolution of cold-deformed GH3625 alloy tube during intermediate annealing. *Trans. Mater. Heat Treat.* **2017**, *38*, 178–184.
17. Ding, Y.; Meng, B.; Gao, Y.; Ma, Y.; Chen, J. Effect of Solution Treatment on Microstructure and Properties of GH3625 Alloy Tubes in Different States. In *High Performance Structural Materials*; Han, Y., Ed.; CMC 2017; Springer: Singapore, 2018.
18. Ding, Y.; Gao, Y.; Dou, Z.; Gao, X.; Liu, D.; Jia, Z. Deformation-induced precipitation behavior of d phase in hot-extruded GH3625 alloy tube. *Acta Metall. Sin.* **2017**, *53*, 695–702.
19. Zhao, Y. Cold deformation of GH625 alloy and its effect on mechanical properties. *J. Mater. Eng.* **2000**, *2*, 36–37. [CrossRef]
20. Wang, Z.; Tian, S.; Kui, H. Grain structure evolution of GH3625 alloy during rolling process. *Met. Funct. Mater.* **2018**, *25*, 45–49.
21. Wang, Z.; Wang, Y.; Tian, S.; Yang, X.; Zhang, P.; Ren, H. Effect of cold drawing deformation on microstructure and properties of GH3625 alloy. *J. Iron Steel Res.* **2011**, *23*, 92–95.
22. Qin, X.W.; Wang, K.; Nai, Q.; Du, L. Effect of heat treatment temperature on microstructure and properties of cold deformed Inconel 625 alloy pipe. *Heat Treat. Met.* **2021**, *46*, 164–169.
23. Gao, Y.; Ding, Y.; Meng, B.; Ma, Y.; Chen, J.; Xu, J. Research progress on the evolution of precipitated phases in Inconel 625 alloy. *J. Mater. Eng.* **2020**, *48*, 13–22.
24. Gao, Y. *Study on Cold Deformation Behavior of GH3625 Alloy Tube*; Lanzhou University of Technology: Lanzhou, China.
25. Floreen, S.; Fuchs, G.E.; Yang, W.J. *The Metallurgy of Alloy 625, Superalloys 718, 625 and Various Derivatives*; Loria, E.A., Ed.; TMS: Warrendale, PA, USA, 1994; pp. 13–37.
26. Shoemaker, L.E. Alloys 625 and 725: Trends in Properties and Applications. *Superalloys* **2005**, *718*, 409–418.
27. Singh, J.B.; Chakravartty, J.K.; Sundararaman, M. Work hardening behaviour of service aged Alloy 625. *Mater. Sci. Eng. A* **2013**, *576*, 239–242. [CrossRef]
28. Meng, B. *Effect of Heat Treatment on Microstructure and Properties of GH3625 Alloy Tube*; Lanzhou University of Technology: Lanzhou, China, 2018.
29. Stasiak, T.; Sow, M.A.; Addad, A.; Touzin, M.; Beclin, F.; Cordier, C. Processing and Characterization of a Mechanically Alloyed and Hot Press Sintered High Entropy Alloy from the Al-Cr-Fe-Mn-Mo Family. *JOM* **2022**, *74*, 971–980. [CrossRef]
30. Helis, L.; Toda, Y.; Hara, T.; Miyazaki, H.; Abe, F. Effect of cobalt on the microstructure of tempered martensitic 9Cr steel for ultra-supercritical power plants. *Mater. Sci. Eng. A* **2009**, *510*, 88–94. [CrossRef]
31. Ding, Y.; Gao, Y.; Dou, Z.; Gao, X.; Jia, Z. Study on Cold Deformation Behavior and Heat Treatment Process of GH3625 Superalloy Tubes. *Mater. Rev.* **2017**, *31*, 7. [CrossRef]
32. Sen, S.K.; Paul, T.C.; Dutta, S.; Hossain, M.N.; Mia, M.N.H. XRD peak profile and optical properties analysis of Ag-doped h-MoO3 nanorods synthesized via hydro-thermal method. *J. Mater. Sci. Mater. Electron.* **2020**, *31*, 1768–1786. [CrossRef]
33. Basak, M.; Rahman, M.L.; Ahmed, M.F.; Biswas, B.; Sharmin, N. The use of X-ray diffraction peak profile analysis to determine the structural parameters of cobalt ferrite nanoparticles using Debye-Scherrer, Williamson-Hall, Halder-Wagner and Size-strain plot: Different precipitating agent approach. *J. Alloys Compd.* **2022**, *895*, 162694. [CrossRef]
34. Guo, J.; Shen, Y. Several issues to pay attention to when calculating grain size using Scherrer's formula. *J. Inn. Mong. Norm. Univ. (Nat. Sci. Chin. Ed.)* **2009**, *38*, 357–358.
35. Stokes, A.R.; Wilson, A.J.C. The diffraction of x rays by distorted crystal aggregates—I. *Proc. Phys. Soc.* **1944**, *56*, 174–181. [CrossRef]
36. Yong, Q. *Second Phase in Steel Materials*; Metallurgical Industry Press: Beijing, China, 2006.
37. Yin, H.; Zhao, J.; Yang, G. The effect of quenching temperature on the microstructure and room temperature strength of COST-FB2 steel used in ultra supercritical thermal power plants. *Mater. Rev.* **2022**, *36*. [CrossRef]

38. Yan, P.; Liu, Z.; Bao, H.; Weng, Y.; Liu, W. Effect of normalizing temperature on the strength of 9Cr–3W–3Co martensitic heat resistant steel. *Mater. Sci. Eng. A* **2014**, *597*, 148–156. [CrossRef]
39. Maruyama, K.; Sawada, K.; Koike, J. Strengthening Mechanisms of Creep Resistant Tempered Martensitic Steel. *ISIJ Int.* **2001**, *41*, 641–653. [CrossRef]

Disclaimer/Publisher's Note: The statements, opinions and data contained in all publications are solely those of the individual author(s) and contributor(s) and not of MDPI and/or the editor(s). MDPI and/or the editor(s) disclaim responsibility for any injury to people or property resulting from any ideas, methods, instructions or products referred to in the content.

Article

Numerical Simulation of Segregation in Slabs under Different Secondary Cooling Electromagnetic Stirring Modes

Daiwei Liu [1,2], Guifang Zhang [1,*], Jianhua Zeng [3] and Xin Xie [3]

[1] Faculty of Metallurgical and Energy Engineering, Kunming University of Science and Technology, Kunming 650093, China; liudaiwei158@163.com
[2] School of Electrical and Information Engineering, Panzhihua University, Panzhihua 617000, China
[3] Pangang Group Research Institute Co., Ltd., Panzhihua 617000, China; zengjianhua68@163.com (J.Z.); xiexin@alu.cqu.edu.cn (X.X.)
* Correspondence: guifangzhang65@163.com

Abstract: Secondary cooling electromagnetic stirring (S-EMS) significantly impacts the internal quality of continuous casting slabs. In order to investigate the effects of S-EMS modes on segregation in slabs, a three-dimensional numerical model of the full-scale flow field, solidification, and mass transfer was established. A comparative analysis was conducted between continuous electromagnetic stirring and alternate stirring modes regarding their impacts on steel flow, solidification, and carbon segregation. The results indicated that adopting the alternate stirring mode was more advantageous for achieving uniform flow fields and reducing the disparity in solidification endpoints, thus mitigating carbon segregation. Specifically, the central carbon segregation index under continuous stirring at 320 A was 1.236, with an average of 1.247, while under alternate stirring, the central carbon segregation index decreased to 1.222 with an average of 1.227.

Keywords: slabs; secondary cooling electromagnetic stirring; alternating stirring; continuous stirring; segregation; numerical simulation

Citation: Liu, D.; Zhang, G.; Zeng, J.; Xie, X. Numerical Simulation of Segregation in Slabs under Different Secondary Cooling Electromagnetic Stirring Modes. *Materials* **2024**, *17*, 2721. https://doi.org/10.3390/ma17112721

Academic Editor: Alexander Yu Churyumov

Received: 15 April 2024
Revised: 25 May 2024
Accepted: 28 May 2024
Published: 3 June 2024

Copyright: © 2024 by the authors. Licensee MDPI, Basel, Switzerland. This article is an open access article distributed under the terms and conditions of the Creative Commons Attribution (CC BY) license (https:// creativecommons.org/licenses/by/ 4.0/).

1. Introduction

The development of steel slab varieties has raised the bar for internal quality standards, particularly emphasizing the homogenization of elemental composition and the refinement of the solidification microstructure [1,2]. However, challenges persist in the production process, such as inadequate isometric crystallinity rates in cast billets and severe segregation of alloying elements, which lead to a diminished steel toughness and plasticity, as well as inconsistent mechanical properties. Therefore, it has become imperative to address elemental segregation in continuous casting production to enhance slab homogenization, ultimately facilitating the production of high-quality steel.

Electromagnetic stirring in the secondary cooling zone during continuous casting (CC) is achieved by installing electromagnetic stirring devices within the secondary cooling zone of a continuous casting machine. The interaction between the stirring magnetic field and the induced electric currents generates Lorentz forces, which compel the flow of liquid metal within the cast slab and facilitate favorable kinetic and thermodynamic conditions conducive to equiaxed crystal growth and solute homogenization. Based on analyses of the carbon content in cast slabs and steel plate properties, Wang et al. [3] optimized the electromagnetic stirring mode and current parameters in the secondary cooling zone, leading to a reduction in the C-type segregation rate in CC slabs. Similarly, Liu et al. [4], through chemical analysis and low-magnification microstructure examination, investigated the effects of electromagnetic stirring frequency parameters in the secondary cooling zone on central segregation and the equiaxed grain rate in CC slabs. Their findings indicated a noticeable alleviation of central segregation in slabs with the adoption of electromagnetic stirring in the secondary cooling zone compared to those without. Dong et al. [5] explored the impact of roller electromagnetic stirring in the secondary cooling zone on the internal quality of Baosteel Q345B, Q550D, and Q345E steel slabs. Industrial experiments revealed significant improvements in the slab's internal quality post implementation of roller electromagnetic stirring in the secondary cooling zone, with the central equiaxed grain rate increasing from an average of 9.7% to 30.3% and the central C-segregation index decreasing from 1.17 to 1.01. Xie et al. [6] investigated the influence of roller electromagnetic stirring in the secondary cooling zone on the internal quality and rolling properties of high-strength steel. Their findings demonstrated that roller electromagnetic stirring improved central segregation from continuous or semi-continuous A- and B-type segregation to punctate C-type segregation, substantially reducing central segregation or central band-like structures in rolled materials.

Through industrial experiments, Li et al. [7] investigated the impact of electromagnetic stirring in the secondary cooling section of continuous casting on the solidification behavior of high-magnetic-induction silicon steel cast slabs. It was observed that with an increase in the S-EMS current intensity, the center-line segregation of carbon and silicon in the molten steel significantly increased. Yao et al. [8] conducted experiments to study the solidification behavior and composition distribution of 22MnB5 high-strength steel slabs under the influence of S-EMS. The intensity of S-EMS currents gradually increased to 100 A, 150 A, and 300 A, leading to a proportional enlargement of the equiaxed crystal zone. At 150 A, the central carbon segregation in the cast slab was weakest. Jiang et al. [9] investigated the transport phenomena of S-EMS during the continuous casting process of slabs. They established a three-dimensional numerical model and validated its effectiveness by measuring the surface temperature of the cast slab and the intensity of magnetic induction. Ma et al. [10] studied the effect of S-EMS on central segregation in slabs through low-magnification grading of the slab and grading of the hot-rolled coil band structure. It was found that while S-EMS could significantly improve central segregation and central looseness in cast slabs and reduce C-type central segregation, it also led to the presence of noticeable white bands. However, the improvement in band structure was not significant.

However, there is currently a lack of research on the impact of electromagnetic stirring modes on the quality of cast slabs. Mei et al. [11], through numerical simulations, demonstrated that compared to continuous stirring, adopting alternating stirring not only induces periodic variations in the flow within the mold but also significantly reduces the level of liquid surface activity in the mold. In industrial experiments conducted by Chun Liu et al. [12], it was found that continuous stirring resulted in a higher equiaxed grain ratio for 65Mn compared to alternating stirring. Li et al. [13] utilized a three-phase solidification model to numerically predict flow patterns and equiaxed grain ratios during the continuous casting process of three-dimensional full-size thin slabs. They analyzed the impact of alternating electromagnetic stirring reversal periods in the secondary cooling zone on the central equiaxed grain ratio of thin slabs. It was observed that with alternating electromagnetic stirring, the secondary cooling zone exhibited a four-vortex or even five-vortex flow pattern. Additionally, as the reversal period increased, the equiaxed grain ratio initially increased and then decreased. However, these studies have yet to address the influence of electromagnetic stirring modes in the secondary cooling zone on segregation.

The continuous casting process of slab in a specific steel plant is investigated in this study. A comprehensive solidification 3D numerical model is established to investigate the coupled phenomena of flow, heat transfer, and segregation. This study aims to compare and analyze the solute transport behavior induced by different electromagnetic stirring modes and elucidate the influence of electromagnetic stirring on central segregation in slab products.

2. Establishment of a Multi-Field Coupling Model

The installation positions and a schematic diagram of electromagnetic stirring in the secondary cooling zone of slab casting are illustrated in Figure 1. Based on actual production data from a specific steel plant, a simplified numerical model was established for grade E355 steel. The average composition of this steel grade was obtained through experimental analysis using ICP (inductively coupled plasma) and a carbon-sulfur analyzer, as shown in Table 1. The actual slab casting speed at the factory ranges from 1.1 m/min to 1.3 m/min, and the superheat ranges from 20 K to 30 K. The model values are set at 1.2 m/min and 25 K, respectively. The first pair of electromagnetic stirring rollers is installed at a distance of 3 m from the meniscus, while the second pair is installed at a distance of 4.7 m. Since the stirrers are positioned far from the submerged entry nozzle, the influence of the magnetic field at the nozzle is not considered. The cross-section of the cast slab is depicted in Figure 1c as 230 mm (thickness) × 1300 mm (width). The diameter of the electromagnetic stirring core is 240 mm, with each electromagnetic stirring roller comprising five coils. The inner three coils have 64 turns each, while the outermost two coils on each side have 32 turns.

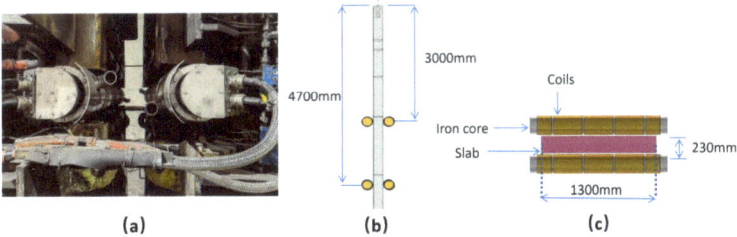

Figure 1. Diagram of the electromagnetic stirring modeling process: (**a**) stirrer field installation; (**b**) roll installation location; (**c**) schematic of a pair of stirring roller models.

Table 1. Main chemical composition of E355 steel (wt. %).

C	Si	Mn	P	S
0.17	0.23	0.55	0.013	0.007

2.1. Model Assumptions

To reasonably simplify the coupling model of flow, heat transfer, solidification, and mass transfer of molten steel during the entire solidification process, the following assumptions are made:

(1) The influence of the molten steel flow on the electromagnetic field is neglected [14], as the magnetic Reynolds number of S-EMS is significantly less than 1 and the electromagnetic field of S-EMS is assumed to be quasi-static due to its low frequency.

(2) The model disregards the Joule heat generated by induced currents, as its magnitude is negligible compared to the latent heat released during steel solidification [15].

(3) The molten steel is treated as an incompressible Newtonian fluid, with constant thermal properties assumed to enhance convergence. Continuous stirring is modeled using time-averaged Lorentz forces rather than transient values.

(4) Turbulent effects in the molten steel are simulated using a low Reynolds number k-ε model, while the mushy zone is treated as a porous medium where the flow satisfies Darcy's law.

(5) The taper and vibration of the mold are neglected [16].

(6) The effects of steel solidification shrinkage and curvature of the caster on molten steel flow and temperature distribution are disregarded [16].

(7) Considering the significant impact of carbon segregation, the model only accounts for the segregation of carbon.

2.2. Governing Equations

To simulate macrosegregation in the slab, it is essential to consider fluid flow, heat transfer, and mass transfer. The main governing equations used in this model are briefly introduced as follows.

(1) Maxwell's equations [17]:

Ampère's law:
$$\nabla \times \vec{H} = \vec{J} + \frac{\partial \vec{D}}{\partial t} \approx \vec{J} \qquad (1)$$

Faraday's law:
$$\nabla \times \vec{E} = -\frac{\partial \vec{B}}{\partial t} \qquad (2)$$

Gauss's law for magnetism:
$$\nabla \cdot \vec{B} = 0 \qquad (3)$$

Ohm's law:
$$\vec{J} = \sigma \vec{E} \qquad (4)$$

The expression for the time-averaged Lorentz force [18]:
$$\vec{F}_e = \frac{1}{2} Re\left(\vec{J} \times \vec{B}^*\right) \qquad (5)$$

where $\vec{H}$ represents the magnetic field intensity in A/m, $\vec{J}$ denotes the current density in A/m², $\vec{E}$ represents the electric field intensity in V/m, $\vec{B}$ represents the magnetic induction in T, σ is the conductivity in S/m, $\vec{F}_e$ represents the Lorentz force in N/m³, Re denotes the real part of a complex number, and $\vec{B}^*$ denotes the complex conjugate of $\vec{B}$.

(2) Continuity equation:
$$\nabla \cdot (\rho u) = 0 \qquad (6)$$

where u represents the velocity of molten steel in m/s and ρ represents the density of molten steel in kg/m³.

(3) Momentum equation:

$$\frac{\partial}{\partial t}(\rho u) + \nabla \cdot (\rho u_i u_j) = -\nabla P + \nabla \cdot \left\{\mu_{eff}\left[\nabla u_i + (\nabla u_j)^T\right]\right\} + \rho g + F_r + \vec{F}_e + S_p \qquad (7)$$

where $\vec{F}_e$ denotes the original term of the Lorentz force in N/m³, μ_{eff} represents the effective viscosity, which is the sum of laminar viscosity μ_l and turbulent viscosity μ_t, and F_r represents the thermal-solutal buoyancy force.

$$F_r = \rho g \beta (T - T_L) \qquad (8)$$

where β represents the thermal expansion coefficient, k⁻¹; T denotes the temperature of the molten steel, k; and S_p signifies the Darcy source term.

$$S_p = \frac{(1-f_l)^2}{\left(f_l^3 + 0.001\right)} A_{mush}(u - u_s) \qquad (9)$$

where f_l denotes the liquid fraction, A_{mush} represents the coefficient of the mushy zone, taken as 2×10^8, and u_s signifies the casting speed of the continuous casting slab.

$$f_l = 1 - f_s = \begin{cases} 0 & T \leq T_s \\ \frac{T-T_s}{T_L-T_s} & T_s < T < T_L \\ 1 & T \geq T_L \end{cases} \quad (10)$$

where T_L and T_s represent the liquidus temperature and solidus temperature, K.

(4) Low Reynolds number k-ε turbulence two-equation model [19]:

$$\nabla \cdot (\rho u \varepsilon) + \frac{\partial}{\partial t}(\rho k) = \nabla \cdot \left[\left(\mu_l + \frac{\mu_t}{\sigma_\varepsilon}\right)\nabla \varepsilon\right] + C_1 f_1 G \rho \frac{\varepsilon}{k} - C_2 f_2 \rho \frac{\varepsilon^2}{k} + \rho E + \frac{(1-f_l)^2}{(f_l^3 + 0.001)} A_{mush} \varepsilon \quad (11)$$

$$\nabla \cdot (\rho u k) + \frac{\partial(\rho \varepsilon)}{\partial t} = \nabla \cdot \left[\left(\mu_l + \frac{\mu_t}{\sigma_k}\right)\nabla k\right] + G + \rho D - \rho \varepsilon + \frac{(1-f_l)^2}{(f_l^3 + 0.001)} A_{mush} k \quad (12)$$

$$G = \mu_t \frac{\partial u_i}{\partial x_j}\left(\frac{\partial u_j}{\partial x_i} + \frac{\partial u_i}{\partial x_j}\right) \quad (13)$$

$$f_2 = 1 - 0.3 \exp\left(-\mathrm{Re}_t^2\right) \quad (14)$$

$$E = 2\frac{\mu_l \mu_t}{\rho}\left(\frac{\partial^2 u_i}{\partial x_j \partial x_k}\right)^2 \quad (15)$$

$$D = 2\mu_l \left(\frac{\partial(\sqrt{k})}{\partial x_j}\right)^2 \quad (16)$$

where k represents turbulent kinetic energy in $m^2 \cdot s^{-2}$, ε represents turbulent dissipation rate in $m^2 \cdot s^{-3}$, μ_l is laminar viscosity, μ_t is turbulent viscosity, C_1 is set to 1.45, f_1 is set to 1.0, and C_2 is set to 2.0.

(5) Energy equation:

$$\frac{\partial(\rho H)}{\partial t} + \nabla \cdot (\rho u H) = \nabla \cdot \left(\left(+\frac{\mu_t}{Pr_t}\right)\nabla T\right) \quad (17)$$

$$H = h_{ref} + f_l L + \int_{T_{ref}}^{T} c_p dT \quad (18)$$

where k_l represents laminar thermal conductivity in $W \cdot m^{-1} \cdot K^{-1}$, Pr_t denotes the Prandtl number, which is set to 0.86, H stands for total enthalpy, h_{ref} is the reference enthalpy, L represents the latent heat of solidification, and c_p is the specific heat of the steel liquid in $J \cdot kg^{-1} \cdot K^{-1}$.

(6) Solutes transport equation [20]:

$$\frac{\partial(\rho c_i)}{\partial t} + \nabla \cdot (\rho u c_i) = \nabla \cdot \left[\left(\rho D_{l,i} + \frac{\mu_t}{Sc_t}\right)\nabla c_i\right] + S_d + S_c \quad (19)$$

$$S_d = \nabla \cdot [\rho f_s D_{s,i} \nabla \cdot (c_{s,i} - c_i)] + \nabla \cdot [\rho f_l D_{l,i} \nabla \cdot (c_{l,i} - c_i)] \quad (20)$$

$$S_c = \nabla \cdot [\rho(u - u_s)(c_{l,i} - c_i)] \quad (21)$$

$$c_i = c_{l,i} f_l + c_{s,i} f_s \quad (22)$$

$$c_{l,i} = \frac{c_i}{1 + f_s(k_i - 1)} \quad (23)$$

$$c_{s,i} = \frac{k_i c_i}{1 + f_s(k_i - 1)} \quad (24)$$

where Sc_t represents the Schmidt number, which is set to 1.0 [21]; S_d denotes the molecular diffusion source term; S_c signifies the convective diffusion source term; $D_{l,i}$ stands for the diffusion coefficient of the solute element in the liquid phase; $D_{s,i}$ represents the diffusion coefficient of the solute element in the solid phase; c_i denotes the concentration of the solute element; $c_{l,i}$ is the local average concentration of the solute element in the liquid phase; $c_{s,i}$ is the local average concentration of the solute element in the solid phase; and k_i denotes the equilibrium distribution coefficient of element between solid and liquid phases.

2.3. The Boundary Conditions

(1) The coil is insulated from the core and the copper mold tube from the slab. Each of the five coil windings of each roller is energized with a two-phase alternating current, with a phase difference of 90° between the currents.

(2) The magnetic field lines are assumed to be parallel to the outer surface of the air unit enveloping the stirrer.

(3) The inlet velocity at the SEN is calculated based on mass conservation by the steelmaking facility. Turbulence-related parameters at the inlet are computed using a semi-empirical equation [22]. The outlet of the computational domain is set as a fully developed boundary.

(4) The SEN walls are assumed to be adiabatic, and the free liquid surface is adiabatic with zero shear stress.

(5) All walls are considered to be no-slip walls, with the wall velocity set to the casting speed. Heat transfer in the mold and secondary cooling zone primarily depends on the position of the slab, as extensively described in our previous work [23].

2.4. Simulation Strategy and Process Parameters

This study divides the computational process into two steps. Firstly, an electromagnetic field numerical model is established using ANSYS APDL, as shown in Figure 2a. Subsequently, a two-phase alternating current is applied to supply power to the five coils of each roller, generating Lorentz forces within the slab. MATLAB interpolation is then used to obtain the Lorentz forces at the fluid calculation grid nodes. Finally, the Lorentz forces are added as momentum source terms to the Navier–Stokes (N-S) equations using a Fluent user-defined function (UDF).

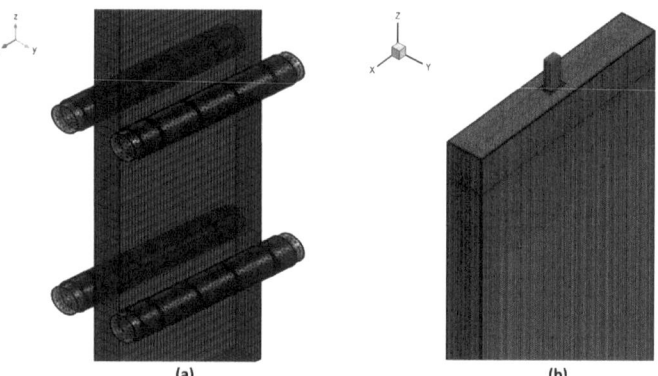

Figure 2. Numerical model mesh diagram: (**a**) mesh of the electromagnetic field model; (**b**) mesh of the full-size model of a slab.

The Fluent computational model, depicted in Figure 2b, includes a 24.8 m long computational domain in the casting direction to ensure complete solidification. Hexahedral structured grids are employed, with grid refinement in the edge regions to obtain accurate solidified shell data. The total number of grids is approximately 3.28 million. The numerical model utilizes the SIMPLE algorithm for iterative calculations, and the momentum equations are discretized using a second-order upwind mode. Convergence criteria are set such that the energy residual is less than 1×10^{-6}, and other variable residuals are less than 1×10^{-4}. The material properties used in the calculations are listed in Table 2. Some of the thermophysical properties were determined using JMatPro10.0 (Sente Software Ltd., Surrey, UK), while others were obtained from reference [24].

Table 2. The material properties [24].

Items	Unit	Value
Permeability of vacuum	H·m^{-1}	1.257×10^{-6}
Relative magnetic permeability of steel, coil, and air	/	1
Relative magnetic permeability of iron core	/	1000
Electrical conductivity of steel	S·m^{-1}	7.14×10^5
Specific heat capacity of steel	J·kg^{-1}·K	680
Thermal conductivity of steel	W·m^{-1}·K^{-1}	29
Viscosity of steel	kg·m^{-1}·s^{-1}	0.0055
Density of steel	kg·m^{-3}	7020
Solidus temperature	K	1763
Liquidus temperature	K	1802
Latent heat of solidification	J·kg^{-1}	270,000
Coefficient of thermal expansion	K^{-1}	1×10^{-4}
Equilibrium partition coefficient of C	/	0.3

According to the specific water-cooling method, the corresponding heat transfer coefficient of the cooling zone can be calculated [25] based on the water flow density. The calculated results for each cooling zone are presented in Table 3.

Table 3. Process parameters of slab continuous casting.

Secondary Cooling Zone	Length (mm)	Heat Transfer Coefficient (Wm^{-2}k^{-1})
Mold	800	1200
Zone 1	405	540
Zone 2	555	763
Zone 3	800	629
Zone 4	1730	557
Zone 5	1927	467
Zone 6	3854	400
Zone 7	5806	306
Zone 8	4485	182

To compare the differences between different stirring modes, simulations and comparative analyses were conducted for two different operating modes of the stirrer, as shown in Table 4. Continuous stirring refers to the continuous application of stirring action, with the current kept stable, ensuring stable Lorentz forces and flow fields. On the other hand, alternate stirring involves the periodic switching of stirring action, resulting in intermittent stirring effects.

Table 4. Different electromagnetic stirring modes.

Mode Number	Roller Operating Mode	Frequency	Period	Current (A)
A	Continuous Stirring	f = 5 Hz	/	320,240,160
B	Alternate Stirring	f = 5 Hz	T = 22 s	320,240,160

By periodically changing the direction of the current supplied to specific coils, it is possible to induce periodic reversal of the Lorentz forces, thereby achieving the transition from the continuous to alternating mode.

3. Model Validation

Model validation was achieved through comparison calculation and measurement of magnetic induction. The center-line magnetic induction of the first pair of roller stirrers was measured at the continuous casting site using a KANETECTM-701 Gauss meter (Kanetec Co., Ltd., Nagano, Japan). Figure 3 shows the comparison between the calculated and measured values of magnetic induction along the centerline of the slab cross-section when the current is 320 amperes and the frequency is 5 hertz. The maximum difference between the two values is less than 5%. When the current is 160 A and 240 A at the same frequency, the differences between the calculated and measured values are also below 5%, indicating a good agreement between the measured and calculated values of magnetic induction.

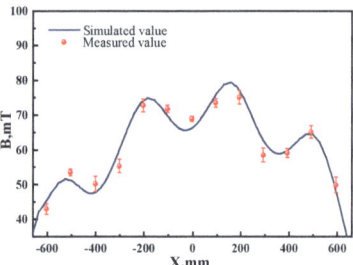

Figure 3. Magnetic induction intensity.

To verify the calculation accuracy of the center segregation model in the slab, carbon segregation was measured without electromagnetic stirring at the center of the slab width. Samples were taken at intervals of 1/2 thickness from the slab center to the surface, with a sampling interval of 20 mm and a drill hole diameter of 5 mm. Six samples were taken for carbon segregation detection. The carbon segregation index (C_i/C_0) predicted by the model was compared with the detected carbon segregation index, as shown in Figure 4. The maximum deviation was 4.8%, which is below 5%. The solute carbon segregation predicted by the model is in good agreement with the actual detection results.

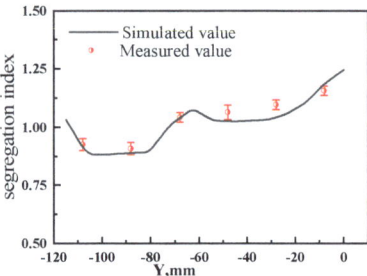

Figure 4. C distribution without S-EMS.

4. Results and Discussion

4.1. Lorentz Force

Figure 5 illustrates the distribution of magnetic induction and Lorentz force on the surface of the slab at a current intensity of 320 A and a frequency of 5 Hz for Mode A. As depicted in the figure, both the magnetic induction (B) and Lorentz force (F) are concentrated within the secondary cooling zone where the stirrer is located, ranging from $Z = -2.5$ m to -5.2 m. On the surface of the slab, each pair of stirring rollers exhibits four peaks in both magnetic induction and Lorentz force, with maximum values of approximately 170 mT and 15,000 N/m^3, respectively. Due to the compact structure of the roller stirrer and its proximity to the slab within the secondary cooling section, where there is minimal gap, magnetic induction is primarily concentrated within a range of 1000 mm around each electromagnetic stirrer.

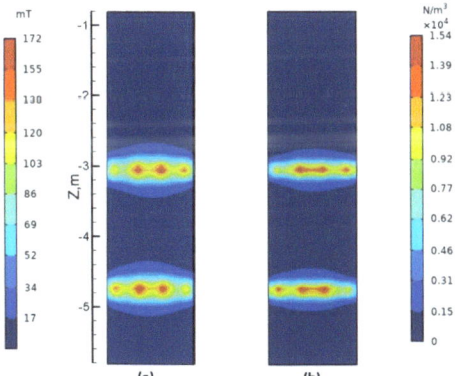

Figure 5. Magnetic induction and Lorentz force distribution on the surface of the slab at a current of 320 A: (**a**) the contour plot of magnetic induction distribution; (**b**) the contour plot of Lorentz force distribution.

The two pairs of rollers on the slab maintain equal but opposite magnetic induction and Lorentz forces from the onset of electrification. Taking the second pair of rollers as an example, the cross-sectional distribution of magnetic induction and Lorentz forces at Z = −4.7 m, with I = 320 A and F = 5 Hz, is depicted in Figure 6. It can be observed that both the magnetic induction (as shown in Figure 6a) and Lorentz force (as shown in Figure 6b) gradually decrease from the surface of the slab to the center. The direction of the Lorentz force is from right to left (pointing towards the negative X-direction), as shown in Figure 6c, and there are four relative peaks of magnetic induction and Lorentz force along the centerline of the slab cross-section. The maximum magnetic field strength is approximately 83 mT, and the maximum Lorentz force is approximately 6200 N/m^3.

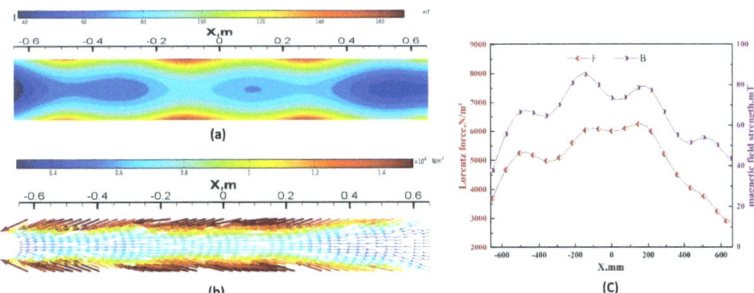

Figure 6. Distribution of B and F in the cross-section (Z = −4.7 m) under continuous stirring with a current of 320 A: (**a**) magnetic induction strength contour plot; (**b**) Lorentz force vector plot; (**c**) numerical values of B and F along the central axis.

Figure 7 illustrates the vector distribution of Lorentz forces at the center cross-section of the slab and a schematic diagram of stirring reversal for Mode B. The magnitude of the Lorentz force remains constant for both pairs of rolls, while its direction periodically changes. Closer to the stirrer, the Lorentz force exerted on the molten steel is greater. At an electrical current intensity of 400 A and a frequency of 5 Hz, the maximum Lorentz force at the central section is approximately 8000 N/m^3. Figure 7a,b depict the vector diagrams of the Lorentz force in the forward and reverse directions along the central longitudinal section. After half a cycle, the Lorentz force reverses, decelerating and then accelerating the molten steel in the opposite direction. The roll-type stirrer switches the stirring direction of the molten steel by altering the phase angle of the alternating currents in the two sets of coils. Figure 7c,d illustrate the schematic of the stirring reversal for the first and second pairs of rolls, respectively. In the alternating stirring mode, the reversal cycle (t) is 22 s. The stirring coils generate a forward Lorentz force for 10 s, followed by a 1 s interruption, and then a reverse Lorentz force for 10 s, followed by another 1 s interruption, repeating this cycle.

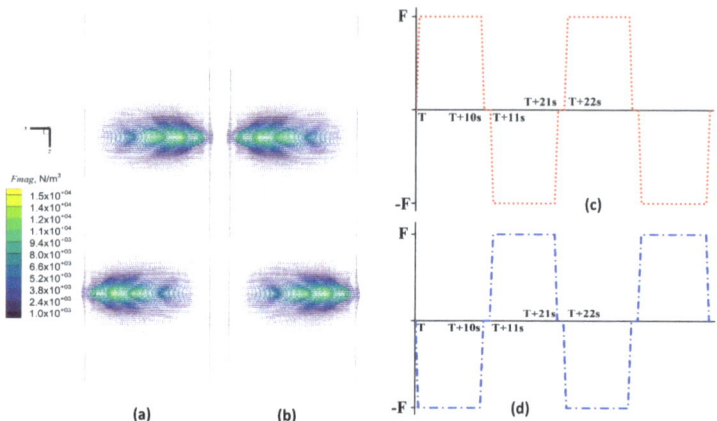

Figure 7. Lorentz force vector distribution and stirring reversal schematics in the central section of the slab at a current of 400 A: (**a**) forward Lorentz force; (**b**) reverse Lorentz force; (**c**) first pair of roll commutation modes; (**d**) second pair of roll commutation modes.

4.2. Flow Field

Figure 8 illustrates the velocity streamlines and flow field contour plots at the center longitudinal section (Y = 0 m) of the slab under different current intensities for Mode A. It is evident that without S-EMS, there is no relative motion between the steel liquid and the solidified shell, resulting in the steel liquid velocity being identical to the casting speed. With the influence of S-EMS, the Lorentz force generated by the electromagnetic stirring roll at −3 m drives the steel liquid to flow from right to left, impacting the narrow face of the mold and ultimately bifurcating into two counter-current flows on either side of the impact point. Due to the Lorentz force induced by the electromagnetic stirring roll at −4.7 m being opposite in direction to that induced by the roll at −3 m, the steel liquid flows from left to right, impacting the narrow face of the mold from the opposite direction. Under the influence of the two pairs of electromagnetic stirring rolls, three vortices are formed, located above the roll installed at −3 m, between the two rolls, and below the roll installed at −4.7 m.

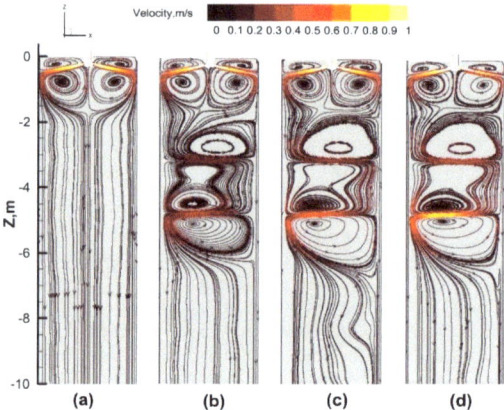

Figure 8. Streamlines and flow field cloud of the longitudinal center section of the slab: (**a**) 0 A; (**b**) 160 A; (**c**) 240 A; (**d**) 320 A.

A comparison of Figure 8a–d reveals that with an increasing current intensity, the velocity of the melt at the same position increases accordingly. This is attributed to the increased Lorentz force with a higher current intensity, which enhances the horizontal flow of the steel liquid due to the lateral Lorentz force, thereby altering the original flow path of the downward-moving steel liquid and intensifying the horizontal flow velocity. This simultaneously promotes heat exchange in the stirring zone, facilitating the reduction in overheating.

Figure 9 illustrates the flow field distribution at a typical moment during one cycle under Mode B with a current intensity of 320 A. From t to t + 5 s, the Lorentz force reverses, exerting a force opposite to the direction of flow, leading to a gradual weakening of the flow intensity. From t + 5 s to t + 11 s, the reverse Lorentz force causes the steel liquid to decelerate to zero before accelerating, resulting in a decrease in flow velocity followed by a reverse increase to maximum velocity. From t + 11 s to t + 16 s, the Lorentz force reverses again, causing a gradual decrease in flow velocity. From t + 16 s to t + 22 s, the reverse Lorentz force leads to a deceleration to zero before accelerating in the opposite direction, with the flow intensity first decreasing and then reversing to accelerate.

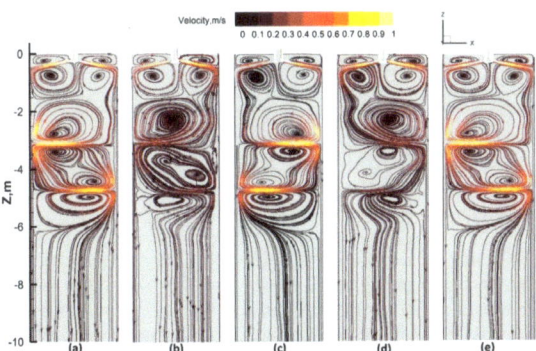

Figure 9. Flow patterns on the cross-section at the slab center during alternating stirring with a current of 320 A: (**a**) t; (**b**) t + 5 s; (**c**) t + 11 s; (**d**) t + 16 s; (**e**) t + 22 s.

In comparison to the continuous stirring mode, the alternating mode results in a continuous and periodic horizontal flow of the steel liquid within the slab. Under the alternating stirring mode, the flow patterns inside the slab exhibit different states, with vortices appearing above and below the stirring rolls.

Figure 10 illustrates the velocity distribution along the centerline of the slab cross-section under different stirring modes (current intensity of 320 A, Z = −4.7 m). In the continuous stirring mode, the steel liquid moves in the negative direction of the X-axis, reaching a maximum velocity of −0.8 m/s at the position of −0.2 m. In the alternating stirring mode, the velocity exhibits periodic variations over one cycle with five time points. At time t, the steel liquid moves in the positive direction of the X-axis, reaching a maximum velocity of 0.8 m/s. From t to t + 5 s, the velocity gradually decreases due to the opposite direction of force relative to the velocity direction, with a maximum velocity of approximately 0.5 m/s. From t + 5 s to t + 11 s, the steel liquid velocity decreases to 0 m/s, then moves in the negative direction of the X-axis, reaching a maximum velocity of −0.8 m/s. From t + 11 s to t + 16 s, due to the change in force direction, which opposes the flow direction, the velocity gradually decreases to −0.5 m/s. From t + 16 s to t + 22 s, the steel liquid velocity decreases to 0 m/s, then moves in the positive direction of the X-axis, reaching a maximum velocity of 0.8 m/s. Comparing the six curves, it can be observed that the average velocity over time in the alternating stirring mode is lower than that in the continuous stirring mode, indicating a lower flow intensity and consequently a lower stirring effect on the steel liquid.

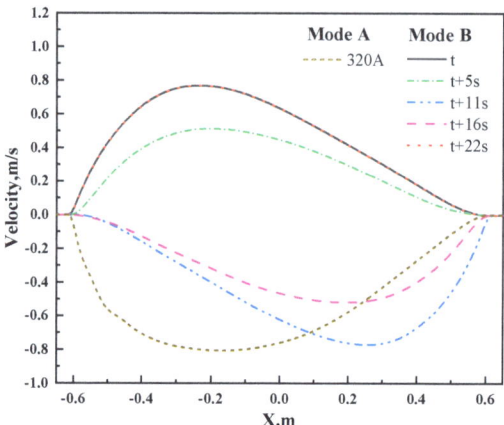

Figure 10. Comparison of transverse velocities at different times for alternating and continuous modes on the centerline of the slab cross-section at Z = −4.7 m with a current of 320 A.

To compare the velocity distributions along the centerline of the slab cross-section under different current intensities (Z = −3 m and Z = −4.7 m) between alternating stirring and continuous stirring, the moment of the maximum velocity in the alternating stirring mode was selected and contrasted with continuous stirring. The results are depicted in Figure 11. For Mode A at Z = −3 m, as the current increases from 160 A to 240 A and then to 320 A, the maximum lateral velocity of the liquid core under the roll decreases from 0.2 m/s to 0.4 m/s and then to 0.5 m/s, respectively. At Z = −4.7 m, the corresponding maximum lateral velocities are −0.4 m/s, −0.55 m/s, and −0.8 m/s as the current increases from 160 A to 240 A and then to 320 A. For Mode B at Z = −3 m, the maximum lateral velocities under the roll decrease from −0.4 m/s to −0.7 m/s and then to −0.8 m/s as the current increases from 160 A to 240 A and then to 320 A, respectively. At Z = −4.7 m, the maximum lateral velocities increase from 0.4 m/s to 0.7 m/s and then to 0.8 m/s after increasing the current from 160 A to 240 A and then to 320 A.

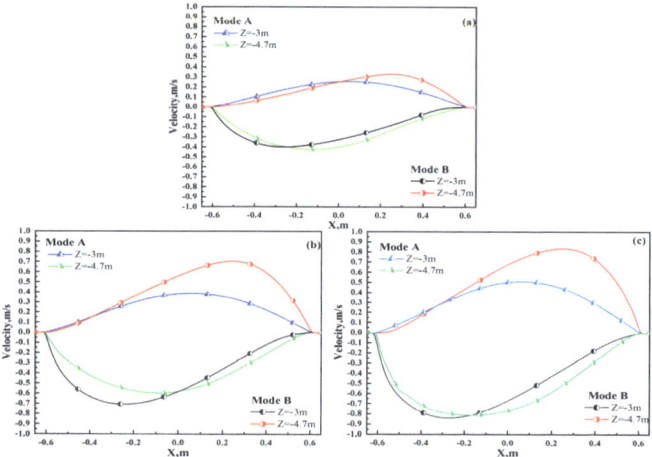

Figure 11. Velocity distribution in the centerline of the cross-section under the roll: (**a**) 160 A; (**b**) 240 A; (**c**) 320 A.

The lateral velocity gradually increases with the current intensity of the stirrer. From Figure 11a, it can be observed that at Z = −3 m, the velocity for Mode B at 160 A is comparable to the maximum velocity at Z = −4.7 m, with a difference of approximately 0.05 m/s, indicating that the horizontal flow velocities of the liquid core under both rolls are symmetrical. Conversely, in continuous mode

(Mode A), as shown in Figure 11c, at 320 A and Z = −3 m, the velocity differs significantly from the maximum velocity at Z = −4.7 m, with a difference of approximately 0.3 m/s, indicating asymmetry in the lateral flow velocities of the liquid core under the two rolls. This asymmetry can be attributed to the first roll being closer to the mold, where the flow field is influenced by both the Lorentz force of the stirrer and the mold's flow field, while the liquid core under the second roll is further away and is not affected by the mold's flow field. Comparing Figure 11a–c, it can be observed that under Mode B conditions, the lateral flow velocities under both rolls are essentially symmetrical, indicating better uniformity in the flow field.

4.3. Carbon Element Segregation

Figure 12 illustrates the distribution of carbon segregation in the longitudinal section of the slab center. In the absence of S-EMS, solute diffusion occurs at the solidification end of the slab with a minimal transverse fluid flow, resulting in a relatively uniform solute distribution. Under continuous stirring, i.e., Mode A, the uneven flow promotes carbon enrichment towards the right narrow edge. The significant influence of the three-ring vortex induced by S-EMS, as shown in Figure 8, affects the carbon concentration distribution in the ingot pool. The second pair of electromagnetic stirring rolls generates a greater flow velocity, flushing away carbon elements near the solidification front on the left narrow edge of the ingot and gradually enriching carbon elements on the right narrow edge of the ingot. Concurrently, with the increase in the S-EMS current intensity, the flow velocity of the steel liquid in the ingot increases, resulting in more carbon elements being flushed away from the solidification front and exacerbating the enrichment phenomenon on the right narrow edge. Under alternating stirring, i.e., Mode B, the periodic variation in the flow field leads to an improvement in the unevenness of carbon elements in the width direction. This indicates that alternating stirring can suppress segregation, and with an increase in current, the uniformity of the element distribution improves. At an alternating stirring current of 320 A, the carbon element uniformity is maximized.

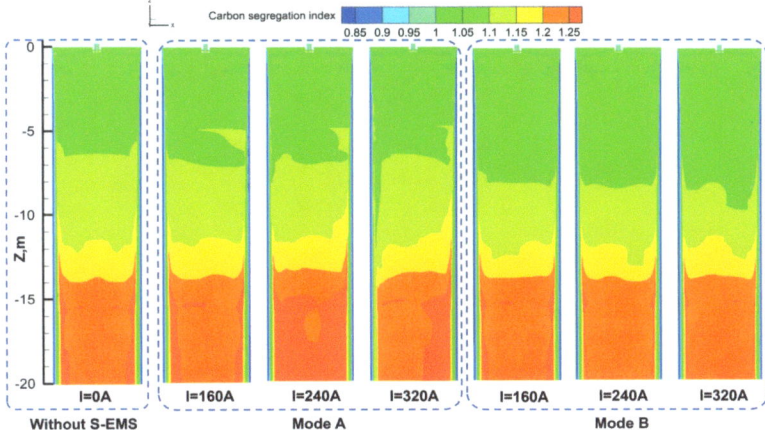

Figure 12. Distribution of carbon concentration at the solidification end of the slab's center longitudinal section under different S-EMS currents.

The distribution of carbon segregation indices under different S-EMS current intensities is depicted in Figure 13. Figure 13a illustrates the carbon segregation distribution in the transverse section of the ingot without electromagnetic stirring, while Figure 13b,c display the distribution of the carbon segregation index along the thickness direction centerline (L1) for Mode A and Mode B, respectively. Notably, significant negative segregation is observed near the subsurface of the slab due to the rapid steel liquid flow velocity and fast solidification at the mold. Without S-EMS (i.e., I = 0 A), the high degree of solute enrichment in the steel liquid eventually solidifies at the center, resulting in significant positive carbon segregation along the centerline of the ingot, particularly evident at the 1/8 centerline of the wide face where the final solidification occurs (as shown in Figure 12). A comparison between Figure 13b,c reveals a more pronounced transient decrease in carbon element segregation indices at approximately X = −0.57 m and X = 0.57 m for Mode A compared to Mode B. This phenomenon can be attributed to the stronger flow intensity and flushing effect in continuous stirring. Specifically, in the absence of S-EMS, the average carbon segregation index along the centerline (L1) is 1.233, with a difference of 0.026 between the maximum and minimum values. At 160 A, the average carbon segregation index increases to 1.236, and at 320 A, it further rises to 1.247, accompanied by an increased fluctuation in segregation indices. Thus, it is evident that increasing the current exacerbates the average carbon segregation and its fluctuation along the ingot centerline, indicating poorer uniformity compared to the absence of electromagnetic stirring.

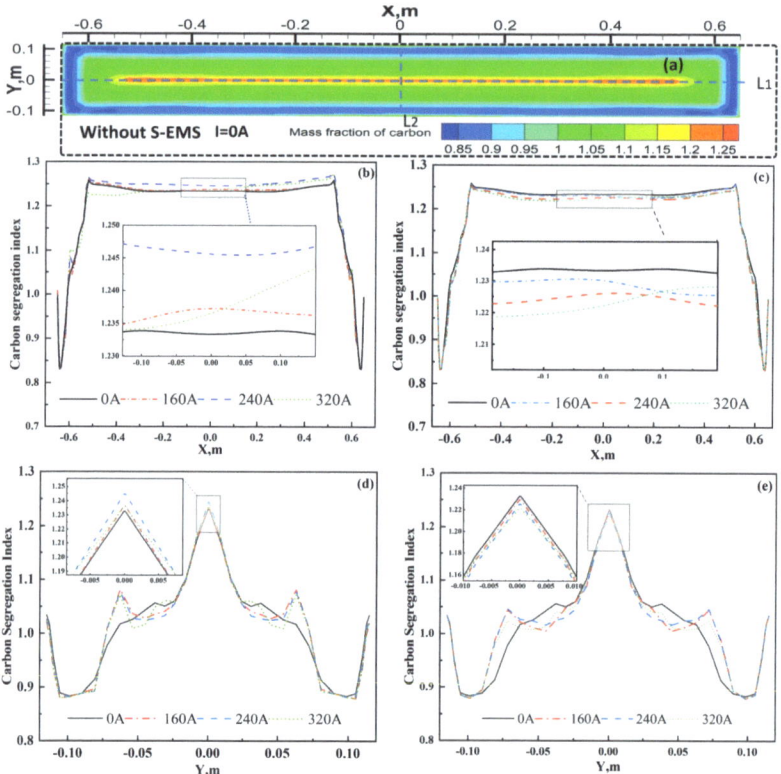

Figure 13. Carbon segregation distribution of centerline in the slab's cross-section and at different S-EMS currents: (**a**) without S-EMS; (**b**) Mode A of line 1; (**c**) Mode B of line 1; (**d**) Mode A of line 2, (**e**) Mode B of line 2.

Under continuous stirring (Mode A), as the S-EMS current intensity increases, the fluctuation of carbon segregation along the ingot centerline intensifies. The degradation of carbon segregation is most severe near the 1/8 centerline length close to the right narrow face of the ingot due to the imbalance in horizontal flow under the upper and lower pairs of rolls and excessive stirring-induced carbon enrichment at the right narrow edge of the ingot. Consequently, employing continuous stirring (Mode A) with electromagnetic stirring installed in the pre-section of the secondary cooling zone, i.e., the first pair of rolls positioned at $Z = -3$ m and the second pair at $Z = -4.7$ m, yields marginal improvements in carbon segregation effects.

Figure 13c demonstrates that under alternating stirring (Mode B), the improvement in carbon segregation is more pronounced with increasing current. The residual of this numerical model is set at 10^{-6}. Between $X = -0.52$ m and $X = 0.52$ m, the average carbon segregation index decreases to 1.232, 1.228, and 1.227 at 160 A, 240 A, and 320 A, respectively, lower than the value of 1.233 at 0 A. The difference between the maximum and minimum segregation indices at 160 A, 240 A, and 320 A is 0.0302, 0.0299, and 0.0311, respectively, which is closer to the difference at 0 A, indicating a favorable improvement in carbon segregation with an increasing current. However, considering the saturation effect of the magnetic induction intensity of the yoke itself, an excessively high current is not advisable.

The distribution of carbon segregation along the centerline (L2) of the narrow face of the slab under different S-EMS currents is illustrated in Figure 13. A comparison between Figure 13d and 13e reveals that, near the center of the L2 line (Y = 0.00 m), the carbon segregation is highest at 240 A in Mode A, at approximately 1.243, while it is lowest at 320 A in Mode B, at approximately 1.221. Between Y = −0.05 m to −0.075 m and Y = 0.05 m to 0.075 m, the maximum carbon segregation index under Mode B, 1.023, is 0.045 lower than that under Mode A, where it is 1.069. Hence, adopting alternating stirring, i.e., Mode B, compared to continuous stirring, i.e., Mode A, can reduce the average carbon segregation index, facilitating carbon homogenization.

In the aforementioned analysis, the alternating stirring mode in the secondary cooling zone achieves continuous reversal of the stirring direction by periodically changing the direction of the Lorentz force. This enhances the symmetry of the transverse flow field under the stirring rolls, thereby improving the carbon segregation in the slab, with significant enhancement effects observed. This offers a new approach to improving the distribution of carbon elements in the production process by employing alternating stirring in the pre-section of the secondary cooling zone.

5. Conclusions

(1) Flow field simulation studies under different stirring modes indicate that in the continuous mode, the lateral velocity of the liquid phase beneath the first and second pairs of rolls is opposite and significantly different. With an increasing current, this difference becomes more pronounced, with a maximum absolute difference in lateral velocity of 0.2 m/s and 0.3 m/s at 160 A and 320 A, respectively. An asymmetry in the internal flow field of the ingot under continuous stirring is observed. In the alternating mode, the maximum lateral velocity of the liquid core beneath the first pair of rolls is approximately equal to that beneath the second pair of rolls, with a maximum absolute difference of approximately 0.05 m/s, indicating a basic symmetry in the lateral velocity of the liquid core beneath the two pairs of rolls.

(2) Simulation studies on the effects of carbon segregation under different stirring modes show that in the continuous mode, the fluctuation in carbon segregation along the centerline of the slab increases with increasing current. Between X = −0.52 m and X = 0.52 m, at 0 A, the average carbon segregation index along the centerline is 1.233, with a difference of 0.026 between the maximum and minimum values, while at 320 A, the average increases to 1.247, with a difference of 0.051 between the maximum and minimum values, indicating an increase in fluctuation. In the alternating mode, the improvement is more significant with an increasing current. Between X = −0.52 m and X = 0.52 m, at 160 A, the average carbon segregation index along the centerline is 1.232, with a difference of 0.0302 between the maximum and minimum values, while at 320 A, the average decreases to 1.227, with a difference of 0.0311 between the maximum and minimum values, indicating minimal fluctuation. Therefore, adopting alternate stirring is advantageous for reducing the average carbon segregation index and promoting carbon homogenization compared to continuous stirring.

Author Contributions: Conceptualization, D.L. and G.Z.; experimental setup design, D.L., G.Z. and X.X.; validation, D.L., J.Z. and X.X.; analysis, G.Z., J.Z. and X.X.; writing, D.L., G.Z., J.Z. and X.X.; project administration, D.L. and G.Z.; funding acquisition, G.Z. All authors have read and agreed to the published version of the manuscript.

Funding: The work was supported by Hunan Zhongke Electric Co., Ltd., grant number "HZ2021F0529A".

Institutional Review Board Statement: Not applicable.

Informed Consent Statement: Not applicable.

Data Availability Statement: The original contributions presented in the study are included in the article, further inquiries can be directed to the corresponding author.

Conflicts of Interest: Authors Jianhua Zeng and Xin Xie are employed by the company Pangang Group Research Institute Co., Ltd. The remaining authors declare that the research was conducted in the absence of any commercial or financial relationships that could be construed as a potential conflict of interest. Besides, the authors declare that this study received funding from Hunan Zhongke Electric Co., Ltd. The funder was not involved in the study design, collection, analysis, interpretation of data, the writing of this article or the decision to submit it for publication.

References

1. Liu, Y.; Hao, W.; Ren, L. Research Progress on Continuous Casting Technology. *Jiangxi Metall.* **2023**, *43*, 484–493.
2. Wang, F. Research on the Preparation Technology of Nickel-Based Corrosion-Resistant Alloy Composite Electromagnetic Continuous Casting. Ph.D. Thesis, Northeastern University, Shenyang, China, 2016.
3. Wang, P.; Wu, W.; Lei, H. Research on Two-Stage Electromagnetic Stirring Process of Wear-Resistant Steel Slab Continuous Casting. *Ind. Heat.* **2021**, *50*, 32–34+38.
4. Liu, Y.; Wang, X. Influence of Electromagnetic Stirring in the Secondary Cooling Zone on Center Segregation in Continuous Casting Slabs. *J. Univ. Sci. Technol. Beijing* **2007**, *29*, 582–585+590.
5. Dong, Z.; Liu, P.; Jia, S.; Han, C.; Jiang, H. Application of Two-Stage Roller-Type Electromagnetic Stirring in Baosteel Slab Continuous Casting Machine. *Spec. Steel* **2011**, *32*, 34–35.
6. Xie, S.; Xiao, A.; Nie, C.; Zhou, J.; Sui, Y.; Liu, P. Effect of Two-Stage Roller-Type Electromagnetic Stirring on Microstructure and Properties of High-Strength Steel. *Contin. Cast.* **2020**, *45*, 46–50+55.
7. Li, X.; Wang, M.; Bao, Y.; Gong, J.; Wang, X.; Pang, W. Solidification Structure and Segregation of High Magnetic Induction Grain-Oriented Silicon Steel. *Met. Mater. Int.* **2019**, *25*, 1586–1592. [CrossRef]
8. Yao, C.; Wang, M.; Ni, Y.; Gong, J.; Xing, L.; Zhang, H.; Bao, Y. Effects of Secondary Cooling Segment Electromagnetic Stirring on Solidification Behavior and Composition Distribution in High-Strength Steel 22MnB5. *JOM* **2022**, *74*, 4823–4830. [CrossRef]
9. Jiang, D.; Zhu, M.; Zhang, L. Numerical simulation of solidification behavior and solute transport in slab continuous casting with S-EMS. *Metals* **2019**, *9*, 452. [CrossRef]
10. Ma, W.; Liu, G.; Li, H. Influence of Electromagnetic Stirring in the Secondary Cooling Zone on the Internal Quality of Slabs. In Proceedings of the 3rd National Conference on Electromagnetic Metallurgy and Strong Magnetic Field Materials Science, Jiaozuo, China, 15–18 August 2016; p. 5.
11. Mei, M.; Chen, M.; Yang, X.; Xuan, M. Influence of Electromagnetic Stirring Modes on Flow Characteristics of Steel in the Crystallizer of Large Slabs. *J. Mater. Metall.* **2023**, *22*, 532–539.
12. Liu, C.; Jin, B.G. The Effect of Electromagnetic Stirring on Steel Strand Quality. *Adv. Mater. Res.* **2012**, *482*, 1699–1702. [CrossRef]
13. Wang, C.; Liu, Z.; Li, B. Effect of Electromagnetic Stirring Reversal Cycle in the Secondary Cooling Zone on Center Equiaxed Crystal Rate of Thin Slabs. *Contin. Cast.* **2023**, *42*, 64–70.
14. Davidson, P.A. Magnetohydrodynamics in materials processing. *Annu. Rev.* **1999**, *31*, 273–300. [CrossRef]
15. Aboutalebi, M.R.; Hasan, M.; Guthrie, R. Coupled turbulent flow, heat, and solute transport in continuous casting processes. *Metall. Mater. Trans. B* **1995**, *26*, 731–744. [CrossRef]
16. Zhang, W.; Luo, S.; Chen, Y.; Wang, W.; Zhu, M.J.M. Numerical simulation of fluid flow, heat transfer, species transfer, and solidification in billet continuous casting mold with M-EMS. *Metals* **2019**, *9*, 66. [CrossRef]
17. Li, Y.; Deng, A.; Li, H.; Yang, B.; Wang, E.J.M. Numerical study on flow, temperature, and concentration distribution features of combined gas and bottom-electromagnetic stirring in a ladle. *Metals* **2018**, *8*, 76. [CrossRef]
18. Zhang, H.; Nagaumi, H.; Zuo, Y.; Cui, J. Coupled modeling of electromagnetic field, fluid flow, heat transfer and solidification during low frequency electromagnetic casting of 7XXX aluminum alloys: Part 1: Development of a mathematical model and comparison with experimental results. *Mater. Sci. Eng. A* **2007**, *448*, 189–203. [CrossRef]
19. Aboutalebi, M.R.; Guthrie, R.I.L.; Seyedein, S.H. Mathematical modeling of coupled turbulent flow and solidification in a single belt caster with electromagnetic brake. *Appl. Math. Model.* **2007**, *31*, 1671–1689. [CrossRef]
20. Barna, M.; Javurek, M.; Wimmer, P. Numeric Simulation of the Steel Flow in a Slab Caster with a Box-Type Electromagnetic Stirrer. *Steel Res. Int.* **2020**, *91*, 2000067. [CrossRef]
21. Song, X.; Cheng, S.; Cheng, Z. Mathematical Modelling of Billet Casting with Secondary Cooling Zone Electromagnetic Stirrer. *Ironmak. Steelmak.* **2013**, *40*, 189–198. [CrossRef]
22. Yang, H.; Zhang, X.; Deng, K.; Li, W.; Gan, Y.; Zhao, L. Mathematical simulation on coupled flow, heat, and solute transport in slab continuous casting process. *Metall. Mater. Trans. B* **1998**, *29*, 1345–1356. [CrossRef]
23. Liu, D.; Zhang, G.; Zeng, J.; Wu, C. Investigation of Solidification Heat Transfer in Slab Continuous Casting Process Based on Different Roll Contact Calculation Methods. *Materials* **2024**, *17*, 482. [CrossRef]
24. Li, S.; Han, Z.; Zhang, J.J.J. Numerical modeling of the macrosegregation improvement in continuous casting blooms by using F-EMS. *Jom* **2020**, *72*, 4117–4126. [CrossRef]
25. Nozaki, T.; Matsuno, J.-I.; Murata, K.; Ooi, H.; Kodama. A secondary cooling pattern for preventing surface cracks of continuous casting slab. *ISIJ* **1978**, *18*, 330–338. [CrossRef]

Disclaimer/Publisher's Note: The statements, opinions and data contained in all publications are solely those of the individual author(s) and contributor(s) and not of MDPI and/or the editor(s). MDPI and/or the editor(s) disclaim responsibility for any injury to people or property resulting from any ideas, methods, instructions or products referred to in the content.

MDPI AG
Grosspeteranlage 5
4052 Basel
Switzerland
Tel.: +41 61 683 77 34

Materials Editorial Office
E-mail: materials@mdpi.com
www.mdpi.com/journal/materials

Disclaimer/Publisher's Note: The title and front matter of this reprint are at the discretion of the Guest Editors. The publisher is not responsible for their content or any associated concerns. The statements, opinions and data contained in all individual articles are solely those of the individual Editors and contributors and not of MDPI. MDPI disclaims responsibility for any injury to people or property resulting from any ideas, methods, instructions or products referred to in the content.

www.ingramcontent.com/pod-product-compliance
Lightning Source LLC
LaVergne TN
LVHW072324090526
838202LV00019B/2349